Springer Proceedings in Mathematics & Statistics

Volume 223

Springer Proceedings in Mathematics & Statistics

This book series features volumes composed of selected contributions from workshops and conferences in all areas of current research in mathematics and statistics, including operation research and optimization. In addition to an overall evaluation of the interest, scientific quality, and timeliness of each proposal at the hands of the publisher, individual contributions are all refereed to the high quality standards of leading journals in the field. Thus, this series provides the research community with well-edited, authoritative reports on developments in the most exciting areas of mathematical and statistical research today.

More information about this series at http://www.springer.com/series/10533

A. Ismael F. Vaz · João Paulo Almeida
José Fernando Oliveira · Alberto Adrego Pinto
Editors

Operational Research

IO2017, Valença, Portugal, June 28–30

Springer

Editors
A. Ismael F. Vaz
School of Engineering
University of Minho
Braga
Portugal

José Fernando Oliveira
INESC TEC—Faculty of Engineering
University of Porto
Porto
Portugal

João Paulo Almeida
Department of Mathematics
and LIAAD-INESC TEC, School
of Technology and Management
Polytechnic Institute of Bragança
Bragança
Portugal

Alberto Adrego Pinto
Department of Mathematics
and LIAAD-INESC TEC
University of Porto
Porto
Portugal

ISSN 2194-1009 ISSN 2194-1017 (electronic)
Springer Proceedings in Mathematics & Statistics
ISBN 978-3-319-71582-7 ISBN 978-3-319-71583-4 (eBook)
https://doi.org/10.1007/978-3-319-71583-4

Library of Congress Control Number: 2017960932

Mathematics Subject Classification (2010): 90Bxx

Printed on acid-free paper

This Springer imprint is published by Springer Nature
The registered company is Springer International Publishing AG
The registered company address is: Gewerbestrasse 11, 6330 Cham, Switzerland

To my family, Monica, Gabriela, Henrique, and Afonso

A. Ismael F. Vaz

To Cristina, Mafalda and Rita

João Paulo Almeida

To São, Beatriz and Mariana

José Fernando Oliveira

To Ana and Maria

Alberto Adrego Pinto

Foreword

It was a privilege to be invited to give a plenary talk at APDIO's IO2017 conference, and it is a privilege to have been invited to write this foreword to the collection of papers.

The papers here give some idea of the variety and scale of work going on in the Portuguese O.R. community. But a good conference—and IO2017 was certainly a good conference—goes beyond the papers.

The conference is an opportunity for early career professionals to begin to experience the excitement and benefits of sharing their work with knowledgeable, supportive peers outside their immediate circle; and for more experienced people to keep in touch with what is going on elsewhere, and influence the collective body of work. It is an opportunity for those informal conversations, whether peer-to-peer or senior-to-junior, that help build the social capital—the network of trust, confidence and exchange—of a profession. It is an opportunity for cross-fertilisation of ideas between people whose paths don't normally cross, for serendipitous encounters that generate unexpected collaborations, for the inspiration that comes from stepping out of the daily routine into other environments.

A good conference has benefits that continue long after the conference has closed, and that spread beyond the people who were present. One of the most important of these is sharing the papers presented to the conference. The scientific papers are the essential backbone of a conference, from which everything else flows. This book represents the tangible output from the conference, that will support the sharing and continuation of its benefits and opportunities.

Valença

Ruth Kaufman, OBE, FORS, FIMA
President, The Operational Research Society

Contents

Chapter 1
New Approach for Optimization of Construction and Demolition Waste Management: Application to the Lisbon Metropolitan Area

António Rebello de Andrade, Marta Castilho Gomes and Joaquim Duque

Abstract The growing concern regarding global sustainability is increasingly evident in the construction sector due to the large amounts of waste produced. This work aims to develop a new approach in order to plan an efficient recycling network for Construction and Demolition Waste (CDW). The approach is based on the methodology of Process Systems Engineering (PSE), allowing an appropriate problem systematization and the definition of flows of materials between the network nodes. The developed mixed-integer linear programming (MILP) model is a tool to support CDW management in assessing the recycling network from a regulatory perspective (minimizing the total cost). Various scenarios can be defined and sensitivity analyses performed, making the aforementioned assessment clear regarding location, types and capacities of the new processes to be installed. The model was applied to the 211 parishes that comprise the Lisbon Metropolitan Area (LMA), in Portugal, but due to its generic formulation can be applied elsewhere at regional or national level. Results show a preference for direct deposition in landfills and the fact that high quality recycling processes are not, at present, economically viable for the LMA.

Keywords Construction and demolition waste · Mixed-integer linear programming · Process systems engineering · Lisbon metropolitan area

---

A. R. d. Andrade
Instituto Superior Técnico, Universidade de Lisboa, Av. Rovisco Pais 1,
1049–001 Lisboa, Portugal
e-mail: a_randrade@hotmail.com

M. C. Gomes (✉)
CERIS, CESUR, Instituto Superior Técnico, Universidade de Lisboa, Av. Rovisco Pais 1,
1049–001 Lisboa, Portugal
e-mail: marta.gomes@tecnico.ulisboa.pt

J. Duque
UAER, Laboratório Nacional de Energia e Geologia, Estr. do Paço do Lumiar, 22,
Lisboa, Portugal
e-mail: joaquim.duque@lneg.pt

A. I. F. Vaz et al. (eds.), *Operational Research*, Springer Proceedings
in Mathematics & Statistics 223, https://doi.org/10.1007/978-3-319-71583-4_1

1.1 Introduction

Construction and demolition waste (CDW) represents 25–30% of the total amount of waste in the EU [7] and its distinct morphological characteristics make managing it a difficult task. Many strategic guidelines have been taken towards reducing the environmental, social and economic impacts generated by this type of waste. European concerns regarding CDW management led to a change in the legal framework, which establishes that by 2020 all member states should have at least 70% of non-hazardous CDW recycled or reused. Portugal, unlike other EU countries, is far from this goal with only 9% of the total CDW production being recycled, 11% reused and 4% incinerated, which means that 76% of this type of waste is landfilled [3].

Therefore, CDW management has become a critical issue in European countries and tools are needed to support decision making in this context. In previous work, [1] collected data from demolition works and developed a novel methodology to estimate demolition waste generation for a region, subsequently applying it to the Lisbon Metropolitan Area. Then, [4] followed the approach of [6] and developed an MILP model for optimizing a CDW recycling network. Using the generated waste data, the model was applied to the LMA at municipal level (18 municipalities).

A novel approach to the problem is herein introduced, using the State Task Network methodology derived from Process Systems Engineering (PSE). Compared to [4], this allowed a more complete representation of all the entities involved and an enhanced design of the network, further detailing the recycling processes and materials' flow between nodes.

The model was applied to the 211 parishes that constitute the LMA and conclusions drawn.

1.2 Optimization Model for Planning a CDW Recycling Network

The optimization model developed for planning a CDW recycling network is based in Process Systems Engineering principles. The model is able to predict the overall behavior of the system and test design alternatives or changes in the processes.

1.2.1 Problem Definition

The new approach considers the breakdown of processes in recycling facilities as presented in Fig. 1.1.

The facility "Recycling Plant" encompasses three processes: screening, production of low quality (LQ) and high quality (HQ) recycled products. The facility "Sorting Plant" includes only the screening process.

Fig. 1.1 Illustrative scheme of processes in CDW recycling facilities

Besides these facilities the MILP model considers:

- Four types of materials: CDW, Intermediate Products (IP), Sold Products (SP), which are recycled products, and Residual Materials (RM).
- Network nodes with attributes that regard: the CDW amount produced; the existing recycling processes and the corresponding capacities; the existing landfills and capacities.
- Costs that comprise the transportation cost of materials (which is a function of the distance between nodes), the cost of landfilling CDW and RM, the cost of processing CDW and IP and the investment cost in new processes (a cost that depends on the corresponding capacity).
- Revenues that correspond to the sales of recycled products (SP).
- An objective function which is the overall cost of the recycling network (to be minimized); this corresponds to a regulatory entity's perspective regarding the system.

Model results regard:

- The installation of new processes, where the type of process (K1, K2, K3 and K4 in (Fig. 1.2)) and the corresponding capacity is defined.
- The flow of materials in the network, i.e. the amount of materials transported between each pair of nodes (which constitute input and output materials for the recycling processes).

Figure 1.2 shows the layout of all processes and materials considered, following the State Task Network methodology developed in [5].

Management options for the CDW (divided into S1 and S2, concrete and undifferentiated material, respectively) are direct landfilling and recycling. The material to be recycled must first be screened either in a sorting plant (process k1) or in a

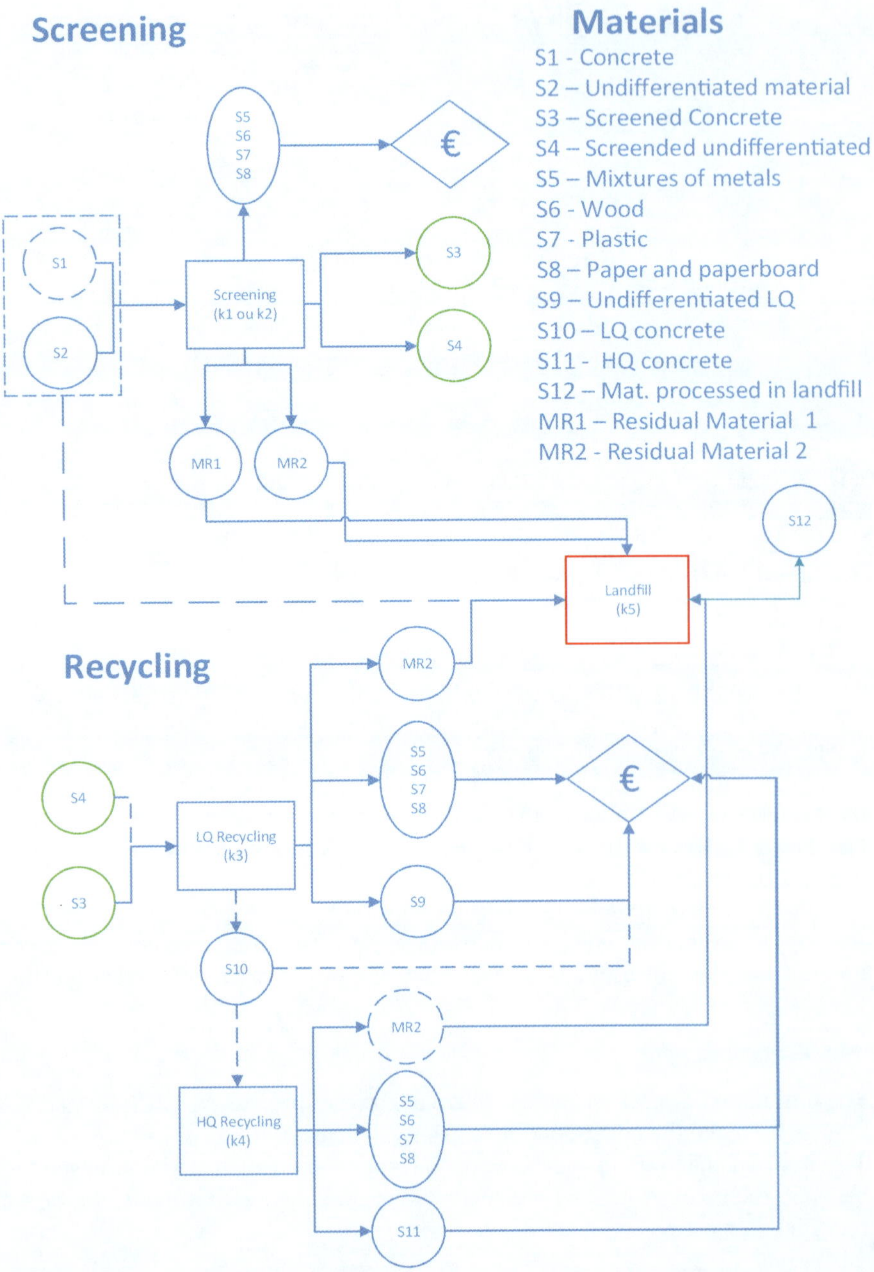

Fig. 1.2 Layout of the CDW network processes. (Note: S1 and S2 materials can only follow the sequences S1→S3→S10→S11 and S2→S4→S9)

recycling plant, which incorporates screening (process k2). This is then followed by the LQ recycling process (k3) and optionally by the HQ recycling process (k4), since the S10 material (LQ concrete) can be sold or continue to be processed into material S11 (HQ concrete), in process K4. Residual materials (RM1 and RM2) and possibly CDW (S1 and S2) are landfilled, which the MILP model considers process K5, whose output is material S12.

1.2.2 Model Description

The objective function to be minimized, total cost, is divided into processes cost and landfilling cost. The processes cost is the sum of operating cost, transportation cost, cost of landfilling residual materials and investment cost of new processes, minus the sales value of recycled products. The landfilling cost is the cost of direct deposition of CDW in landfills.

Model constraints can be summarized as follows:

- Pre-existing recycling processes cannot be closed and only one process, of each type, is allowed to be installed in a given network node.
- The processing capacities of sorting and recycling plants have an upper limit.
- Furthermore, a process capacity must be chosen from a predefined set of capacity values.
- The sale of materials s5–s11 cannot surpass the existing demand at each node.
- The flow of materials between network nodes is determined by mass balance equations at each node and process.
- To initialize mass balances, a virtual process is considered (k0).

1.3 Results and Discussion

The MILP model was programmed in GAMS modeling language and implemented in an Intel©Core^{TM} i7 CPU, 4700MQ, with 2.40 GHz and 8 GB of RAM. It was solved to optimality with CPLEX 12.6.0.1.

At the validation stage, the model was solved for a representative 10 parish sample of the LMA. Then, the LMA case study with its 211 parishes was approached (the parish distribution considered is the one before the administrative reorganization in 2013). As CDW production at nodes (parishes) the demolition waste amounts estimated by [1] were used. The composition of material flows and the processes considered were gathered from [2]. Regarding CDW composition, 30% is concrete material (S1) and 70% undifferentiated material (S2). Other data was collected specifically for this work, namely by contacting companies of the CDW sector.

A [211×211] distance matrix between parishes was computed with ArcGIS software, based on the road network (Fig. 1.3). The distance between each pair of parishes

Fig. 1.3 LMA map displaying parish centroids, parish limits and the roads considered

was computed considering the parish centroids and the minimum cost route between them.

Table 1.1 presents a summary of results (all solutions reported are optimal). Scenarios[1] A-E* correspond to a comparison of model solutions with and without the legal imposition of recycling at least 70% of the amount of CDW produced. Scenarios F-Q correspond to a sensitivity analysis of the critical parameters of the model, where the parameter value was increased and decreased by 20%.

As an example, the analysis of scenarios B and B* is presented in detail in Fig. 1.4, where the different material flows are displayed. These scenarios consider that there are no pre-existing processes in LMA and only new plants can operate.

In scenario B (left hand side of Fig. 1.4), all CDW produced is landfilled (recycling rate is zero) and recycling processes are not installed - this management option is not economically viable. When considering the legal enforcement that at least 70% of the CDW must be recycled (scenario B* - right hand side of Fig. 1.4), the installation of processes k2 and k3 takes place in node i171 (300.000 ton/year) and process k2 in node i22 (150.000 ton/year). The recycling rate is 70%.

Analysis of results for scenarios A to E (without legal enforcement) leads to conclude that the direct disposal of CDW in landfills is preferable to CDW recycling. When legal enforcement is applied, the solution that minimizes costs is also the option of minimum recycling (70%). S2 material, which represents the major flow of CDW,

[1]The symbol "*" means that the legal enforcement of at least 70% of recycled waste was imposed.

Table 1.1 Summary of results

Scenario	Parameters/constraints	No. of processes (by type)				Total cost (M €)	Total cost (%)	Processing cost (M€)	Cost of direct disposal (M€)	Direct disposal rate (%)	Recycling rate (%)
		k1	k2	k3	k4						
A	Current conditions	7	6	6	0	12.27	100	0.62	11.65	93.6	6.4
A*	Current conditions	7	6	6	0	17.16	139.85	13.25	3.91	30	70
B	No pre-existing processes	0	0	0	0	12.41	101.14	0	12.41	100	0
B*	No pre-existing processes	0	2	1	0	19.89	162.10	15.87	4.02	30	70
C	K4 installation	7	6	6	1	13.15	107.17	1.5	11.65	93.6	6.4
C*	K4 installation	7	6	6	1	18.04	147.03	14.13	3.91	30	70
D	Capacity reduction (pre-exist. processes)	7	6	6	0	12.27	100	0.62	11.65	93.6	6.4
D*	Capacity reduction (pre-exist. processes)	7	7	7	0	17.99	146.62	14.08	3.91	30	70
E	Composition of CDW: 50% S1, 50% S2	7	6	6	0	11.04	89.98	1.03	10.01	89.34	10.66
E*	Composition of CDW: 50% S1, 50% S2	7	6	6	0	13.91	113.37	10.02	3.89	30	70
F	(+20%) transportation costs	7	6	6	0	13.25	107.99	0.68	12.56	93.55	6.45
G	(−20%) transportation costs	7	6	6	0	11.29	92.01	0.53	10.76	94.01	5.99
H	(+20%) demand of materials	7	6	6	0	12.27	100	0.62	11.65	93.6	6.4
I	(−20%) demand of materials	7	6	6	0	12.27	100	0.62	11.65	93.6	6.4
J	(+20%) landfill costs (S1,S2)	7	6	6	0	13.68	111.49	0.70	12.98	92.72	7.28

(continued)

Table 1.1 (continued)

Scenario	Parameters/constraints	No. of processes (by type)				Total cost (M €)	Total cost (%)	Processing cost (M€)	Cost of direct disposal (M€)	Direct disposal rate (%)	Recycling rate (%)
		k1	k2	k3	k4						
K	(−20%) landfill costs (S1,S2)	7	6	6	0	10.85	88.43	0.52	10.33	94.65	5.35
L	(+20%)%) landfill costs (MR1, MR2)	7	6	6	0	12.36	100.73	0.57	11.79	95.05	4.95
M	(−20%) landfill costs (MR1, MR2)	7	6	6	0	12.14	98.94	0.78	11.36	90.31	9.69
N	(+20%) sales value of materials	7	6	6	0	12.19	99.34	0.73	11.46	91.46	8.54
O	(−20%) sales value of materials	7	6	6	0	12.33	100.49	0.57	11.76	94.74	5.26
P	(+20%) investment costs	7	6	6	0	12.27	100	0.62	11.65	93.6	6.4
Q	(−20%) investment costs	7	6	6	0	12.27	100	0.62	11.65	93.6	6.4

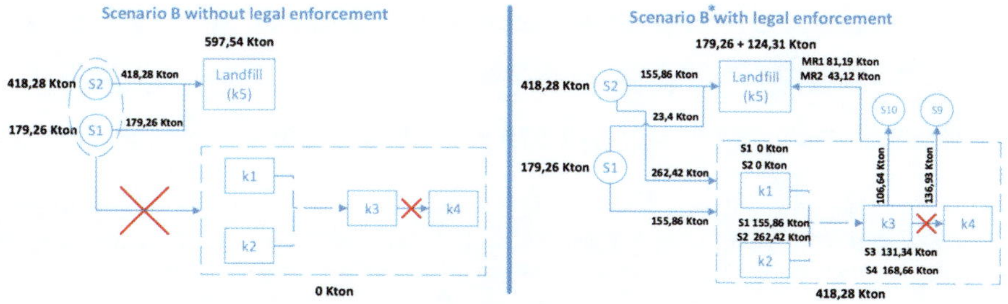

Fig. 1.4 Material flows in scenarios B and B*

is not worthwhile to recycle due to the high costs involved when compared to S1 material. The high capacity of the existing processes in LMA make the installation of new processes unattractive.

Process k4 is not economically feasible, a fact emphasized by scenarios C and C*, where despite imposing the installation of this process, the S10 material is sold instead of entering the HQ recycling process. This reflects what actually occurs in LMA as there is no process of this type installed.

In the remaining scenarios (F-Q), results are consistent with a positive or negative variation of 20% in the parameters' value. The minimum cost solution is the one of scenario K (1085M€), in which the landfilling costs of S1 and S2 had a 20% decrease, and the one of maximum cost (1368M€) is the one in scenario J, that corresponds to an increase of 20% in the same parameter. So, landfill cost is the model parameter with the higher influence on the total cost of the recycling network.

1.4 Conclusion

In this work, an MILP model for optimization of a CDW recycling network was developed. The model considers the breakdown of processes in recycling facilities following an approach based on PSE concepts, and enables decision-making regarding the installation of new processes in the network: where to locate them, what are the process types and the processing capacities. For the LMA case study and regarding the minimization of total cost of the recycling network, direct deposition in landfills is the CDW management option which implies lower costs, and as such is preferred to recycling.

In many countries the increase of disposal costs in landfills proved to be an effective measure to foster the recycling rate, as confirmed in this work in the sensitivity analysis of this model parameter. In order to meet the goal set by the EU for 2020, the analysis of various scenarios was considered for solutions with and without the legal enforcement of recycling at least 70% of the CDW produced. This legal requirement may be applied by a regulatory authority, which can steadily rise the landfilling cost in order to achieve the 2020 objective in a more sustainable way. An additional conclusion is that the HQ recycling process in the LMA does not contribute to the reduction of total network cost.

A future development of this work is to take the perspective of transforming entities (recycling operators) into account, translate it into an objective function and solve the model (with the same constraints). Feasibility of waste management and operation by private entities may thus be evaluated. Another issue to address is the inclusion of the environmental assessment component of global sustainability, as presented in [4]. Moreover, the time dimension may be included, which allows the model to simultaneously optimize network design, process scheduling and materials transportation.

References

1. M. Bernardo, M.C. Gomes, J. Brito, Demolition waste generation for development of a regional management chain model. Waste Manag. **49**, 156–169 (2016)
2. A. Coelho, Análise de viabilidade de implantação de centrais de reciclagem de resíduos da construção e demolição em portugal - PARTE III - análise de viabilidade de uma central de reciclagem. Technical report, Universidade Técnica de Lisboa, Instituto Superior Técnico (2012)
3. A. Coelho, J. Brito, Economic viability analysis of a construction and demolition waste recycling plant in Portugal - part I: location, materials, technology and economic analysis. J. Clean. Prod. **39**, 338–352 (2012)
4. M.N. Correia, M.C. Gomes, J. Duque, Optimization of construction and demolition waste management: application to the Lisbon metropolitan area, *Operations Research and Big Data* (Springer, Berlin, 2015), pp. 33–40
5. J. Duque, A. Barbosa-Póvoa, A.Q. Novais, Design and planning of sustainable industrial networks: application to a recovery network of residual products. Ind. Eng. Chem. Res. **49**, 4230–4248 (2010)
6. M. Hiete, J. Stengel, J. Ludwing, F. Schultmann, Matching construction and demolition waste supply to recycling demand: a regional management chain model. Build. Res. Inf. **39**, 333–351 (2011)
7. M. Mália, J. de Brito, M. Bravo, Indicadores de resíduos de construção e demolição para construções residenciais novas. Ambiente Construído, Porto Alegre **11**, 117–130 (2011)

Chapter 2
Existence of Nash Equilibria on Integer Programming Games

Margarida Carvalho, Andrea Lodi and João Pedro Pedroso

Abstract We aim to investigate a new class of games, where each player's set of strategies is a union of polyhedra. These are called integer programming games. To motivate our work, we describe some practical examples suitable to be modeled under this paradigm. We analyze the problem of determining whether or not a Nash equilibria exists for an integer programming game, and demonstrate that it is complete for the second level of the polynomial hierarchy.

Keywords Integer programming games · Nash equilibria · Computational complexity

2.1 Introduction

Game theory is a generalization of decision theory that takes into account the interaction of multiple decision makers which are concerned about finding their "best" strategies subject to the fact that each controls some, but not all, actions that can take place. See Fudenberg and Tirole [6] for an overview in this field.

Our goal is to study a particular class of games called *integer programming games* (IPG), namely, the existence of "solutions" which in this context are called *Nash equilibria* (NE). We highlight three contributions concerning IPGs: the computational

M. Carvalho (✉)
IVADO Fellow, Canada Excellence Research Chair, École Polytechnique de Montréal, Montreal, QC, Canada
e-mail: margarida.carvalho@polymtl.ca

A. Lodi
Canada Excellence Research Chair, École Polytechnique de Montréal, Montreal, QC, Canada
e-mail: andrea.lodi@polymtl.ca

J. P. Pedroso
Faculdade de Ciências Universidade do Porto and INESC TEC, Porto, Portugal
e-mail: jpp@fc.up.pt

A. I. F. Vaz et al. (eds.), *Operational Research*, Springer Proceedings in Mathematics & Statistics 223, https://doi.org/10.1007/978-3-319-71583-4_2

complexity study of the problem of deciding the existence of a pure NE and of a NE, and the determination of sufficient conditions to guarantee the existence of NE.

Our paper is structured as follows. Section 2.1.1 fixes notation and covers the game theory background. Section 2.1.2 presents examples of integer programming games highlighting the importance of studying the existence of NE. Section 2.2 provides literature review. Section 2.3 classifies the computational complexity of the problems related with the existence of NE to IPGs and states sufficient conditions for NE to exist. Finally, we conclude and discuss further research directions in Sect. 2.4.

2.1.1 Definitions and Notation

We consider games with a finite set of decision makers $M = \{1, 2 \ldots, m\}$, called *players*. Each player $p \in M$ has the set of feasible strategies X^p. We denote the set of all players' strategies combinations by $X = \prod_{p \in M} X^p$ and the operator $(\cdot)^{-p}$ to denote $(\cdot)$ for all players except player p. We call each $x^p \in X^p$ and $x \in X$ a player p's *pure strategy* and a *pure profile of strategies*, respectively. Let player p's payoff function be $\Pi^p(x^p, x^{-p})$. Our investigation focuses on simultaneous *non-cooperative complete information* games, i.e., players play simultaneously, they are self-interested and have full information of each other payoffs and strategies. Each player aims to maximize her payoff, which is influenced by other participants' decisions. In other words, each player p's goal is to select her *best response* against the opponents' strategies x^{-p} by solving the following mathematical programming problem:

$$\underset{x^p \in X^p}{\text{maximize}} \quad \Pi^p(x^p, x^{-p}). \tag{2.1}$$

A pure profile of strategies $x \in X$ that solves the optimization problem (2.1) for all players is called *pure equilibrium*. A game may fail to have pure equilibria and, therefore, a broader solution concept for a game must be introduced. To that end, we recall some basic concepts of measure theory. Let Δ^p denote the space of Borel probability measures over X^p and $\Delta = \prod_{p \in M} \Delta^p$. Each player p's expected payoff for a profile of strategies $\sigma \in \Delta$ is

$$\Pi^p(\sigma) = \int_{X^p} \Pi^p(x^p, x^{-p}) d\sigma. \tag{2.2}$$

A *Nash equilibrium* (NE) is a profile of strategies $\sigma \in \Delta$ such that

$$\Pi^p(\sigma) \geq \Pi^p(x^p, \sigma^{-p}), \qquad \forall p \in M \qquad \forall x^p \in X^p. \tag{2.3}$$

In a NE each player p's expected payoff from σ cannot be improved by unilaterally deviating to a different strategy.[1]

The *support of a strategy* $\sigma^p \in \Delta^p$, denoted as $\text{supp}(\sigma^p)$, is the set of player p's strategies played with positive probability, i.e., $\text{supp}(\sigma^p) = \{x^p \in X^p : \quad \sigma^p(x^p) > 0\}$. Given $\sigma \in \Delta$, if each player's support size is 1, then it is a pure profile of strategies, otherwise, we call it (strictly) mixed. For the sake of simplicity, whenever the context makes it clear, we use the term (strategy) profile to refer to a pure one.

A game is called *continuous* if each player p's strategy set X^p is a nonempty compact metric space and the payoff $\Pi^p(x^p, x^{-p})$ is continuous.

A *separable game* is a continuous game with the payoff functions taking the form

$$\Pi^p(x^p, x^{-p}) = \sum_{j_1=1}^{k_1} \dots \sum_{j_m=1}^{k_m} a^p_{j_1 \dots j_m} f^1_{j_1}(x^1) \dots f^m_{j_m}(x^m), \tag{2.4}$$

where $a^p_{j_1 \dots j_m} \in \mathbb{R}$ and the f^p_j are real-valued continuous functions.

A game is *finite* if the X^p are finite and the $\Pi^p(x^p, x^{-p})$ are arbitrary. Stein et al. [18] state that finite games are special cases of separable games. *Normal-form games* (or strategic-form games) are finite games represented through a multidimensional matrix with an entry for each pure profile of strategies $x \in X$, where that entry is an m dimensional vector of the players' payoffs associated with x.

Based on the definition presented by Köppe et al. [9], we define an *integer programming game* (IPG) as a game with $X^p = \{x^p : A^p x^p \leq b^p, \quad x^p_i \in \mathbb{N} \text{ for } i = 1, \dots, B_p\}$, where A^p is a $r_p \times n_p$ matrix (with $n_p \geq B_p$), b^p a column vector of dimension r_p, and the payoff functions $\Pi^p(x^p, x^{-p})$ are continuous and can be evaluated in polynomial time. Note that IPGs contain mathematical programming problems in the special case of a single player.

Observe that any finite game can be modeled as an IPG: associate a binary variable for each player's strategy (which would model the strategy selected), add a constraint summing the decision variables up to one (this ensures that exactly one strategy is selected) and formulate the players' payoffs according to the payoff values for combinations of the binary variables[2]; moreover, these payoff functions are continuous, since we can endow X with the discrete metric, which makes any function automatically continuous. In an IPG, strategies' sets can be unbounded and thus not compact, which makes the IPG class not contained in that of continuous games. If the strategies' sets in an IPG are nonempty and bounded, then the X^p are finite unions of convex polyhedra which are compact metric spaces and thus, the IPG is a continuous game. Figure 2.1 displays the connections between these game classes.

[1] The equilibrium conditions (2.3) only reflect a player p deviation to strategy in X^p and not in Δ^p, because a strategy in Δ^p is a convex combination of strategies in X^p, and thus cannot lead to a better payoff than one in X^p.

[2] In specific, a player's payoff is a summation over all pure profiles of strategies, where each term is the product of the associated binary variables and the associated payoff.

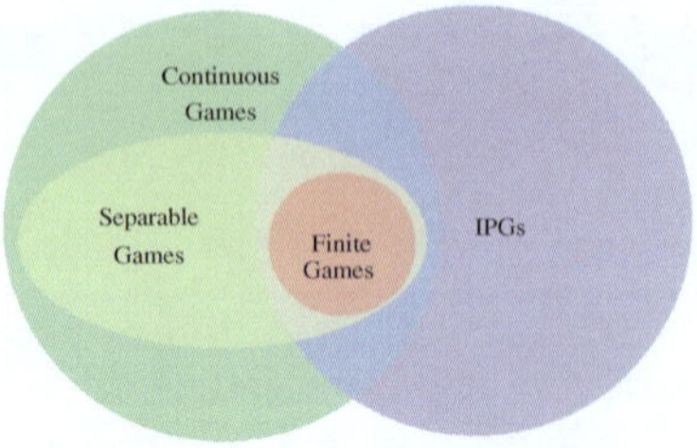

Fig. 2.1 Game classes

2.1.2 Examples

Next, we describe two games: the *knapsack game* which is the simplest purely integer programming game that one could devise, the *competitive lot-sizing game* which has practical applicability in production planning.

Knapsack Game

A simple and natural IPG would be one with linear payoff functions for the players. Under this setting, each player p aims to solve

$$\max_{x^p \in \{0,1\}^n} \sum_{i=1}^{n} v_i^p x_i^p + \sum_{k=1,k\neq p}^{m} \sum_{i=1}^{n} c_{k,i}^p x_i^p x_i^k \tag{2.5a}$$

$$\text{s. t.} \quad \sum_{i=1}^{n} w_i^p x_i^p \leq W^p, \tag{2.5b}$$

where the parameters v_i^p, $c_{k,i}^p$, w_i^p and W^p are real numbers. This model can describe situations where m entities aim to decide in which of n projects to invest such that each entity budget constraint (2.5b) is satisfied and the associated payoffs are maximized (2.5a). The second term of the payoff (2.5a) reflects the opponents' influence: if player p and player k select project i ($x_i^p = x_i^k = 1$) then, player p earns $c_{k,i}^p > 0$ or loses $c_{k,i}^p < 0$.

In Carvalho [2] mathematical programming tools are used to compute some refined equilibria of this game.

Competitive Lot-sizing Game

The competitive lot-sizing game is a Cournot competition played through T periods by a set of firms (players) that produce the same good; see [2] for details. Each player has to plan her production as in the lot-sizing problems (see [17]) but, instead of satisfying a known demand in each period of the time horizon, the the market price depends on the total quantity of the good that is sold in the market. Each player p has to decide how much will be produced in each time period t (production variable x_t^p) and how much will be placed in the market (variable q_t^p). For producing a positive quantity, player p must pay a fixed and a proportional amount (setup and variable costs, respectively). A producer can build inventory (variable h_t^p) by producing in advance and incurring in an inventory cost. In this way, we obtain the following model for each player (producer) p:

$$\max_{y^p \in \{0,1\}^T, x^p, q^p, h^p} \quad \sum_{t=1}^{T} P_t(q_t) q_t^p - \sum_{t=1}^{T} F_t^p y_t^p - \sum_{t=1}^{T} C_t^p x_t^p - \sum_{t=1}^{T} H_t^p h_t^p \tag{2.6a}$$

$$\text{s. t.} \quad x_t^p + h_{t-1}^p = h_t^p + q_t^p \quad \text{for } t = 1, \ldots, T \tag{2.6b}$$

$$0 \leq x_t^p \leq M_t^p y_t^p \quad \text{for } t = 1, \ldots, T \tag{2.6c}$$

where F_t^p is the setup cost, C_t^p is the variable cost, H_t^p is the inventory cost and M_t^p is the production capacity for period t; $P_t(q_t) = a_t - b_t \sum_{j=1}^{m} q_t^j$ is the unit market price. The payoff function (2.6a) is player p's total profit; constraints (2.6b) model product conservation between periods; constraints (2.6c) ensure that the quantities produced are non-negative and whenever there is production ($x_t^p > 0$), the binary variable y_t^p is set to 1 implying the payment of the setup cost F_t^p. In this game, the players influence each other through the unit market price function $P_t(q_t)$.

2.2 Literature Review

Nash [13] defined the most widely accepted concept of solution for non-cooperative games, the Nash equilibrium. From the definition given in the previous section, a NE associates a probability distribution to each player's set of strategies such that no player has incentive to unilaterally deviate from that NE if the others play according with the equilibrium. In other words, in an equilibrium, simultaneously each player is maximizing her expected payoff given the equilibrium strategies of the other players. In a pure NE only a single strategy of each player has positive probability assigned (i.e., probability 1).

The state-of-the-art game theory tools are confined to "well-behaved" continuous games, where payoff functions and strategy sets meet certain differentiability and concavity conditions, and normal-form games.

The class of continuous games contains a wide range of relevant games. Glicksberg [8] proved that every continuous game has a NE. However, literature focuses in continuous games for which the strategies sets are convex and payoffs are quasi-concave, since by Debreu, Glicksberg and Fan's famous theorem there is a pure NE which computation can be reduced to a constrained problem by the application of the Karush–Kuhn–Tucker (KKT) conditions to each player's optimization problem. In this context, an outlier is the work in [18], where the support size of general separable games is studied. Note that the tools mentioned above are not valid, for separable games since they may fail to satisfy concavity conditions.

Finite games have received wide attention in game theory. Nash [13] proved that any finite game has a NE. Daskalis et al. [5] proved that computing a NE is

PPAD-complete, which is believed to be a class of hard problems, since it is unlikely that $PPAD$ (Polynomial Parity Arguments on Directed graphs) is equal to the polynomial time complexity class P. Nisan et al. [14] describe general algorithms to compute Nash equilibria, which failed to run in polynomial time. We refer the interested reader to the surveys and state-of-art algorithms collected in von Stengel [19]. Currently, some of those algorithms are available on GAMBIT [12], the most up-to-date software for the computation of NE for normal-form games.

On the other hand, enumerating all players' feasible strategies (as in finite games) for an IPG can be impractical, or the players' strategies in an IPG might lead to non well-behaved games, for example where the player's maximization problems are non-concave. This shows that the existent tools and standard approaches for finite games and convex games are not directly applicable to IPGs.

The literature in IPGs is scarce and often focus in the particular structure of specific games. Kostreva [10] and Gabriel et al. [7] propose methods to compute NE for IPGs, however it lacks a computational time complexity guarantee and a practical validation through computational results. Köppe et al. [9] present a polynomial time algorithm to compute pure NE under restrictive conditions, like number of players fixed and sum of the number of player's decision variables fixed, to name few. There are important real-world IPGs, in the context of e.g., electricity markets [16], production planning [11], health-care [3]; this highlights the importance of exploring such game models.

2.3 Existence of Nash Equilibria

It can be argued that players' computational power is bounded and thus, since the space of pure strategies is simpler and contained in the space of mixed strategies – i.e., the space of Borel probability measures – pure equilibria are more plausible outcomes for games with large sets of pure strategies. In this way, it is important to understand the complexity of determining a pure equilibrium to an IPG.

According with Nash famous theorem [13] any finite game has a Nash equilibrium. Since a purely integer bounded IPGs is a finite game, it has a NE. However, Nash theorem does not guarantee that the equilibrium is pure, which is illustrated in the following example.

Example 2.1 (*No pure Nash equilibrium*) Consider the two-player game such that player A solves

$$\underset{x^A}{\text{maximize}} \quad 18x^A x^B - 9x^A \quad \text{subject to} \quad x^A \in \{0, 1\}$$

and player B:

$$\underset{x^B}{\text{maximize}} \quad -18x^A x^B + 9x^B \quad \text{subject to} \quad x^B \in \{0, 1\}.$$

Under the profile $(x^A, x^B) = (0, 0)$ player B has incentive to change to $x^B = 1$; for the profile $(x^A, x^B) = (1, 0)$ player A has incentive to change to $x^A = 0$; for the profile $(x^A, x^B) = (0, 1)$ player A has incentive to change to $x^A = 1$; for the profile $(x^A, x^B) = (1, 1)$ player B has incentive to change to $x^B = 0$. Thus there is no pure NE. The (mixed) NE is $\sigma^A = \sigma^B = (\frac{1}{2}, \frac{1}{2})$ with expected payoff of zero for both players.

In Sect. 2.3.1, we classify both the computational complexity of deciding if there is a pure and a mixed NE for an IPG. It will be shown that even with linear payoffs and two players, the problem is Σ_2^p-complete. Then, in Sect. 2.3.2, we state sufficient conditions for the game to have finitely supported Nash equilibria.

2.3.1 Complexity of the Existence of NE

The complexity class Σ_2^p contains all decision problems that can be written in the form $\exists x \forall y P(x, y)$, that is, as a logical formula starting with an existential quantifier followed by a universal quantifier followed by a Boolean predicate $P(x, y)$ that can be evaluated in polynomial time; see Chap. 17 in Papadimitriou's book [15].

Theorem 2.1 *The problem of deciding if an IPG has a pure NE is* Σ_2^p*-complete.*

Proof The decision problem is in Σ_2^p, since we are questing if there is a solution in the space of pure strategies such that for any unilateral deviation of a player, her payoff is not improved (and evaluating the payoff value for a profile of strategies can be done in polynomial time).

It remains to prove Σ_2^p-hardness, that is all problems in Σ_2^p can be reduced in polynomial time to the problem of deciding if an IPG has a pure NE. The following problem is Σ_2^p-complete (see Caprara et al. [1]) and we will reduce it to a problem of deciding if an IPG has a pure NE:

DeNegre bilevel Knapsack Problem - DN

INSTANCE Non-negative integers n, A, B, and n-dimensional non-negative integer vectors a and b.

QUESTION Is there a binary vector x such that $\sum_{i=1}^{n} a_i x_i \leq A$ and for all binary vectors y with $\sum_{i=1}^{n} b_i y_i \leq B$, the following inequality is satisfied

$$\sum_{i=1}^{n} b_i y_i (1 - x_i) \leq B - 1?$$

Our reduction starts from an instance of DN. We construct the following instance of IPG.

- The game has two players, $M = \{Z, W\}$.
- Player Z controls a binary decision vector z of dimension $2n + 1$; her set of feasible strategies is

$$\begin{aligned}\sum_{i=1}^{n} a_i z_i &\leq A \\ z_i + z_{i+n} &\leq 1 \quad i = 1, \ldots, n \\ z_{2n+1} + z_{i+n} &\leq 1 \quad i = 1, \ldots, n.\end{aligned}$$

- Player W controls a binary decision vector w of dimension $n + 1$; her set of feasible strategies is

$$Bw_{n+1} + \sum_{i=1}^{n} b_i w_i \leq B. \tag{2.7}$$

- Player Z's payoff is $(B - 1)w_{n+1}z_{2n+1} + \sum_{i=1}^{n} b_i w_i z_{i+n}$.
- Player W's payoff is $(B - 1)w_{n+1} + \sum_{i=1}^{n} b_i w_i - \sum_{i=1}^{n} b_i w_i z_i - \sum_{i=1}^{n} b_i w_i z_{i+n}$.

We claim that in the constructed instance of IPG there is an equilibrium if and only if the *DN* instance has answer YES.

(Proof of if). Assume that the *DN* instance has answer YES. Then, there is x satisfying $\sum_{i=1}^{n} a_i x_i \leq A$ such that $\sum_{i=1}^{n} b_i y_i(1 - x_i) \leq B - 1$. Choose as strategy for player Z, $\widehat{z} = (x, \overbrace{0, \ldots, 0}^{n}, 1)$ and for player W $\widehat{w} = (\overbrace{0, \ldots, 0}^{n}, 1)$. We will prove that $(\widehat{z}, \widehat{w})$ is an equilibrium. First, note that these strategies are guaranteed to be feasible for both players. Second, note that none of the players has incentive to deviate from $(\widehat{z}, \widehat{w})$:

- Player Z's payoff is $B - 1$, and $B - 1 \geq \sum_{i=1}^{n} b_i w_i$ holds for all the remaining feasible strategies w of player W.
- Player W's has payoff $B - 1$ which is the maximum possible given $\widehat{z}$.

(Proof of only if). Now assume that the IPG instance has answer YES. Then, there is a pure equilibrium $(\widehat{z}, \widehat{w})$.

If $\widehat{w}_{n+1} = 1$, then, by (2.7), $\widehat{w} = (\overbrace{0, \ldots, 0}^{n}, 1)$. In this way, since player Z maximizes her payoff in an equilibrium, $\widehat{z}_{2n+1} = 1$, forcing $\widehat{z}_{i+n} = 0$ for $i = 1, \ldots, n$. The equilibrium inequalities (2.3), applied to player W, imply that, for any of her feasible strategies w with $w_{n+1} = 0$, $B - 1 \geq \sum_{i=1}^{n} b_i w_i (1 - \widehat{z}_i)$ holds, which shows that *DN* is a YES instance with the leader selecting $x_i = \widehat{z}_i$ for $i = 1, \ldots, n$.

If $\widehat{w}_{n+1} = 0$, under the equilibrium strategies, player Z's payoff term $(B - 1)\widehat{w}_{n+1}z_{2n+1}$ is zero. Thus, since in an equilibrium player Z maximizes her payoff, it holds that $\widehat{z}_{i+n} = 1$ for all $i = 1, \ldots, n$ with $\widehat{w}_i = 1$. However, this implies that player W's payoff is non-positive given the profile $(\widehat{z}, \widehat{w})$. In this way, player W would strictly improve her payoff by unilaterally deviating to $w = (\overbrace{0, \ldots, 0}, 1)$.

In conclusion, w_{n+1} is never zero in a pure equilibrium of the constructed game instance. □

Extending the existence property to mixed equilibria would increase the chance of an IPG to have a NE, and thus, a solution. Next theorem shows that the problem remains Σ_2^p-complete.

Theorem 2.2 *The problem of deciding if an IPG has a NE is* Σ_2^p*-complete.*

Proof Analogously to the previous proof, the problem belongs to Σ_2^p.

It remains to show that it is Σ_2^p-hard. We will reduce the following Σ_2^p-complete to it (see [1]):

Dempe Ritcht Problem - DR

INSTANCE Non-negative integers n, A, C and C', and n-dimensional non-negative integer vectors a and b.

QUESTION Is there a value for x such that $C \leq x \leq C'$ and for all binary vectors satisfying $\sum_{i=1}^{n} b_i y_i \leq x$, the following inequality holds

$$Ax + \sum_{i=1}^{n} a_i y_i \geq 1?$$

Our reduction starts from an instance of DR. We construct the following instance of IPG.

- The game has two players, $M = \{Z, W\}$.
- Player Z controls a non-negative variable z and a binary decision vector $(z_1, \ldots, z_{n+1})$; her set of feasible strategies is

$$\begin{aligned} \sum_{i=1}^{n} b_i z_i &\leq z \\ z_i + z_{n+1} &\leq 1, \qquad\qquad i = 1, \ldots, n \\ z &\leq C'(1 - z_{n+1}) \\ z &\geq C(1 - z_{n+1}). \end{aligned}$$

- Player W controls a non-negative variable w and binary decision vector $(w_1, \ldots, w_n)$.
- Player Z's payoff is $Az + \sum_{i=1}^{n} a_i z_i w_i + z_{n+1}$.
- Player W's payoff is $z_{n+1} w + \sum_{i=1}^{n} b_i w_i z_i$.

We claim that in the constructed instance of IPG there is an equilibrium if and only if the DR instance has answer YES.

(Proof of if). Assume that the *DR* instance has answer YES. Then, there is x such that $C \leq x \leq C'$ and $Ax + \sum_{i=1}^{n} a_i y_i \geq 1$ for a y satisfying $\sum_{i=1}^{n} b_i y_i \leq x$. As strategy for player Z choose $\widehat{z} = C'$ and $(\widehat{z}_1, \ldots, \widehat{z}_n, \widehat{z}_{n+1}) = (y_1, \ldots, y_n, 0)$; for player W choose $\widehat{w} = 0$ and $(\widehat{w}_1, \ldots, \widehat{w}_n) = (y_1, \ldots, y_n)$. We prove that $(\widehat{z}, \widehat{w})$ is an equilibrium. First, note that these strategies are guaranteed to be feasible for both players. Second, note that none of the players has incentive to deviate from $(\widehat{z}, \widehat{w})$:

- Player Z's payoff cannot be increased, since it is equal or greater than 1 and for $i = 1, \ldots, n$ such that $\widehat{z}_i = 0$ the payoff coefficients are zero.
- Analogously, player W's payoff cannot be increased, since for $i = 1, \ldots, n$ such that $\widehat{w}_i = 0$ the payoff coefficients are zero and the payoff coefficient of $\widehat{z}_{n+1}\widehat{w}$ is also zero.

(Proof of only if). Assume that *DR* is a NO instance. Then, for any x in $\left[C, C'\right]$ the leader is not able to guarantee a payoff of 1. This means that in the associated IPG, player Z has incentive to choose $z = 0$ and $(z_1, \ldots, z_n, z_{n+1}) = (0, \ldots, 0, 1)$. However, this player Z's strategy leads to a player W's unbounded payoff. In conclusion, there is no equilibrium. □

In the proof of Theorem 2.2, it is not used the existence of a mixed equilibrium to the constructed IPG instance. Therefore, it implies Theorem 2.1. The reason for presenting these two theorems is because in Theorem 2.1, the reduction is a game where the players have finite sets of strategies, while in Theorem 2.2, in the reduction, a player has an unbounded set of strategies.

2.3.2 Conditions for the Existence of NE

Glicksberg [8] and Stein et al. [18] provide results on the existence and characterization of equilibria for continuous and separable games (recall the definitions in Sect. 2.1.1), which we will apply to IPGs. Their proofs rely on the fact that the payoff functions have the form (2.4), enabling to describe the "degree"of interdependence among players.

In an IPG, each player p's strategy set X^p is a nonempty compact metric space if X^p is bounded and nonempty. This together with the fact that in Sect. 2.1.1 we assumed that each player's payoff is continuous, allow us to conclude the following:

Lemma 2.1 *Every IPG such that X^p is nonempty and bounded is a continuous game.*

Given that every continuous game has a NE [8],

Theorem 2.3 *Every IPG such that X^p is nonempty and bounded has a Nash equilibrium.*

Applying Stein et al. [18] results, we obtain the following:

Theorem 2.4 *For any Nash equilibrium σ of a separable IPG, there is a Nash equilibrium τ such that each player p mixes among at most $k_p + 1$ pure strategies and $\Pi^p(\sigma) = \Pi^p(\tau)$.*

Proof Apply Theorem 2.8 of [18] to a separable IPG. □

If in an IPG each player's set of strategies X^p is bounded and the payoff takes the form (2.4), IPG is separable. Assuming that these two conditions are satisfied (so that Theorems 2.3 and 2.4 hold) is not too strong when modeling real-world applications. In other words, the players' strategies are likely to be bounded due to limitations in the players' resources, which guarantees that an IPG has an equilibrium (Theorem 2.3). For instance, recall the knapsack game and the competitive lot-sizing game from Sect. 2.1.2 in which each player's set of strategies is bounded. In the knapsack game, payoffs are linear, thus by Theorem 2.4, we deduce that the bound on the equilibria supports for each player is $n + 1$.

Interesting IPGs, the competitive lot-sizing game (recall Sect. 2.1.2), has quadratic payoff functions that can be written in the form (2.4).

Corollary 2.1 *Let IPG be such that X^p is nonempty and bounded, and*

$$\Pi^p(x^p, x^{-p}) = c^p x^p + \sum_{k \in M} (x^k)^\intercal Q_k^p x^p, \tag{2.8}$$

where $c^p \in \mathbb{R}^{n_p}$ and Q_k^p is a $n_k \times n_p$ real matrix. Then, for any Nash equilibrium σ there is a Nash equilibrium τ such that each player p mixes among at most $1 + n_p + \frac{n_p(n_p-1)}{2}$ pure strategies and $\Pi^p(\sigma) = \Pi^p(\tau)$.

Proof In order to write player p's payoff in the form (2.4), there must be a function $f_{j_p}^p(x^p)$ for $1, x_1^p, \ldots, x_{n_p}^p, x_1^p x_1^p, x_1^p x_2^p, \ldots, x_1^p x_{n_p}^p, x_2^p x_2^p, \ldots, x_{n_p}^p x_{n_p}^p$; thus, $k_p = 1 + n_p + \frac{n_p(n_p-1)}{2}$ in Theorem 2.4.

The thesis [2] presents an algorithmic approach that uses the fact that we can restrict our investigations to finitely supported NE.

2.4 Conclusions and Further Directions

Literature in non-cooperative game theory lacks the study of games with diverse sets of strategies with practical interest. This paper is a first attempt to address the computational complexity and existence of equilibria to integer programming games.

We classified the game's complexity in terms of existence of pure and mixed equilibria. For both cases, it was proved that the problems are Σ_2^p-complete. However, if the players' set of strategies is bounded, the game is guaranteed to have an equilibrium. Chen et al. [4] proved that computing a NE for a finite game is PPAD-complete even with only two players. Thus, recalling Fig. 2.1, computing a NE to a separable IPG is PPAD-hard. Even when there are equilibria, the computation of one is a PPAD-hard problem, which is likely to be a class of hard problems. Furthermore, the PPAD class does not seem to provide a tight classification of the computational complexity of computing an equilibrium in IPGs. In fact, the PPAD class has its root in finite games that are an easier class of games, in comparison with general IPGs. Note that for IPGs, verifying if a profile of strategies is an equilibrium implies solving each player's best response optimization, which can be a NP-complete problem, while for finite games this computation can be done efficiently. In this context, it would be interesting to explore the definition of a "second level PPAD" class, that is, a class of problems for which a solution could be verified in polynomial time if there was access to a NP oracle.

In this paper, we also determined sufficient conditions for the existence of equilibria on IPGs. Moreover, these theoretical results enabled us to conclude that the support of a NE is finite. This is a key result in the correctness of the algorithm that computes an equilibrium for an IPG presented in [2]. Future work in this context should address the question of determining all equilibria, computing an equilibrium satisfying a specific property (e.g., computing the equilibrium that maximizes the social welfare, computing a non-dominated equilibrium) and equilibria refinements or new solution concepts under a games with multiple equilibria. From a mathematical point of view, the first two questions embody a big challenge, since it seems to be hard to extract problem structure to the general IPG class of games. The last question raises another one, which is the possibility of considering different solution concepts to IPGs.

Acknowledgements Part of this work was performed while the first author was in the Faculty of Sciences University of Porto and INESC TEC. The first author thanks the support of Institute for data valorisation (IVADO), the Portuguese Foundation for Science and Technology (FCT) through a PhD grant number SFRH/BD/79201/2011 and the ERDF European Regional Development Fund through the Operational Programme for Competitiveness and Internationalisation - COMPETE 2020 Programme within project POCI-01-0145-FEDER-006961, and National Funds through the FCT (Portuguese Foundation for Science and Technology) as part of project UID/EEA/50014/2013. We thank the referees for comments and questions that helped clarifying the presentation.

References

1. A. Caprara, M. Carvalho, A. Lodi, G.J. Woeginger, A study on the computational complexity of the bilevel knapsack problem. SIAM J. Optim. **24**(2), 823–838 (2014)
2. M. Carvalho, Computation of equilibria on integer programming games, Ph.D. thesis, Faculdade de Ciências da Universidade do Porto (2016)
3. M. Carvalho, A. Lodi, J.P. Pedroso, A. Viana, Nash equilibria in the two-player kidney exchange game. Math. Programm. **161**(1–2), 1–29 (2016). https://doi.org/10.1007/s10107-016-1013-7. ISSN 1436-4646
4. X. Chen, X. Deng, Settling the complexity of two-player nash equilibrium. in *47th Annual IEEE Symposium on Foundations of Computer Science, 2006. FOCS '06* (2006), pp. 261–272. https://doi.org/10.1109/FOCS.2006.69

5. C. Daskalakis, P. Goldberg, C. Papadimitriou, The complexity of computing a nash equilibrium. SIAM J. Comput. **39**(1), 195–259 (2009). https://doi.org/10.1137/070699652
6. D. Fudenberg, J. Tirole, *Game Theory* (MIT Press, Cambridge, 1991)
7. S.A. Gabriel, S. Ahmad Siddiqui, A.J. Conejo, C. Ruiz, Solving discretely-constrained nash-cournot games with an application to power markets. Netw. Spat. Econ. **13**(3), 307–326 (2013). https://doi.org/10.1007/s11067-012-9182-2. ISSN 1566-113X
8. I.L. Glicksberg, A further generalization of the kakutani fixed point theorem, with application to nash equilibrium points. Proc. Am. Math. Soc. **3**(1), 170–174 (1952). https://doi.org/10.2307/2032478. ISSN 00029939
9. M. Köppe, C. Thomas Ryan, M. Queyranne, Rational generating functions and integer programming games. Op. Res. **59**(6), 1445–1460 (2011). https://doi.org/10.1287/opre.1110.0964. ISSN 0030-364X
10. M.M. Kostreva, Combinatorial optimization in nash games. Comput. Math. Appl. **25**(10–11), 27–34 (1993), URL http://www.sciencedirect.com/science/article/pii/0898122193902784. https://doi.org/10.1016/0898-1221(93)90278-4. ISSN 0898-1221
11. H. Li, J. Meissner, Competition under capacitated dynamic lot-sizing with capacity acquisition. Int. J. Prod. Econ. **131**(2), 535–544 (2011), URL http://www.sciencedirect.com/science/article/pii/S0925527311000375. https://doi.org/10.1016/j.ijpe.2011.01.022. ISSN 0925-5273.
12. R.D. Mckelvey, A.M. Mclennan, T.L. Turocy, Gambit: software tools for game theory. Version 16.0.0 (2016), http://www.gambit-project.org
13. J. Nash, Non-cooperative games. Ann. Math. **54**(2), 286–295 (1951)
14. N. Nisan, T. Roughgarden, E. Tardos, V.V. Vazirani, *Algorithmic Game Theory* (Cambridge University Press, New York, 2007)
15. C.H. Papadimitriou, *Computational Complexity* (Addison-Wesley, Reading, 1994). ISBN 0201530821
16. M.V. Pereira, S. Granville, M.H.C. Fampa, R. Dix, L.A. Barroso, Strategic bidding under uncertainty: a binary expansion approach. IEEE Trans. Power Syst. **20**(1), 180–188 (2005). https://doi.org/10.1109/TPWRS.2004.840397. ISSN 0885-8950
17. Y. Pochet, L.A. Wolsey, *Production Planning by Mixed Integer Programming*, Springer Series in Operations Research and Financial Engineering (Springer, New York, 2006). ISBN 0387299599
18. N.D. Stein, A. Ozdaglar, P.A. Parrilo, Separable and low-rank continuous games. Int. J. Game Theory **37**(4), 475–504 (2008). https://doi.org/10.1007/s00182-008-0129-2. ISSN 0020-7276
19. B. von Stengel, Special issue of on computation of nash equilibria in finite games. Econ. Theory **42**(1), 1–7 (2010). https://doi.org/10.1007/s00199-009-0452-2

Chapter 3
The Reverse Logistics of Unsold Medications in Pharmacies in Campania, Italy

Rosekelly Araújo Costa, Teresa Pereira and Isabel Cristina Lopes

Abstract This paper is a study in Reverse Logistics (RL) that aims to analyse the reverse flow of medications with expired dates, in the pharmacies of the Campania region in Italy. The main objective is to analyse the final destination of medications that are not sold and are collected in pharmacies. The analysis of how the company responsible for the collection of the medications works was made using semi-structured interviews, and a subsequent factor analysis of the collected data. The pharmacies of the main cities of this region were investigated, in order to understand their importance in this process, as well as to understand their main difficulties and challenges. A statistical analysis of the data allowed us to verify how pharmacies are accustomed to the current legislation and are aware of the importance of their role in the RL of the medications that are not sold due to expired date. It was observed that pharmacies are very satisfied with the company responsible for the collection and referral of medications and their materials to an adequate final destination. Both of them work in tune, respond well to current legislation and respect the environment.

Keywords Reverse logistics · Expired date medications · Pharmacies · Region of campania - Italy · Factor analysis

---

R. A. Costa
ISCAP - Accounting and Business School, Polytechnic of Porto, Porto, Portugal
e-mail: rosekelly12@hotmail.com

T. Pereira
ISEP - School of Engineering, and CIDEM - Centro de Investigação e Desenvolvimento em Engenharia Mecânica, Polytechnic of Porto, Porto, Portugal
e-mail: mtp@isep.ipp.pt

I. C. Lopes (✉)
LEMA - Mathematical Engineering Lab, and CEOS.PP, ISCAP - Accounting and Business School, Polytechnic of Porto, Porto, Portugal
e-mail: cristinalopes@iscap.ipp.pt

A. I. F. Vaz et al. (eds.), *Operational Research*, Springer Proceedings in Mathematics & Statistics 223, https://doi.org/10.1007/978-3-319-71583-4_3

3.1 Introduction

This research consists of a case study in the field of Reverse Logistics (RL), analysing the reverse flow of medications with expired dates in pharmacies in the region of Campania, Italy. The final destination of non-sold medications collected in the pharmacies is an important issue, because they are chemical substances that directly affect our planet, causing a strong environmental degradation and a threat in the quality of life of the natural and social ecosystem.

Decisions related to the reuse of non-renewable resources or the disposal of waste negatively affect the environment. Sustainable development is often translated by the use of methods such as recycling, reuse, recovery and waste management [18]. Lately, consumers and governments have pushed companies hard to reduce environmental impact on their processes and production [1]. To make these methods feasible (collection, treatment, recycling and environmentally safe), companies launch them into the reverse distribution channels that are incorporated into the business cycle, or production cycle, goods that are already in the post-sale or post-consumption phase. It is then important to plan, operate and control these flows and inverse information, so that there is aggregation of value, not only economic, but also ecological, social, global and logistic [8].

The main objectives of this work is to analyse the RL context in which the collection of expired date medications is inserted, to understand how the system of operations in the case study works, to compare it with the RL theory, and finally to understand what influences the pharmacies to practice RL. The methodology used is the case study, using as instruments of data collection the semi-structured interview with a company that operates in reverse logistics and the survey sent to pharmacies. The data gathered with the survey was subject to a statistical treatment with a principal component analysis (PCA) and factor analysis (FA). According to Govindan [5], survey based analysis may lead to introduce new variables and constraints in decision making models for reverse supply chains, and therefore the contribution of this work to the Operations Research literature is the identification of important issues to be accounted for when forecasting demand and revenues in the pharmacies sector.

Section 3.2 reviews the literature on RL concerning pharmacies supply chain. In Sect. 3.3, the methodology of this work is described in detail and in Sect. 3.4 the results of the interview are presented, describing the RL process. The discussion of the results of the survey to the pharmacies is in Sect. 3.5 and its conclusions are presented in Sect. 3.6.

3.2 Literature Review

In recent years, reverse logistics (RL) has become a field of importance for all organizations due to growing environmental concerns, legislation, corporate social responsibility and sustainable competitiveness [1]. Bazan et al. [4] emphasise that the responsible management of product return flows in production and inventory environments is a rapidly increasing requirement for companies.

Ballou [3] refers that logistics is composed of important activities that add value to utility, time and place, such as: Transportation, Inventory Management, Order Processing, Getting the Product, Product Programming, Protection Pack, Material Handling, Storage and Information Maintenance. The supply chain (SC) is defined by Ballou [3] as a set of cross-functional activities (transport, inventory control, etc.) that are repeated numerous times throughout the channel by which to convert raw materials into finished products.

Reverse logistics includes all of the activities that are mentioned in the above definition. The difference is that reverse logistics encompasses these activities and the information related as they operate in reverse. Therefore, reverse logistics is defined by Hawks [7] as the process of planning, implementing, and controlling the efficient, cost effective flow of raw materials, in-process inventory, finished goods and related information from the point of consumption to the point of origin for the purpose of recapturing value or proper disposal. Re-manufacturing and refurbishing activities may also be included in the definition of reverse logistics. Reverse logistics is more than reusing containers and recycling packaging materials. Redesigning packaging to use less material, or reducing the energy and pollution from transportation are important activities, but they might be secondary to the real importance of overall reverse logistics. Reverse logistics contemplates important stages in the life cycle of a product, such as the repairs, reuse, recycling, recovery and wear of the product. The management of all these phases is fundamental to the sustainability of companies [10].

Reverse logistics also includes processing returned merchandise due to damage, seasonal inventory, restock, salvage, recalls, and excess inventory. It also includes recycling programs, hazardous material programs, as is the case of pharmacies' RL, obsolete equipment disposition, and asset recovery. Agrawal et al. [1] emphasise that RL refers to the sequence of activities required to collect the used product from the customers for either reuse or repair or re-manufacture or recycle or dispose of it.

Concerning RL on pharmacies, the authors only found works related to RL and reverse supply chain (RSC) in the pharmaceutical industry on a basis of improvement management and/or on the coordination of processes and involved third parties, in a competitive and regulation-driven environmental sector. A mathematical model for optimizing the supply chain with reverse flows with planning and scheduling levels is proposed in [2], but it is applied to the pharmaceutical industry at an international business markets level, and not to the pharmacies level. In the pharmaceutical industry and its supply chain, leftover medications that have not been properly disposed not only damage the environment but also might turn into a peril to people's health if being redistributed illegally in undeveloped countries [19]. For this industry, concerning this supply chain, the RL is usually not owned by a single company. The coordination of the recover process of unsold or improper medications and medical waste at customer zones is an important issue [18]. Characteristics of these products have an influence on transport restrictions, storage possibilities and safety measures [17]. According to Kumar et al. [12], this type of activities, based on strict regulations, are usually externalised and handled through specialized third-party logistics (3PL).

3.3 Methodology

This research was conducted in Italy, in the Campania region of Naples.

The case study was carried out at SmaltEco, located in Arzano, 15 km from Naples. SmaltEco is a 3PL company that deals with the return of expired medications that have not been sold by pharmacies.

A qualitative approach was adopted through a semi-structured interview at SmaltEco to understand the role of RL concerning the pharmaceutical SC and the mandatory regulamentation.

The quantitative method was also used, sending questionnaires to pharmacies in the main cities of the Campania region in Italy, as a way of identifying their role in RL, as well as identifying their main difficulties and challenges.

This research seeks to identify the important factors for pharmacies to be encouraged to practice RL, as well as to identify if they operate according to the current legislation, and find out the level of satisfaction of these with the company that deals with the RL of the medications.

In the development of this study, will be considered as overdue medications those medications that remain for a long time on the shelves of pharmacies and are not sold until their expiration date.

3.3.1 The Semi-structured Interview for the RL Company

A script was prepared for the semi-structured interview, with the aim of systematizing the fundamental questions for the collection of information. Two visits were made to SmaltEco, which is responsible for the RL of overdue medications in that region. On the first visit, a semi-structured interview was carried out, and on the second visit, a visit to the warehouse to observe the work of the company.

The first part of the interview, block 1, is designed with the purpose of knowing the company and the interviewer job position and decision making power.

In block 2, the concern is to know the process and the information flow between the company and the pharmacies in the Campania region, and the criteria for selecting business partners (eg. incinerators, recycling points, etc.), and also the storage and separation criteria.

In block 3, it is sought to know the regulations and specifications that should be fulfilled concerning EU and country laws, and also which own resources and processes or if subcontracted ones are used.

3.3.2 The Survey for the Pharmacies

The survey was developed in four blocks of questions, with all questions on a Likert scale of 1 to 5, where 1 corresponds to "totally disagree" and 5 corresponds to "totally agree" (see Table 3.1). In Block 1, the 7 questions are related to medications provisioning and stock management. In Block 2, the 3 questions are related to the

Table 3.1 Questions in the survey to the pharmacies, measured in Likert scale from 1 (*totally disagree*) to 5 (*totally agree*). The original survey was conducted in italian language

Block I - The Procurement
Q5 - Pharmacies play a key role in Reverse Supply Chain Logistics
Q6 - Replenishment, according to the demand, contributes to the decrease of the rate of Reverse Logistics
Q7 - Having low stocks, adjusted to demand, decreases Reverse Logistics
Q8 - Working with automatic systems to control stocks, decreases Reverse Logistics
Q9 - The existence of forecast models for the group of pharmacies, contributes to Reverse Logistics
Q10 - I consider it important for pharmacies to create an umbrella association that allows the use of technologies for the forecasting, management and continuous replenishment of the associated pharmacies
Q11 - I consider it important for pharmacies to work in Co-competitiveness (cooperation + competitiveness)
Block II - The Flow of Information
Q12 - The flow of information directly interferes with Reverse Logistics
Q13 - An information system, where a wholesale market has access to stocks of pharmacies, to replace them when necessary, decreases the rate of Reverse Logistics
Q14 - The use of management software between the collection company and the pharmacy would help to manage the return process of expired drugs
Block III - The Collection
Q15 - I consider the current collection service to be efficient
Q16 - I am satisfied with the current frequency of collection (3 to 4 times a year)
Q17 - If the periodic collection of medicinal products is reduced to monthly, it will help reduce storage space
Q18 - If the periodic collection is reduced to monthly, it will improve or contribute to a greater commitment of pharmacies to Reverse Logistics
Block IV - The Legislation
Q19 - The current legislation helps to encourage pharmacies to practice RL
Q20 - Some medicines can not be returned. Implementing a law, in which it would make these returns possible, would lower the rate of return of these drugs
Q21 - According to the legislation in force, some drugs are obliged to be in the stocks of pharmacies, this causes RL rate to increase

flow of information, communication between the pharmacy and the wholesale market and the company responsible for collecting the expired medications. In Block 3, the 4 questions are all related to the company responsible for collecting the medications, to see if pharmacies are satisfied with this service and how it can be improved. In Block 4, the 3 questions are related to the current legislation on the return of medications and seeks to understand if the current legislation helps to encourage pharmacies to practice RL.

The Campania region has 1674 pharmacies. Surveys were first sent by email in the first week of May 2016; after having only 5% of responses in two months, we then opted for the personal delivery of the survey in the pharmacies. Then, 124 inquiries were retrieved from a total of 500 that were sent / delivered, corresponding to a valid response rate of 24.8%.

3.3.3 Statistical Analysis of the Survey to Pharmacies

Statistical analysis was performed using IBM SPPS software toolpack, version 22. For the qualitative data, it was used the Principal Components Analysis (PCA) method and Factor Analysis (FA) model. PCA and FA are exploratory multivariate analysis techniques that allow to reduce the information of multiple correlated variables into a smaller set of independent variables, using linear combinations of the original variables, known as components and factors, representing most of the information present in the original variables [15, 16]. To use these techniques, Guadagnoli and Velicer [6] refer a minimum sample size of 100–200 observations, which is also recommended by several authors. MacCallum et al. [13] define that, as a rule for the sample size, a ratio of valid responses per existing variables should be greater than 5 (in this case, it was greater than 7).

In PCA we wish to form projections $\mathbf{z} = (z_1, z_2, \ldots, z_q)$ of the standardised data $\mathbf{x} = (x_1, x_2, \ldots, x_p)$ such that $z_i = a_{i1}x_1 + a_{i2}x_2 + \cdots + a_{ip}x_p$ where the coefficients form a $q \times p$ matrix A, and we hope that we can form an adequate representation of the data with $q \ll p$. The principal components are the uncorrelated z_i indices which are measuring different dimensions of the data, ordered by decreasing variances: $Var(z_1) > Var(z_2) > \cdots > Var(z_q)$. The variances of the principal components are the eigenvalues λ_i of the sample covariance matrix $\mathbf{S}$, or of the correlation matrix $\mathbf{R}$. The constants $a_{i1}, a_{i2}, \ldots, a_{ip}$, called loadings, are the coordinates of the corresponding eigenvector $\mathbf{e_i}$ multiplied by $\sqrt{\lambda_i}$, that is, scaled so that the sum of squares of its coordinates is unitary. When doing a PCA, there is always hope that the variances of most of the indices are negligible [14]. The proportion of the total variance explained by the k-th principal component is $\frac{\lambda_k}{\lambda_1+\cdots+\lambda_p}$. If most of the total population variance can be attributed to the first components, than these can replace the original p variables without much loss of information [11].

After performing the matrix rotation, PCA becomes FA. Both of these techniques are usually seen as data reduction methods but, while PCA is a deterministic method, FA is actually a model that includes an error component.

Altough PCA is usually applied to quantitative variables, SPSS has implemented an optimal scaling procedure that assigns a numerical value to each category of the ordinal variables and creates corresponding quantitative variables with metric properties, enabling principal component analysis to be performed on categorical variables [15].

It is important that the variables are in some sense comparable, because if variables are recorded on widely differing scales, a principal component analysis of the covariance matrix will largely reflect the variables with the numerically greatest variance [9]. As our variables of interest were all measured in the same scale, they didn't need to be standardized.

3.4 Case Study

With the interviews, the RL operational process of the company which deals with the RL of expired medications and which have not been sold by the pharmacies in Campania was understood and described. In every pharmacy throughout the country, there are collection points where consumers deposit the medications that have expired their date. The urban cleaning company of each city is responsible for collecting the expired medicinal products that the final consumers deposit in the pharmacies. According to the normatives of Assinde, in each region in Italy there is a company, subcontracted by this organization, responsible for the collection and return flow of expired medications. The SmaltEco company is the one responsible for the collection of these medications in Campania region, but it also does the preparation for the destination (collection, separation, recycling, incineration). It performs the collection of medications in pharmacies three to four times a year. In this whole process, SmaltEco is subject to the Assinde system, an arrangement of pharmaceutical laboratories in Italy, which ensures the pharmacies the return of medications with expired date, but also guarantees a refund between 65–95% of the medications value if the medications belong to the Assinde agreement. For medicinal products that do not belong to the group Assinde, the pharmacies pay a percentage of 6% to the company responsible for the collection, because according to the existing Italian legislation, the pharmacies are responsible for end destination of medications with the expired validity date. The medications are stored before and after the separation. Before the separation they are stored without any criterion, since the company has not a CER code (European waste code) for the storage, therefore at this stage, hazardous and non-hazardous medications are stored all together, still in their boxes. After the separation, they are stored and identified according to the CER code. The medications from laboratories not belonging to the Assinde agreement must also have an appropriate final disposal, and in this case, they are forwarded to the incinerator. 10 tons are processed weekly by SmaltEco and, after the separation, near 9.2 tons are forwarded to central incineration. In the region of Campania, there is only one incinerator, which is State-owned, located in Pantano, 21 km from Naples. It is not used due the high costs per kilo: it costs the double of the used ones, including shipping. Hence, the most used incinerator is in Pisa, 550 km from Naples. In relation to the cost-benefit principle, another option is the private incinerator, located in Milan, 780 km from Naples. As SmaltEco operates in almost all Italy, a routing vehicle plan is made to reduce costs.

Recycling materials, such as plastics, paper, glass, etc., represent around 750 kg of the weekly tons received and are targeted to the licensed recycling centres, in Naples, Rome, Pomezia and Milan. The shipment and the total cost of transporting these products to recycling centres are SmaltEco responsibility as they are already included in the 6% charged price to pharmacies. The recycling center chosen is the one located in the vehicle routing plan of the medications collection. SmaltEco has an economic policy on RL that is emphasised on the company's incineration transportation costs and RL external cost, but it can be improved by the company and the government, to contribute for a greener RL activity in this RSC.

In the survey to the pharmacies, the first two questions are the identification of the pharmacy and its location. Most of the pharmacies surveyed were concentrated in Naples, with 46.77%, followed by Caserta with 16.94%, Avellino with 13.71%, Benevento with 12.9% and Salerno with 9.68%. The employees responsible for answering the questionnaire were mostly pharmacists, corresponding to 46%, followed by the pharmaceutical director with 41.1%, and the owners, a small percentage of 10.5%. The pharmacies surveyed have between 1 and 6 employees, and the mode is to have 3 employees (40.3%), followed by 29.8% of pharmacies with 2 employees.

3.5 Results and Discussion

Regarding the survey to the pharmacies, the 17 questions in the survey were measured in a Likert scale from 1 to 5 (Table 3.1), and the corresponding ordinal qualitative variables were analysed with principal components analysis. In Table 3.2 the Spearman correlation coefficients between the 17 variables are presented. The matrix highlights the strong correlation between Q6 and the other questions.

The Bartlett sphericity test presented a very expressible result $\chi^2 \simeq 1919.7$ with 136 degrees of freedom, for which corresponds a p-value smaller than 0.001. Whereby we reject H0, concluding that the variables are significantly correlated.

The Kaiser-Meyer-Olkin (KMO) measure produced a value of 0.801, which is considered good, according to [16] (see Table 3.3), proving the adequacy of the application of principal factor analysis technique to this sample.

Figure 3.1 shows the scree plot. According to [11], one should look for an elbow in this graph and retain the number of principal components of the point at which the remaining eigenvalues are relatively small and all about the same size. This graph suggests the retention of a number of components between two and six.

Table 3.4 shows the percentage explained by each factor before and after the rotation. Taking into account the retention rule that states that factors with values greater than 1 should be retained, we then considered only the first four factors. The first factor explains 49.255% of the variance, the second factor explains 11.329%, the third factor explains 8.361% and the fourth explains 7.224%. Cumulatively they explain 76.159% of the variability of the original seventeen variables. While the total variance explained by the four factors (76.159%) does not vary with the rotation, the same does not happen with the explained variance for each factor.

Table 3.2 Spearman correlation matrix

Spearman's rho	Q5	Q6	Q7	Q8	Q9	Q10	Q11	Q12	Q13	Q14	Q15	Q16	Q17	Q18	Q19	Q20	Q21
Q5																	
Q6	0.801[b]																
Q7	0.578[b]	0.701[b]															
Q8	0.602[b]	0.543[b]	0.880[b]														
Q9	0.441[b]	0.507[b]	0.537[b]	0.481[b]													
Q10	0.293[b]	0.402[b]	0.550[b]	0.616[b]	0.706[b]												
Q11	0.339[b]	0.418[b]	0.619[b]	0.545[b]	0.640[b]	0.523[b]											
Q12	0.492[b]	0.613[b]	0.397[b]	0.340[b]	0.484[b]	0.543[b]	0.492[b]										
Q13	0.499[b]	0.531[b]	0.690[b]	0.668[b]	0.197[a]	0.343[b]	0.392[b]	0.341[b]									
Q14	0.400[b]	0.474[b]	0.371[b]	0.341[b]	0.327[b]	0.391[b]	0.600[b]	0.680[b]	0.436[b]								
Q15	0.653[b]	0.555[b]	0.373[b]	0.368[b]	0.459[b]	0.262[b]	0.216[a]	0.391[b]	0.370[b]	0.217[a]							
Q16	0.559[b]	0.527[b]	0.466[b]	0.425[b]	0.460[b]	0.302[b]	0.258[b]	0.359[b]	0.372[b]	0.161	0.797[b]						
Q17	0.384[b]	0.493[b]	0.664[b]	0.629[b]	0.295[b]	0.424[b]	0.514[b]	0.495[b]	0.517[b]	0.344[b]	0.058	0.251[b]					
Q18	0.281[b]	0.251[b]	0.516[b]	0.574[b]	0.144	0.338[b]	0.399[b]	0.372[b]	0.361[b]	0.226[a]	−0.011	0.186[a]	0.725[b]				
Q19	0.480[b]	0.514[b]	0.447[b]	0.372[b]	0.371[b]	0.117	0.301[b]	0.172	0.541[b]	0.268[b]	0.683[b]	0.758[b]	0.120	-0.024			
Q20	0.498[b]	0.505[b]	0.289[b]	0.191[a]	0.608[b]	0.360[b]	0.410[b]	0.618[b]	0.151	0.475[b]	0.544[b]	0.535[b]	0.249[b]	0.097	0.363[b]		
Q21	0.218[a]	0.300[b]	0.295[b]	0.297[b]	0.071	0.301[b]	0.042	0.176	0.581[b]	0.216[a]	0.413[b]	0.334[b]	0.035	−0.065	0.467[b]	0.143	

[a]Correlation is significant at the 0.05 level (2-tailed)
[b]Correlation is significant at the 0.01 level (2-tailed)

Table 3.3 Kayser-Meyer-Olkin statistics. *Source*:[16]

KMO	Factor analysis
0.9–1	Very good
0.8–0.9	Good
0.7–0.8	Average
0.6–0.7	Reasonable
0.5–0.6	Bad
<0.5	Unacceptable

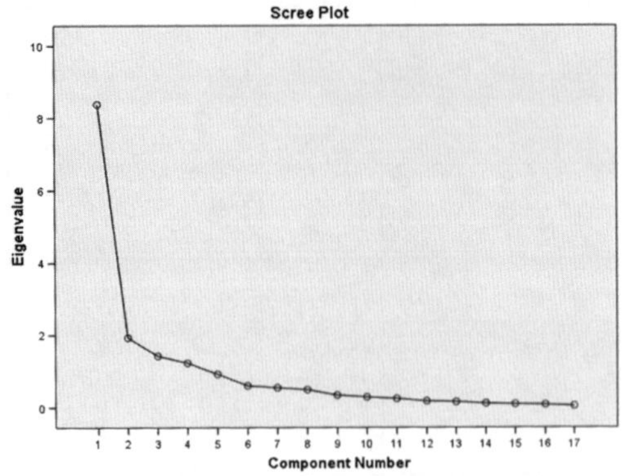

Fig. 3.1 Scree Plot: the eigenvalues are represented in relation to the number of components

Table 3.4 Eigenvalues of principal component analysis

Component	Initial eigenvalues			Rotated sums of squared loadings		
	Total	% of variance	Cumulative %	Total	% of variance	Cumulative %
1	8.373	49.255	49.255	4.937	29.044	29.044
2	1.926	11.329	60.585	3.080	18.120	47.164
3	1.421	8.361	68.945	2.605	15.322	62.486
4	1.226	7.214	76.159	2.324	13.673	76.159
5	0.920	5.411	81.571			
6	0.600	3.529	85.100			
7	0.545	3.206	88.306			
8	0.491	2.890	91.196			
9	0.337	1.984	93.180			
10	0.285	1.676	94.855			
11	0.240	1.413	96.268			
12	0.173	1.017	97.285			
13	0.151	0.886	98.170			
14	0.110	0.649	98.820			
15	0.088	0.520	99.339			
16	0.073	0.430	99.769			
17	0.039	0.231	100			

Table 3.5 Component matrix of principal components analysis (before rotation). Communalities and eigenvalues are also presented. Weights over 0.5 are highlighted

Questions	Component				Communalities
	1	2	3	4	
Q5	**0.872**	−0.262	0.055	−0.100	0.843
Q6	**0.875**	−0.194	0.054	0.003	0.806
Q7	**0.914**	0.072	−0.003	0.003	0.841
Q8	**0.810**	0.226	−0.424	0.069	0.893
Q9	0.230	0.172	0.256	**0.654**	0.576
Q10	**0.598**	0.457	−0.153	0.209	0.633
Q11	**0.667**	0.288	0.271	0.117	0.615
Q12	**0.689**	0.167	0.354	0.099	0.637
Q13	**0.817**	0.073	−0.447	0.118	0.886
Q14	**0.763**	−0.005	0.179	0.296	0.702
Q15	**0.760**	−0.465	0.093	−0.125	0.817
Q16	**0.707**	−0.269	0.116	−0.446	0.784
Q17	**0.593**	**0.566**	−0.066	−0.314	0.776
Q18	0.463	**0.629**	−0.013	−0.451	0.813
Q19	0.635	**−0.518**	−0.076	−0.148	0.699
Q20	**0.701**	−0.179	0.497	0.071	0.776
Q21	0.485	−0.323	**−0.661**	0.269	0.848
Eigenvalues	8.373	1.926	1.421	1.226	

The component matrix, in Table 3.5, presents the coefficients that correlate the variables with the components before the rotation, that is, the loadings. It allows to verify which factor best explains each one of the variables. Note that the sum of the squares of the coefficients of the variables for each factor (column sum) is the eigenvalue of the components, and the sum of the squares of the coefficients of each variable (row sum) is the communality of each variable. The communalities are also presented in Table 3.5. It is possible to observe that all variables have a strong correlation with the extracted factors, since the percentage of common variance of the variables in the extracted factors is greater than 50% for all variables, explaining at least 57.6% of the total variance.

The coefficients in Table 3.5 can be interpreted as a correlation between variables and factors, because they vary between −1 and 1. Thus, the correlation between *low stocks decreases RL* (Q7) and factor 1 is 0.914, while the correlation between *legislation that obliges medications to be in stock increases RL rate* (Q21) with factor 3 is of −0.661. It should be noted that the affectation of the variable *Change of periodicity of collection* (Q17) to a given factor (in this case factors 1 and 2) is unclear, since there is a high coefficient for both factors. In this case we will have to analyse this variable through the use of rotations. The purpose of rotation is to maximize the coefficient values, so that each variable is associated with only one

Table 3.6 Rotated component matrix. Weights over 0.5 are highlighted

Questions	Component			
	1	2	3	4
Q5	**0.806**	0.232	0.308	0.211
Q6	**0.735**	0.240	0.338	0.307
Q7	**0.593**	0.464	0.376	0.363
Q8	0.292	**0.527**	**0.694**	0.218
Q9	−0.078	−0.072	0.087	**0.746**
Q10	0.043	**0.533**	0.404	0.429
Q11	0.337	0.455	0.062	**0.539**
Q12	0.451	0.375	0.007	**0.542**
Q13	0.363	0.391	**0.747**	0.208
Q14	0.489	0.196	0.268	**0.594**
Q15	**0.863**	0.035	0.240	0.115
Q16	**0.823**	0.306	0.071	−0.088
Q17	0.169	**0.847**	0.141	0.099
Q18	0.101	**0.896**	−0.010	−0.014
Q19	**0.767**	−0.043	0.327	−0.041
Q20	**0.708**	0.121	−0.084	**0.503**
Q21	0.252	−0.120	**0.878**	−0.003

factor. The rotated components using VARIMAX orthogonal method with Kaiser normalization was performed and is presented in Table 3.6. The rotated component matrix is advantageous because it highlights the meaning of the factors.

1. Factor 1 can be highlighted by the assertion of blocks 1, 3 and 4, and can be referred as **RL role and legislation**;
2. Factor 2 is highly influenced by the statement "if periodic collection were reduced to monthly, it would improve or contribute to a greater commitment of pharmacies to RL"from block 3, and can be named **RL collection periodicity**;
3. Factor 3 is strongly related to the **minimum required stock and process automation**;
4. Factor 4 is highlighted by the question "The existence of forecast models for the group of pharmacies contributes to Reverse Logistics", referring to procurement, in block 1, and it could be named **Cooperative planning and forecast**.

3.6 Conclusions

In this work, the RL process of the pharmacies in Campania region was analysed with an interview to a 3PL company that operates at a national level and which is responsible for RL collection of unsold and expired medications, storage, separation and final destination. This allowed to understand how the reverse supply chain for unwanted medications operates and contributes for a greener RL. Also, a survey was developed for the pharmacies itself to understand the impact of the RL process on the pharmacies inventory policy and business performance, and how it can be improved. The data was analysed using PCA and FA, and 4 factors were extracted, explaining 76.2% of the original data.

From the SmaltEco interview, the RSC was identified. Although the business is according to the regulamentation, there are actions that can be improved in terms of RL environmental sustainability, such as prices of local incineration, transportation costs and ecological footprint.

From the survey, it is possible to conclude that, as regards the question of investigation, the factors that most influence the practice of RL are: Pharmacies and its legislation play a key role in Reverse Supply Chain; If the periodic collection were reduced to monthly, it would improve or contribute to a greater commitment of pharmacies to RL; Process automation decreases RL due to decrease of inventory levels, and minimum stocks of some medications required by law increases RL rate; Cooperative planning and forecast models contribute to RL.

Of all the factors, the least important was "I consider it important for pharmacies to create an umbrella association that allows the use of technologies for the forecasting, management and continued replenishment of associated pharmacies." This makes it clear that pharmacies have no interest in forming a summit, thus demonstrating their complete independence.

The subject discussed in this paper may raise awareness in the O.R. community for the issues that are specific to the reverse logistics practice in the pharmacies sector.

Acknowledgements We acknowledge the financial support of CIDEM, R&D unit funded by the FCT-Portuguese Foundation for the Development of Science and Technology, Ministry of Science, Technology and Higher Education, under the Project UID/EMS/0615/2016.

References

1. S. Agrawal, R.K. Singh, Q. Murtaza, A literature review and perspectives in reverse logistics. Resour. Conserv. Recycl. **97**, 76–92 (2015)
2. A.C.S. Amaro, A.P.F.D. Barbosa-Póvoa, Planning and scheduling of industrial supply chains with reverse flows: a real pharmaceutical case study. Comput. Chem. Eng. **32**(11), 2606–2625 (2008)
3. R.H. Ballou, *Business Logistics Management*, 4th edn. (Prentice Hall, New Jersey, 2006)
4. E. Bazan, M.Y. Jaber, S. Zanoni, A review of mathematical inventory models for reverse logistics and the future of its modelling: an environmental perspective. Appl. Math. Modell. **40**(5–6), 4151–4178 (2016)
5. K. Govindan, H. Soleimani, D. Kannan, Reverse logistics and closed-loop supply chain: a comprehensive review to explore the future. Eur. J. Oper. Res. **240**(3), 603–626 (2015)
6. E. Guadagnoli, W.E. Velicer, Relation of sample size to the stability of component patterns. Psychol. Bull. **103**(2), 265–275 (1988)

7. K. Hawks, *VP Supply Chain Practice, Navesink* (Reverse Logistics Magazine, Winter/Spring, 2006)
8. B.T. Hazen, R.E. Overstreet, D.J. Hall, J.R. Huscroft, J.B. Hanna, Antecedents to and outcomes of reverse logistics metrics. Ind. Mark. Manag. **46**, 160–170 (2015)
9. P. Hewson. Multivariate statistics with R (2009), https://local.disia.unifi.it/rampichini/analisi_multivariata/Hewson_notes.pdf
10. S.K. Jakhar, Performance evaluation and flow allocation decision model for a sustainable supply chain of an industry. J. Clean. Prod. **87**, 391–413 (2015)
11. R. Johnson, D. Wichern, *Applied Multivariate Statistical Analysis*, 6th edn. (Pearson Prentice Hall, New Jersey, 2017)
12. S. Kumar, E. Dieveney, A. Dieveney, Reverse logistic process control measures for the pharmaceutical industry supply chain. Int. J. Prod. Perform. Manag. **58**(2), 188–204 (2009)
13. R.C. MacCallum, K.F. Widaman, K.J. Preacher, S. Hong, Sample size in factor analysis: the role of model error. Multivar. Behav. Res. **36**(4), 611–637 (2001)
14. B. Manly, J. Alberto, *Multivariate Statistical Methods: A Primer*, 4th edn. (CRC Press, US, 2016)
15. J. Maroco. *Análise Estatística com utilização do SPSS*. Edições Sílabo, 2ª edn. (2003)
16. M.H. Pestana, J.N. Gageiro, *Análise de dados para ciências sociais - A complementaridade do SPSS*. Edições Sílabo, 5ª edn. (2008)
17. A. Rolewicz-Kalińska, Logistic constraints as a part of a sustainable medical waste management system. Transp. Res. Procedia **16**, 473–482 (2016)
18. D. Weraikat, M.K. Zanjani, N. Lehoux, Coordinating a green reverse supply chain in pharmaceutical sector by negotiation. Comput. Ind. Eng. **93**, 67–77 (2016a)
19. D. Weraikat, M.K. Zanjani, N. Lehoux, Two-echelon pharmaceutical reverse supply chain coordination with customers incentives. Int. J. Prod. Econ. **176**, 41–52 (2016b)

Chapter 4
A Penalty Approach for Solving Nonsmooth and Nonconvex MINLP Problems

M. Fernanda P. Costa, Ana Maria A. C. Rocha and Edite M. G. P. Fernandes

Abstract This paper presents a penalty approach for globally solving nonsmooth and nonconvex mixed-integer nonlinear programming (MINLP) problems. Both integrality constraints and general nonlinear constraints are handled separately by hyperbolic tangent penalty functions. Proximity from an iterate to a feasible promising solution is enforced by an oracle penalty term. The numerical experiments show that the proposed oracle-based penalty approach is effective in reaching the solutions of the MINLP problems and is competitive when compared with other strategies.

Keywords MINLP · Penalty function · DIRECT · Oracle

4.1 Introduction

In this paper, we address the solution of nonsmooth and nonconvex mixed-integer nonlinear programming (MINLP) problems by a penalty approach. It is assumed that the problem is in the form

$$\begin{aligned} \underset{x \in X \subset \mathbb{R}^n}{\text{glob min}} \quad & f(x) \\ \text{subject to} \quad & g_j(x) \le 0,\ j = 1, \ldots, p \\ & h_l(x) = 0,\ l = 1, \ldots, m \\ & x_i \in \mathbb{R} \text{ for } i \in I_c \subseteq I \equiv \{1, \ldots, n\} \\ & x_j \in \mathbb{Z} \text{ for } j \in I_d \subseteq I \end{aligned} \tag{4.1}$$

M. F. P. Costa (✉)
Centre of Mathematics, University of Minho, Campus de Gualtar,
4710-057 Braga, Portugal
e-mail: mfc@math.uminho.pt

A. M. A. C. Rocha · E. M. G. P. Fernandes
Algoritmi Research Centre, University of Minho, Campus de Gualtar,
4710-057 Braga, Portugal
e-mail: arocha@dps.uminho.pt

E. M. G. P. Fernandes
e-mail: emgpf@dps.uminho.pt

A. I. F. Vaz et al. (eds.), *Operational Research*, Springer Proceedings
in Mathematics & Statistics 223, https://doi.org/10.1007/978-3-319-71583-4_4

where $f, g_j, h_l : \mathbb{R}^n \to \mathbb{R}$ are continuous possibly nonlinear functions in a compact subset of $\mathbb{R}^n$, herein defined as $X = \{x : -\infty < lb_i \le x_i \le ub_i < \infty, i = 1, \dots, n\}$, $I_c \cap I_d = \emptyset$ and $I_c \cup I_d = I$. Thus, I_c is the index set of the continuous variables and I_d consists of the indices of the integer variables. Here, integer variables include binary variables. Let C be the following subset of $\mathbb{R}^n$, $C = \{x \in \mathbb{R}^n : g_j(x) \le 0, j = 1, \dots, p,\ h_l(x) = 0, l = 1, \dots, m\}$, and let $W_c = C \cap X$ be a closed set. Consider the set D, which is the cartesian product of the sets D_j, $j \in I_d$, where

$$D_j = \{d \in \mathbb{Z} : lb_j \le d \le ub_j\}, j \in I_d, \tag{4.2}$$

let $\mathscr{I}$ be defined by $\mathscr{I} = \left\{x \in X : x_j \in \mathbb{Z} \text{ for } j \in I_d \subseteq I\right\}$ and let $W = C \cap \mathscr{I}$ be the nonempty feasible region of the problem (4.1). When a continuous relaxation of the integer variables is applied, $W \equiv W_c$. A continuous relaxation means that the integer variables can be treated as continuous variables, and all function (f, g and h) values can be computed for $x_j \in \mathbb{R}$, $j \in I_d$ (instead of $x_j \in \mathbb{Z}$, $j \in I_d$). The MINLP problem (4.1) is said to be convex if f and $g_1(x), \dots, g_p(x)$ are convex functions and $h_1(x), \dots, h_m(x)$ are affine functions over X. This means that by relaxing the integrality constraint on x_j, $j \in I_d$, a convex program is obtained (minimizing a convex function over a convex set). Otherwise, the MINLP is said to be nonconvex.

Most techniques available in the literature require the definition and the use of convex model functions and the continuous relaxations of the integer variables. However, some real-life MINLP problems that emerge in mechanical, electrical and chemical engineering applications involve nonsmooth and nonconvex functions and the specific integer variables cannot be relaxed [1]. Most exact methods for nonconvex MINLP are based on the branch-and-bound (BB) technique. Effective examples are the spatial-BB algorithm [2, 3], branch-and-reduce type algorithms [2, 4] and the α-BB algorithm [5].

Heuristics for nonconvex MINLP are also available in the literature. A heuristic approach extension of the boundary tracking optimization is presented in [6]. In [7], a variable neighborhood search heuristic is proposed and in [8], two heuristics are analyzed: the first aims to obtain an initial feasible solution, the second one searches for an improved solution within the neighborhood of a given point.

Extensions of the feasibility pump algorithm to nonconvex MINLP are available in [9]. A derivative-free method that relies on two search procedures, a line search strategy for the continuous variables and a local search for the discrete ones, is presented in [10]. Recently, penalty-based algorithms aiming to penalize integrality violation are available in the literature [11–13].

Metaheuristics are nowadays very popular and aim to compute fast and good approximations to optimal solutions of nonconvex MINLP problems. A mixed-integer hybrid differential evolution (MIHDE) [14] has been successfully applied to mixed-integer optimization problems and a particle swarm optimization is presented in [15]. A parameter free penalty approach with a genetic algorithm (GA) [16] and a filter technique combined with a GA [17] are analyzed when solving nonconvex MINLP. In [18], the BBMCSFilter method, which relies on a BB framework and a derivative-free methodology to solve nonsmooth and nonconvex NLP, is presented.

Two extended versions of the ant colony optimization framework are available in [19] and a new version of the firefly algorithm (FA), that uses four preference rules to select solutions that are feasible or have the least objective function values, is tested in [20]. Review on MINLP techniques and applications are available in [2, 21, 22]. A brief overview of the start-of-the-art in software for the solution of MINLP problems can be found in [23].

In this study, a penalty continuous formulation of the MINLP problem (4.1) is used. First, a penalty function has been selected from a class of penalty functions that are applied to general integer problems [11–13]. Second, two other penalty functions have been constructed in order to penalize the general constraints violation as well as to enforce convergence to a solution, denoted by the oracle, that is feasible and has the least function value found so far. Thus, after relaxing the integrality constraints on the variables and adding a particular penalty term to the objective function, $P_d(x; \varepsilon_d)$, aiming to penalize the integrality constraint violation, as well as by adding another penalty term, $P_c(x; \varepsilon_c)$, to penalize the general constraints violation, the following continuous bound constrained nonlinear programming (BCNLP) problem emerges

$$\begin{aligned} &\operatorname*{glob\,min}_{x\in X} \ \Psi(x; \varepsilon_d, \varepsilon_c) \equiv f(x) + P_d(x; \varepsilon_d) + P_c(x; \varepsilon_c) \\ &\text{subject to } \ x_i \in \mathbb{R}, \ i = 1, \ldots, n, \end{aligned} \tag{4.3}$$

where $\varepsilon_d, \varepsilon_c \in \mathbb{R}^+$ are positive penalty parameters [24]. The motivation is that problem (4.1) is equivalent to the continuous BCNLP problem, in the sense that they have the same global minimizers. The optimal solution of the BCNLP problem can then be easily obtained by well-established and known solvers.

In the sequel, the herein presented work adds a new penalty term to the objective function in problem (4.3), aiming to enforce convergence to the oracle, represented by o^*, and defined as the best found feasible solution, aiming to predict a global optimum. The goal of the oracle penalty is to penalize solutions that move away from o^*. The new proposed algorithm is tested and compared with other nonconvex MINLP strategies.

Thus, our contribution in this article is directed to the combination of three penalty terms aiming to penalize the integrality violation, the nonlinear inequality and equality constraints violation and the distance to the oracle o^*. The penalty term for the integrality constraints is based on the hyperbolic tangent function, as proposed in [11], and the equality and inequality constraints are dealt with penalties also defined by the hyperbolic tangent function [24]. Similarly, the new penalty imposed on the distance of the current solution to the oracle is also based on the hyperbolic tangent function. The motivation for the use of the hyperbolic tangent function is that its boundedness property makes the BCNLP penalty problem easier to solve than with some of its competitors. The solution of the BCNLP problem is then obtained using the DIRECT algorithm [25], a deterministic and derivative-free algorithm for finding global solutions inside hyperrectangles. We illustrate the performance of the proposed penalty approach on a well-known set of MINLP test problems.

The remainder of the paper proceeds as follows. Section 4.2 introduces the penalty methodology and Sect. 4.3 addresses the implementation of the penalty terms and investigates the use of the penalty parameters and the oracle parameter. Section 4.4 contains the results of all the numerical experiments and the conclusions are summarized in Sect. 4.5.

4.2 Penalty Approaches

The following equivalence result based on a penalty approach will be used [11–13].

Property 4.1 *Assuming that W and W_c are compact sets, there exists a value $\bar{\varepsilon} > 0$ such that, for any $\varepsilon_d \in (0, \bar{\varepsilon}]$, the problems*

$$\min f(x), \quad \textit{subject to } x \in W$$

and

$$\min F(x; \varepsilon_d) \equiv f(x) + P_d(x; \varepsilon_d), \quad \textit{subject to } x \in W_c \tag{4.4}$$

where

$$P_d(x; \varepsilon_d) = \frac{1}{\varepsilon_d} \sum_{j \in I_d} \min_{d \in D_j} \tanh(|x_j - d|) \tag{4.5}$$

are equivalent in the sense that they have the same minimizers.

This property is a consequence of Property 2.5 in [11]. The below presented assumptions (A1)–(A3) on f and on the penalty $P_d(x; \varepsilon_d)$ (see (4.5)) are required to prove Property 4.1.

(A1) Function f is bounded on W_c and there exists an open set $A \supset W$ and real numbers $\alpha, L > 0$ such that for all $x, y \in A$, f satisfies

$$|f(x) - f(y)| \leq L\|x - y\|^{\alpha}.$$

(A2) For all $x, y \in W$ and for all $\varepsilon_d \in \mathbb{R}^+$,

$$P_d(x; \varepsilon_d) = P_d(y; \varepsilon_d).$$

(A3) There exists an $\bar{\varepsilon}$, and for all $z \in W$ there exists a neighborhood $S(z)$ such that

$$P_d(x; \varepsilon_d) - P_d(z; \varepsilon_d) \geq \bar{L}\|x - z\|^{\alpha}, \text{ for all } x \in S(z) \cap (W_c \backslash W), \varepsilon_d \in (0, \bar{\varepsilon}],$$

where $\bar{L} > L$ and α is chosen as in (A1). Furthermore, let $S = \bigcup_{z \in W} S(z)$, $\exists \bar{x} \notin S$ such that

$$\lim_{\varepsilon_d \to 0} (P_d(\bar{x}; \varepsilon_d) - P_d(z; \varepsilon_d)) = +\infty \text{ for all } z \in W,$$

$$P_d(x; \varepsilon_d) \geq P_d(\bar{x}; \varepsilon_d) \text{ for all } x \in W_c \backslash S \text{ and for all } \varepsilon_d > 0.$$

Problem (4.4) comes out by relaxing the integer constraints on the variables and adding a particular penalty term to the objective function f.

Let $P_c(\cdot; \varepsilon_c) : \mathbb{R}^n \to \mathbb{R}$ be a penalty term, that aims to penalize general equality and inequality constraints violation, defined by

$$P_c(x; \varepsilon_c) = \frac{1}{\varepsilon_c} \left(\sum_{j=1}^{p} \tanh(g_j^+(x)) + \sum_{l=1}^{m} \tanh(|h_l(x)|) \right), \tag{4.6}$$

where $g_j^+(x) = \max\{g_j(x), 0\}$ and $\varepsilon_c \in \mathbb{R}^+$ is the penalty parameter. We note that $P_c(x; \varepsilon_c) = 0$ when $x \in C$ and $P_c(x; \varepsilon_c) > 0$ when $x \notin C$. Generally speaking, under suitable assumptions on the objective function F of problem (4.4) and on the penalty $P_c(\cdot; \varepsilon_c)$, the problems

$$\min \Psi(x; \varepsilon_d, \varepsilon_c) \equiv F(x; \varepsilon_d) + P_c(x; \varepsilon_c), \quad \text{subject to } x \in X,$$

and (4.4) are equivalent (see Theorem 2.1 in [12]).

For the sake of simplicity, we define

$$P(x; \varepsilon_d, \varepsilon_c) = P_d(x; \varepsilon_d) + P_c(x; \varepsilon_c). \tag{4.7}$$

Both penalty terms in $P(x; \varepsilon_d, \varepsilon_c)$ are based on the hyperbolic tangent function, $\tanh : \mathbb{R} \to [-1, 1] \subset \mathbb{R}$, an odd function which is differentiable, strictly increasing on $\mathbb{R}$, and satisfies $\tanh(t) = 0$ iff $t = 0$ and

$$\lim_{t \to 0^+} \frac{\tanh(t)}{t} = 1, \quad \lim_{t \to +\infty} \tanh(t) = 1 \text{ and } \lim_{t \to +\infty} \frac{d \tanh(t)}{dt} = 0.$$

Under some suitable assumptions on f and $P(x; \varepsilon_d, \varepsilon_c)$ (see Theorem 2.1 in [12], as well as Property 2.5 in [11] in the context of the hyperbolic tangent function) we may remark the following.

Remark 4.1 Under suitable assumptions on f and $P(x; \varepsilon_d, \varepsilon_c)$, let W and X ($W \subseteq X \subset \mathbb{R}^n$) be compact sets. Then, $\exists \tilde{\varepsilon} \in \mathbb{R}^+$ such that for all $\varepsilon_d, \varepsilon_c \in (0, \tilde{\varepsilon}]$, the problems (4.1) and (4.3) have the same global minimizers.

4.3 Oracle-Based Penalty Algorithm

The extension of the above presented penalty approach to solve MINLP problems is investigated.

We note here that the term $P_d(x; \varepsilon_d)$ (see (4.5)) penalizes the distance from x to a point z (in terms of the components $i \in I_d$) that satisfies $z := [x]_r \in \mathscr{I} \subset X$ where $z_i \in \mathbb{Z}, i \in I_d$ results from rounding x_i to the nearest integer and $z_l = x_l$ for $l \in I_c$, thus compelling x to come near z. However, since z may not be a global minimizer, our proposal considers a new penalty term that aims to reduce the distance from x to a very promising solution, o^* (ideally a global optimizer), that satisfies $o^* \in W$ and has an objective function value not greater than $f(z)$. The o^* is a parameter vector, herein also denoted by the oracle, likewise it is used in [26], due to its predictive nature. Although the original idea of the oracle penalty method corresponds to a transformation of the objective function f into an additional equality constraint $h_{m+1}(x) = f(x) - \gamma = 0$, where γ is the oracle parameter [26], our proposal is equivalent to having an extra equality constraint that aims to enforce the proximity of the current solution to the oracle. Thus, we add a new penalty term to P, measuring proximity from x to o^*, with the aim of finding a solution near the oracle with a lower objective function value $f(x) < f(o^*) \leq f(z)$

$$q(x; o^*) = \sum_{i=1}^{n} \tanh(|x_i - o_i^*|). \tag{4.8}$$

Remark 4.2 We note that, in the context of incorporating the function 'tanh' in the penalty terms, this corresponds to adding new equality constraints $x_i = o_i^*$ to the problem (4.1) and that the feasible set of the "new problem" is now $W_o = \{x \in W : x_i = o_i^*, i = 1, \ldots, n\}$. When the oracle parameter o^* is a global minimizer to the problem (4.1), a feasible solution to the "new problem" ($x \in W_o$) is the global solution of the MINLP problem.

Thus, the new proposed BCNLP problem for finding a global solution to a MINLP problem like (4.1) is

$$\begin{aligned} &\operatorname*{glob\,min}_{x \in X} \; \Psi(x; \varepsilon_d, \varepsilon_c, o^*) \equiv f(x) + P(x; \varepsilon_d, \varepsilon_c, o^*) \\ &\text{subject to } \; x_i \in \mathbb{R},\; i = 1, \ldots, n, \end{aligned} \tag{4.9}$$

where the oracle penalty function reads as follows:

$$P(x; \varepsilon_d, \varepsilon_c, o^*) = P_d(x; \varepsilon_d) + P_c(x; \varepsilon_c) + \frac{1}{\varepsilon_c} q(x; o^*). \tag{4.10}$$

When there is a guess about the global minimizer, this information may be used to speed the convergence of the algorithm. To apply the oracle penalty function when there is no guess about the global minimizer, some modifications are required to make the method more robust regarding the oracle parameter selection. We assume that the two following conditions hold:

- $f(o^*) > f(x^*)$;
- there exists at least one $z^* \in W$ such that $f(o^*) = f(z^*) \geq f(x^*)$.

Thus, the oracle vector o^* should be updated whenever a solution better than o^* is produced, i.e., if a solution $z \in \mathscr{I}$ is found such that $f(z) \leq f(o^*)$ and $\Theta(z) \leq \Theta(o^*)$, where

$$\Theta(x) = \max_{j=1,\dots,p;l=1,\dots,m} \left\{ g_j^+(x), |h_l(x)| \right\} \tag{4.11}$$

represents the maximum general constraints violation, then the new value for the oracle is the following $o^* = z$.

The algorithm based on the proposed oracle penalty function, denoted by oracle-based penalty algorithm (ObPA), is shown in Algorithm 1. To initialize the oracle, we set $o^* = [x^0]_r$, where the initial approximation, x^0, is randomly generated in X.

Input: $x^0 \in X, \varepsilon > 0, \delta > 0, \eta > 0, \mu > 0, \varepsilon_d^1 > \varepsilon, \varepsilon_c^1 > \varepsilon, \delta^1 > \delta, \eta^1 > \eta, \mu^1 > \mu$;
Set $k = 1$;
Initialize the oracle as $o^* = z^0 = [x^0]_r$;
while *the stopping rule defined in (4.14) does not hold* **do**
  if $\Theta(z^{k-1}) \leq \Theta(o^*)$ *and* $f(z^{k-1}) \leq f(o^*)$ **then**
    Set $o^* = z^{k-1}$;
  end
  if $\Theta(o^*) \leq \eta^k$ **then**
    Compute x^k, an approximation to the solution of problem (4.9) such that

$$\Psi(x^k; \varepsilon_d^k, \varepsilon_c^k, o^*) \leq \Psi(x; \varepsilon_d^k, \varepsilon_c^k, o^*) + \delta^k \text{ for all } x \in X \tag{4.12}$$

  else
    Compute x^k, an approximation to the solution of problem (4.3) such that

$$\Psi(x^k; \varepsilon_d^k, \varepsilon_c^k) \leq \Psi(x; \varepsilon_d^k, \varepsilon_c^k) + \delta^k \text{ for all } x \in X \tag{4.13}$$

  end
  Set $z^k = [x^k]_r$;
  if $\|x^k - z^k\|_\infty > \mu^k$ **then**
    $\varepsilon_d^{k+1} = \max\{0.1\varepsilon_d^k, \varepsilon\}$; $\mu^{k+1} = \mu^k$; $\delta^{k+1} = \delta^k$;
  else
    $\varepsilon_d^{k+1} = \varepsilon_d^k$; $\mu^{k+1} = \max\{0.1\mu^k, \mu\}$; $\delta^{k+1} = \max\{0.9\delta^k, \delta\}$;
  end
  if $\Theta(x^k) > \eta^k$ **then**
    $\varepsilon_c^{k+1} = \max\{0.1\varepsilon_c^k, \varepsilon\}$; $\eta^{k+1} = \eta^k$; $\delta^{k+1} = \delta^k$;
  else
    $\varepsilon_c^{k+1} = \varepsilon_c^k$; $\eta^{k+1} = \max\{0.1\eta^k, \eta\}$; $\delta^{k+1} = \max\{0.9\delta^k, \delta\}$;
  end
  Set $k = k + 1$;
end

Algorithm 1: ObPA

In addition to forcing the integer variables to take integer values, another important issue is to reduce the overall general constraint violation measured by Θ. The ObPA has the ability to select the penalty objective function for the BCNLP problem. Either penalty (4.10) or (4.7) is used according to the general constraint feasibility level of the oracle. At iteration k, if $\Theta(o^*) \leq \eta^k$ then it is worth to penalize $|x_i - o_i^*|$ componentwise, so that an approximation near to the oracle is computed (and penalty (4.10) is used); otherwise, an approximation in the vicinity of the oracle is not of the upmost importance and the penalty (4.7) is used instead.

Besides the penalty parameters and the feasibility tolerance η^k, another parameter, μ^k, is required to check the level of integrality violation at the current solution x^k. Furthermore, the parameter δ^k represents the error bound which reflects the accuracy required for the current approximation x^k to the solution of the BCNLP problem.

Simple rules to control the reduction of parameters ε_d^k, ε_c^k, η^k, μ^k and δ^k are used and lower bounds are imposed to prevent the BCNLP problems of becoming very hard to solve. The penalty parameters ε_d^k and ε_c^k are reduced, using $\varepsilon_d^{k+1} = \max\{0.1\varepsilon_d^k, \varepsilon\}$ and $\varepsilon_c^{k+1} = \max\{0.1\varepsilon_c^k, \varepsilon\}$ respectively, when the corresponding violation measures ($\|x^k - z^k\|_\infty$ and $\Theta(x^k)$) at the computed approximation x^k are not satisfactory; otherwise, they are maintained.

The ObPA stops when an approximation x^k, which has a sufficiently small general constraints feasibility measure and is within an error of δ (in relative terms) of the known global solution, is computed. Thus, the stopping conditions are

$$\Theta(x^k) \leq \eta \text{ and } \frac{|f(x^k) - f^*|}{\max\{1, |f^*|\}} \leq \delta, \tag{4.14}$$

where η and δ are very small positive tolerances.

Remark 4.3 The use of the known global solution to stop the algorithm, during these preliminary tests, aims to analyze its effectiveness. In case f^* is not available, the second condition in (4.14) is replaced by the relative difference between the function values of two consecutive iterations less than or equal to the specified error tolerance, δ.

Finally, we now briefly elaborate on the global optimization method to solve the BCNLP problems formulated in (4.9) and (4.3). The deterministic algorithm DIRECT [25] is used. The problems to be addressed by DIRECT are defined in (4.9) and (4.3) in such a way that conditions (4.12) and (4.13) respectively are satisfied. The method does not require any derivative information and has been originally proposed to solve BCNLP problems, by producing finer and finer partitions of the hyperrectangles generated from X, and evaluating Ψ at their centers. The algorithm is a modification of the standard Lipschitzian approach that eliminates the need to specify the Lipschitz constant [25]. To perform a balance between global and local search, the algorithm makes use of two important concepts: potentially optimal hyperrectangle and grouping according to size. The center, c_i, the objective function value at the center point, $\Psi(c_i; \cdot)$, and the size, d_i, of each hyperrectangle i are used to define the groups of hyperrectangles, to select the potentially optimal hyperrectangles and to divide them into smaller ones, until a convergence condition is satisfied [27]. In the context of Algorithm 1, three stopping criteria were considered for DIRECT: (i) an error tolerance on the BCNLP objective penalty function value, δ^k, (ii) a maximum number of iterations, or (iii) a maximum number of function evaluations.

4.4 Numerical Experiments

To make a preliminary evaluation of the practical behavior of the proposed ObPA for solving nonconvex MINLP problems, we use a set of benchmark problems, identified as f1 to f29 in the subsequent tables (see [4, 17, 28]). The algorithm is implemented in Matlab™ (registered trademark of the MathWorks, Inc.) programming language. The algorithmic parameters are set as follows: $\eta = 1E-04$, $\delta = 1E-03$, $\mu = 1E-04$, $\varepsilon = 1E-05$, $\varepsilon_d^1 = 1$, $\varepsilon_c^1 = 0.1$, $\eta^1 = 0.1$, $\mu^1 = 0.1$. However, if the stopping conditions (4.14) do not hold for the given η and δ, ObPA is allowed to run for 30 iterations.

At each iteration k, when DIRECT is used to solve the BCNLP problems (4.9) or (4.3), by imposing the conditions (4.12) or (4.13) respectively, the error tolerance on the penalty function value is δ^k. We note that the parameter δ^1 is set to one, slowly decreases from one iteration to the other, until it reaches the value $\delta = 1E-03$. The maximum number of iterations is made to depend on the number of variables ($5n$ for f7; $10n$ for f3, f4, f8, f12, f14, f16, f18, f19, f24 and f26; $20n$ for f1, f5, f11 and f20; $50n$ for f9 and f17; $70n$ for f2, f22, f23, f28 and f29; $100n$ for f6, f13, f21 and f25; $150n$ for f15; $250n$ for f27; $300n$ for f10) and the maximum number of function evaluations is set to 50,000.

First, we compare the results produced by ObPA, as presented in Algorithm 1, with those obtained by a variant that does not use the oracle penalty, i.e., the BCNLP problem (4.3) is always solved in all iterations. See Table 4.1. The table shows the name of the problem, P, the best known optimal solution available in the literature, f^*, the solution produced by the algorithm, f_{sol}, the number of function evaluations required to achieved the reported solution, nfe, the number of iteration, nit, and the CPU time in seconds, T. From the results, it is possible to conclude that the proposed ObPA was able to find the global optimum for 20 of the 29 problems (according to the stopping conditions shown in (4.14) with $\eta = 1E-04$ and $\delta = 1E-03$). For the remaining nine problems, the algorithm run for 30 iterations. From the table, we may also conclude that the solutions obtained by the variant without the oracle penalty have been greatly deteriorated in three problems (f5, f7, f9) and slightly deteriorated in two (f11 and f12). The solutions for all the other problems are comparable, being f19 the only one with a slight improvement. Overall the results obtained by the proposed ObPA are superior to those of the tested variant.

Second, the results produced by ObPA are compared with those obtained by the BBMCSFilter, a BB-based multistart coordinate search filter method published in [18] and the results reported in [17], where a filter-based genetic algorithm (FGA) is presented. Table 4.2 shows the name of the problem, being the set f1–f12 also

Table 4.1 Numerical results produced by Algorithm 1 and by the variant without the oracle penalty

P	f^*	Algorithm 1				Variant without the oracle penalty			
		f_{sol}	nfe	nit	$T(sec.)$	f_{sol}	nfe	nit	T (sec.)
f1	2	2.000456	589	2	$1.29E-01$	2.000472	509	2	$1.10E-01$
f2	2.124	2.124481	5433	2	$2.54E+00$	2.124481	4891	2	$2.39E+00$
f3	1.07654	1.076392	1423	3	$5.96E-01$	1.076534	1233	3	$5.35E-01$
f4	99.239637	99.244695	629	2	$2.69E-01$	99.244695	523	2	$2.29E-01$
f5	3.557463	3.701380	103,049	30	$6.47E+01$	5.225669	87,569	30	$6.42E+01$
f6	4.579582	4.579600	88,843	3	$5.10E+01$	4.579600	77,111	3	$4.45E+01$
f7	−17	−16.691358	2039	30	$5.61E-01$	−10.333333	1757	30	$5.63E-01$
f8	−32217.4	−32215.640357	56,685	2	$4.39E+01$	−32215.640357	56,685	2	$4.62E+01$
f9	7.6671801	7.667232	20,523	4	$1.02E+01$	8.240213	198,977	30	$1.02E+02$
f10	−2.4444	−2.438023	354,975	30	$8.95E+01$	−2.438023	308,273	30	$7.97E+01$
f11	3.2361	3.236034	1417	2	$7.06E-01$	3.260172	21,901	30	$1.10E+01$
f12	1.125	1.125301	263	2	$5.94E-02$	1.132343	6911	30	$1.55E+00$
f13	87.5	89.500017	707,913	30	$3.22E+02$	89.500051	591,301	30	$2.74E+02$
f14	−6.666667	−6.666514	241	2	$8.68E-02$	−6.666514	223	2	$1.10E-01$
f15	−5.6848	−5.684732	14,315	3	$7.53E+00$	−5.684732	12,789	3	$6.79E+00$
f16	2.000	2.000119	1873	2	$8.24E-01$	2.000356	1549	2	$7.01E-01$
f17	3.4455	3.445514	5941	3	$1.19E+00$	3.445514	5235	3	$1.08E+00$
f18	2.2000	2.200032	5097	4	$2.28E+00$	2.200198	1445	2	$6.54E-01$
f19	6.00972	6.548438	39,871	30	$2.36E+01$	6.424818	35,395	30	$2.09E+01$
f20	−17.0000	−16.999953	16,871	6	$7.87E+00$	−16.999953	14,627	6	$7.03E+00$

(continued)

Table 4.1 (continued)

P	f^*	Algorithm 1				Variant without the oracle penalty			
		f_{sol}	nfe	nit	T (sec.)	f_{sol}	nfe	nit	T (sec.)
f21	−4.514202	−4.514198	25,999	4	$1.46E+01$	−4.514154	23,529	4	$1.39E+01$
f22	−13.401904	−13.401855	67,081	4	$3.60E+01$	−13.401855	36,605	3	$1.90E+01$
f23	−1.08333	−1.078680	206,889	30	$9.65E+01$	−1.078667	204,335	30	$9.44E+01$
f24	−0.94347	−0.664913	300,055	30	$1.98E+02$	−0.664913	267,527	30	$1.73E+02$
f25	189.3116	189.375606	14,855	4	$7.52E+00$	189.375388	13,161	4	$6.80E+00$
f26	31	31.000339	777	4	$1.85E-01$	31.001016	685	4	$1.60E-01$
f27	−32	−31.998899	34,169	4	$8.32E+00$	−31.998628	30,237	4	$6.90E+00$
f28	73.0353	78.769766	1425,125	30	$9.96E+02$	78.769766	1423,437	30	$1.00E+03$
f29	−1.923	−0.913446	991,839	30	$9.19E+02$	−0.913446	1451,577	30	$1.12E+03$

Table 4.2 Numerical results produced by the Algorithm 1, the BBMCSFilter in [18] and the FGA in [17]

P	$(\|I_c\|,\|I_d\|)$	Algorithm 1		BBMCSFilter †			FGA §		
		f_{sol}	nfe	f_{avg}	SD	nfe_{avg}	f_{avg}	SD	nfe_{avg}
f1	(1,1)	2.000456	589	2.000817	$3.6E-04$	3530	2.0000	$1.6E-06$	4530
f2	(1,1)	2.124481	5433	2.124590	$1.4E-06$	1259	2.1852	$6.1E-02$	3799
f3	(2,1)	1.076392	1423	1.081640	$8.1E-03$	5274	1.0769	$3.8E-04$	5752
f4	(2,1)	99.244695	629	99.239635	$1.0E-07$	670	99.5784	$3.4E-01$	9854
f5	(3,4)	3.701380	103,049	3.560848	$2.0E-03$	76,775	3.6822	$1.2E-01$	11,492
f6	(3,4)	4.579600	88,843	4.582322	$9.3E-04$	75,413	4.8048	$2.3E-01$	9937
f7	(1,1)	−16.691358	2039	−16.998054	$2.3E-03$	4296	−16.8267	$1.7E-01$	4147
f8	(3,2)	−32215.640357	56,685	−32217.428	$0.0E+00$	18,051	−32217	$2.7E-02$	6609
f9	(2,3)	7.667232	20,523	7.667583	$9.5E-04$	28,090	7.7472	$8.0E-02$	11,480
f10	(1,1)	−2.438023	354,975	−2.444444	$0.0E+00$	2736	−2.444	$4.4E-04$	4125
f11	(1,2)	3.236034	1417	3.236121	$8.7E-05$	41,635	3.3395	$1.0E-01$	5028
f12	(1,1)	1.125301	263	1.125115	$2.9E-04$	7770	1.125	$1.4E-06$	4757
f13	(2,2)	89.500017	707,913	87.507043	$1.7E-02$	41,852	–	–	–
f14	(1,1)	−6.6665143	241	−6.666131	$1.8E-04$	1122	–	–	–
f15	(1,2)	−5.684732	14,315	−5.651952	$2.6E-02$	393,345	–	–	–
f16	(2,2)	2.000119	1873	2.000000	$0.0E+00$	29,847	–	–	–
f17	(1,1)	3.445514	5941	3.445808	$2.1E-04$	5469	–	–	–
f18	(1,3)	2.200032	5097	2.200000	$0.0E+00$	11,182	–	–	–
f19	(4,2)	6.548438	39,871	6.010714	$6.6E-04$	37,132	–	–	–
f20	(2,3)	−16.999953	16,871	−16.994605	$5.5E-03$	27,149	–	–	–
f21	(1,3)	−4.514198	25,999	−4.513448	$6.8E-04$	50,146	–	–	–

(continued)

Table 4.2 (continued)

		Algorithm 1		BBMCSFilter [†]			FGA [§]		
P	$(\|I_c\|,\|I_d\|)$	f_{sol}	nfe	f_{avg}	SD	nfe_{avg}	f_{avg}	SD	nfe_{avg}
f22	(2,4)	−13.401855	67,081	−13.401930	$3.6E-04$	84,790	–	–	–
f23	(2,2)	−1.078680	206,889	−1.083245	$5.4E-05$	2458	–	–	–
f24	(3,8)	-0.664913	300,055	–	–	–	–	–	–
f25	(2,1)	189.375606	14,855	–	–	–	–	–	–
f26	(0,2)	31.000339	777	–	–	–	–	–	–
f27	(1,1)	−31.998899	34,169	–	–	–	–	–	–
f28	(6,5)	78.769766	1425,125	–	–	–	–	–	–
f29	(5,3)	−0.913446	991,839	–	–	–	–	–	–

[†] The NLP relaxation is stopped after 10 sample points are generated in the multistart algorithm and 30 runs are executed

[§] The algorithm stops when a solution with error $1E-3$ is found or the number of function evaluations reaches 10,000; $P_s = 20$, $R = 50$

Table 4.3 Other numerical comparisons

P	Algorithm 1		EXP-MIP		4-rule FA ♭		MIHDE §		ACOmi †		PSO ‡		pen-GA ♮	
	f_{sol}	$nfe\ (nit)$	f_{exp}	# nod.	f_{avg}	nfe_{avg}	f_{avg}	nfe_{avg}	f_{avg}	nfe_{avg}	% suc.	nfe_{avg}	% suc.	nfe^{suc}_{avg}
f1	2.000011	1589 (2)	–	–	2.0000	3409	–	13,104	–	–	–	–	84	172
f2	2.124476	13,449 (2)	–	–	2.7149	5253	–	29,166	–	–	100	3500	85	64
f3	1.076392	1423 (3)	1.076	0	1.0767	5178	–	28,455	1.1459	4250	–	–	43	18,608
f4	99.244695	629 (2)	–	–	–	–	–	60,950	–	–	100	4000	59	7447
f5	3.701662	38,287 (11)	–	–	–	–	–	12,375	–	–	–	–	41	3571
f6	4.579600	33,859 (3)	4.579	2	4.7758	12,157	–	–	4.5796	731	100	30,000	–	–
f7	−16.998720	1501 (2)	−17	1	−16.9998	3243	–	983	−17	307	–	–	–	–
f8	−32215.640357	56,685 (2)	–	–	–	–	–	50,976	–	–	–	–	100	100
f9	7.667232	20,523 (4)	7.667	2	8.0695	8622	–	–	7.6672	363	–	–	–	–
f10	−2.438023	12,395 (3)	–	–	−2.4380	3501	–	–	−2.4444	270	–	–	–	–
f11	3.236034	1397 (2)	–	–	3.2361	4405	–	–	23.475	1180	–	–	–	–
f24	−0.686926	62,391 (3)	−0.912	1	–	–	–	–	–	–	–	–	93	258
f26	31.000339	801 (4)	31	1	–	–	–	–	–	–	–	–	–	–
f29	−1.393493	36,191 (5)	–	–	–	–	–	–	–	–	88	40,000	–	–

♭ Termination conditions: $|f^k - f^*|/|f^*| \leq 1E-04$ and violation $\leq 1E-03$; $P_s = 20$, $R = 30$
§ Termination condition: $|f^{k+20} - f^k| < 1E-05$ or a maximum of 2000 iterations; $P_s = 3$, $R = 10$
† Algorithm stops when a solution with error $1E-03$ is reached or a maximum of 10,000 function evaluations is attained; $P_s = 20$, $R = 30$
‡ Termination conditions: $|f^{k+50} - f^k| < 1E-05$ or a maximum of 200 iterations; $P_s = 50$, $R = 100$
♮ Termination conditions: $|f^k - f^*| \leq 1E-02$ or a maximum of 200 iterations; $P_s = 10n$, $R = 100$

used in [17] and the set f1–f23 used in [18]. In the second column of the table, the pair inside parenthesis corresponds to $(|I_c|,|I_d|)$. The remaining columns contain: the solution produced by ObPA, f_{sol}, and the number of function evaluations, nfe, the average value of the objective function values produced by all the executed runs (with BBMCSFilter and FGA), f_{avg}, the standard deviation of the function values, SD, and the average number of function evaluations (over all the runs), nfe_{avg}. The character '–' in the tables means that the information is not available in the cited papers, 'P_s' is the size of the population and 'R' gives the number of independent executed runs. From the comparison, we may conclude that the produced solutions are of good quality. For most problems, the number of required function evaluations is moderate when compared with the numbers produced by the other algorithms, with the exception in nine problems where it is much higher. In seven of these problems, the algorithm reached 30 iterations since one of the conditions in (4.14) was not satisfied. Thus, from the comparison with the BBMCSFilter and FGA, the ObPA proves to be competitive either in terms of the quality of the found solutions or in the number of function evaluations.

Finally, using a small subset of the problems, we compare our results with those reported by other strategies. Table 4.3 reports the solution produced by Algorithm 1, f_{sol}, the number of function evaluations, nfe, and the number of iterations, nit. The algorithm is made to stop when a solution with an error of $1E-03$ is reached or a maximum of $5000n$ function evaluations is attained. The other results in the table are collected from the exact penalty for mixed-integer programs (EXP-MIP) in [13], the 4-rule FA in [20], the MIHDE in [14], the extended version of the ant colony optimization (ACOmi) in [19], the particle swarm optimization (PSO) in [15] and the penalty GA (pen-GA) in [16]. The table also shows the solution found by EXP-MIP, f_{exp}, and the number of nodes (corresponding to the number of branch and reduce iterations), '# nod.'.

As far as the stochastic heuristics are concerned, Table 4.3 shows: the average of the objective function values (over all the executed runs), f_{avg}, the average number of function evaluations, nfe_{avg}, the percentage of successful runs (according to the stopping condition based on the proximity of f to f^*), % suc., and the average number of function evaluations of the successful runs alone, nfe^{suc}_{avg}. From the results we may conclude that the proposed ObPA performs reasonably well.

4.5 Conclusions

In this paper, an oracle-based penalty approach for solving nonsmooth and nonconvex MINLP problems is proposed. A continuous reformulation BCNLP problem is solved by the deterministic DIRECT solver. The penalty function to be optimized involves a combination of penalty terms to penalize the integrality constraints, the equality and inequality constraints and the distance to the oracle, based on hyperbolic tangent penalty functions. The numerical experiments show that the proposed algorithm gives competitive results when compared with other methods in the literature.

Future developments will be directed to improve the efficiency of the oracle-based penalty algorithm, in terms of the number of function evaluations, by using an alternative deterministic and derivative-free global optimizer to solve the continuous BCNLP problems.

Acknowledgements The authors would like to thank two anonymous referees for their valuable comments and suggestions to improve the paper.

This work has been supported by COMPETE: POCI-01-0145-FEDER-007043 and FCT - Fundação para a Ciência e Tecnologia, within the projects UID/CEC/00319/2013 and UID/MAT/00013/2013.

References

1. O. Exler, T. Lehmann, K. Schittkowski, A comparative study of SQP-type algorithms for nonlinear and nonconvex mixed-integer optimization. Math. Program. Comput. **4**(4), 383–412 (2012)
2. S. Burer, A.N. Letchford, Non-convex mixed-integer nonlinear programming: a survey. Surv. Oper. Res. Manag. Sci. **17**(2), 97–106 (2012)
3. S. Lee, I.E. Grossmann, A global optimization algorithm for nonconvex generalized disjunctive programming and applications to process systems. Comput. Chem. Eng. **25**(11), 1675–1697 (2001)
4. H.S. Ryoo, N.V. Sahinidis, Global optimization of nonconvex NLPs and MINLPs with applications in process design. Comput. Chem. Eng. **19**(5), 551–566 (1995)
5. C.S. Adjiman, I.P. Androulakis, C.A. Floudas, Global optimization of mixed-integer nonlinear problems. AIChE J. **46**(9), 1769–1797 (2000)
6. V.K. Srivastava, A. Fahim, An optimization method for solving mixed discrete-continuous programming problems. Comput. Math. Appl. **53**(10), 1481–1491 (2007)
7. L. Liberti, G. Nannicini, N. Mladenović, A good recipe for solving MINLPs, in *Matheuristics: Hybridizing Metaheuristics and Mathematical Programming*, vol. 10, ed. by V. Maniezzo, T. Stützle, S. Voß (Springer, US, 2010), pp. 231–244
8. G. Nannicini, P. Belotti, Rounding-based heuristics for nonconvex MINLPs. Math. Program. Comput. **4**(1), 1–31 (2012)
9. C. D'Ambrosio, A. Frangioni, L. Liberti, A. Lodi, A storm of feasibility pumps for nonconvex MINLP. Math. Program. **136**(2), 375–402 (2012)
10. G. Liuzzi, S. Lucidi, F. Rinaldi, Derivative-free methods for bound constrained mixed-integer optimization. Comput. Optim. Appl. **53**(2), 505–526 (2012)
11. M.F.P. Costa, A.M.A.C. Rocha, R.B. Francisco, E.M.G.P. Fernandes, Firefly penalty-based algorithm for bound constrained mixed-integer nonlinear programming. Optimization **65**(5), 1085–1104 (2016)
12. S. Lucidi, F. Rinaldi, Exact penalty functions for nonlinear integer programming problems. J. Optim. Theory Appl. **145**(3), 479–488 (2010)
13. S. Lucidi, F. Rinaldi, An exact penalty global optimization approach for mixed-integer programming problems. Optim. Lett. **7**(2), 297–307 (2013)
14. Y.-C. Lin, K.-S. Hwang, F.-S. Wang, A mixed-coding scheme of evolutionary algorithms to solve mixed-integer nonlinear programming problems. Comput. Math. Appl. **47**(8–9), 1295–1307 (2004)
15. L. Yiqing, Y. Xigang, L. Yongjian, An improved PSO algorithm for solving non-convex NLP/MINLP problems with equality constraints. Comput. Chem. Eng. **31**(3), 153–162 (2007)
16. K. Deep, K.P. Singh, M. L. Kansal, C. Mohan, A real coded genetic algorithm for solving integer and mixed integer optimization problems. Appl. Math. Comput. **212**(2), 505–518 (2009)

17. A. Hedar, A. Fahim, Filter-based genetic algorithm for mixed variable programming. Numer. Algebr. Control Optim. **1**(1), 99–116 (2011)
18. F.P. Fernandes, M.F.P. Costa, E.M.G.P. Fernandes, Branch and bound based coordinate search filter algorithm for nonsmooth nonconvex mixed-integer nonlinear programming problems, in *Computational Science and Its Applications – ICCSA 2014, Part II*, LNCS, vol. 8580, ed. by B. Murgante, S. Misra, A.M.A.C. Rocha, C. Torre, J.G. Rocha, M.I. Falcão, D. Taniar, B.O. Apduhan, O. Gervasi (Springer, Berlin, 2014), pp. 140–153
19. M. Schlüter, J.A. Egea, J.R. Banga, Extended ant colony optimization for non-convex mixed integer nonlinear programming. Comput. Oper. Res. **36**(7), 2217–2229 (2009)
20. M.F.P. Costa, A.M.A.C. Rocha, R.B. Francisco, E.M.G.P. Fernandes, Extension of the firefly algorithm and preference rules for solving MINLP problems, in *International Conference of Numerical Analysis and Applied Mathematics (ICNAAM 2016)*, AIP. Conf. Proc. **1863**, 270003-1–270003-4 (2017)
21. P. Belotti, C. Kirches, S. Leyffer, J. Linderoth, J. Luedtke, A. Mahajan, Mixed-integer nonlinear optimization. Acta Numer. **22**, 1–131 (2013)
22. F. Boukouvala, R. Misener, C.A. Floudas, Global optimization advances in mixed-integer nonlinear programming, MINLP, and constrained derivative-free optimization, CDFO. Eur. J. Oper. Res. **252**(3), 701–727 (2016)
23. M.R. Bussieck, S. Vigerske, MINLP solver software, in *Wiley Encyclopedia of Operations Research and Management Science*, ed. by J.J. Cochran, L.A. Cox, P. Keskinocak, J.P. Kharoufeh, J.C. Smith (Wiley, New York, 2011)
24. A.M.A.C. Rocha, M.F.P. Costa, E.M.G.P. Fernandes, Solving MINLP problems by a penalty framework, in *Proceedings of XIII Global Optimization Workshop*, ed. by A.M. Rocha, M.F. Costa, E. Fernandes, (2016), pp. 97–100
25. D.R. Jones, C.D. Perttunen, B.E. Stuckman, Lipschitzian optimization without the Lipschitz constant. J. Optim. Theory Appl. **79**(1), 157–181 (1993)
26. M. Schlüter, M. Gerdts, The oracle penalty method. J. Glob. Optim. **47**(2), 293–325 (2010)
27. D.E. Finkel, DIRECT Optimization Algorithm User Guide, Center for Research in Scientific Computation. (CRSC-TR03-11, North Carolina State University, Raleigh, NC 27695-8205, March 2003)
28. C.A. Floudas, P.M. Pardalos, C. Adjiman, W.R. Esposito, Z.H. Gümüs, S.T. Harding, J.L. Klepeis, C.A. Meyer, C.A. Schweiger, *Handbook of Test Problems in Local and Global Optimization*, Nonconvex Optimization and its Applications (Springer Science & Business Media, Dordrecht, 1999)

Chapter 5
A Biased-Randomized Heuristic for the Home Healthcare Routing Problem

Manuel Eliseu, M. Isabel Gomes and Angel A. Juan

Abstract The home healthcare routing problem (HHRP) refers to the problem of allocating and routing caregivers to care-dependent people at their homes. It has been mostly tackled in the literature as a rich vehicle routing problem with time windows. This paper proposes a biased-randomized heuristic, based on the well-known savings heuristic, to solve the HHRP. The algorithm is tested in small but real-case instances where patients' visits may occur more than once a day and, in such cases, all the visits have to be performed by the same caregiver. The results show the algorithm provides good quality results in reasonably low computing times.

Keywords Home healthcare · Vehicle routing problem with time windows Biased-randomized heuristic · Real case instances

5.1 Introduction

The increase in average life expectancy, as a result of new developments in medicine, along with the decrease of the birth rate in developed countries is making the so called "modern society" to grow older [12]. The decrease of informal care of the elderly is leading families to seek for institutionalization solutions, uprooting their relatives from the environment they are so deeply attached. These services may vary from social support, palliative care, personal care and/or food supply. The main benefits of home healthcare services include people's preference of remaining at home [5],

---

M. Eliseu
Faculty of Sciences and Technology, Nova University of Lisbon, Lisbon, Portugal
e-mail: m.eliseu@campus.fct.unl.pt

M. I. Gomes (✉)
Centro de Matemática e Aplicações, Faculty of Sciences and Tecnology,
Nova University of Lisbon, Lisbon, Portugal
e-mail: mirg@fct.unl.pt

A. A. Juan
IN3 - Computer Science Deptartment, Open University of Catalonia, Barcelona, Spain
e-mail: ajuanp@uoc.edu

A. I. F. Vaz et al. (eds.), *Operational Research*, Springer Proceedings
in Mathematics & Statistics 223, https://doi.org/10.1007/978-3-319-71583-4_5

preventing social isolation [15] and a better cost-efficiency ratio when compared to the provision of these services in institutions [16]. The Portuguese home healthcare services are mostly provided by private companies or charity organizations, with the latter considerably outnumbering the former. One of the major problems faced by home healthcare service providers is staff assignment and scheduling. Too often, these tasks are performed manually, thus requiring a huge amount of time and, given the complexity of such decisions, leading to poor quality scheduling and routing plans.

In this work we address the home healthcare routing problem (HHRP) faced by a non-profit organization operating in the Lisbon region. The service is managed by a social worker who is in charge of planning the tasks of 6 caregivers, who are working in teams of two. Given the nearness of the patients to be visited, the caregivers walk between patients' homes and the Parish Day Center. Every week, the social worker needs to provide each team with a list of patients and the visiting plan, so that all patients have their needs fulfilled. All the planning is done with pen and paper, and although she knows a more efficient planning can be done, she lacks the tools and the knowledge to develop them. This paper presents the first step to create a decision support tool for solving the HHRP. We propose an approach based on a biased-randomized version of a well-known routing heuristic, which can be easily embedded into a spreadsheet for facilitating managerial use.

This paper will develop as follows. In the next section a short literature review is presented focusing on the heuristic and meta-heuristic approaches that have been used to solve the HHRP problem so far. Next, an illustrative case will be introduced and compared with the traditional vehicle routing problem with time windows (VRPTW). In Sect. 5.4, details on the solving methodology are provided. Results will be presented and discussed in Sect. 5.5. Lastly, some conclusions and future work are given.

5.2 Literature Review

The HHRP fits within the resource planning and allocation problem [13]. Its operational level of decision has been mostly tackled in the literature as a rich VRPTW, as shown in the recent review of Fikar and Hirsch [8]. This is a very well known problem that has been deeply studied by the academia. However, the existing models do not cover some of the particularities one finds in the HHRP: continuity of care, nurses' skills that have to match patients' needs, and work regulations, among others.

The first works concerning the HHRP were published between 1998 and 2006. They addressed the problems in national context and proposed decision support systems (DSS) that integrated GIS technology. The first one was published in 1998 by Begur et al. [2]. These authors developed a DSS for the Visiting Nurse Association, in USA, to help them planning the allocation of nurses to patients and determine the daily visits sequence for each nurse. This DSS routing software is based on a well-known routing heuristic and provides simultaneously the assignment of patients and the routing for each nurse that minimizes the total travel time. Later in 2006, Bertels and Fahle [4] combined different meta-heuristics and exact approaches to address the nurse rostering problem and routing decisions taking into account patients' and nurses' preferences, legal aspects, nurses' qualifications, ergonomics, and other aspects. The developed algorithms were embedded into a DSS, which according to the authors can handle most real-life HHRPs. In the same year, Eveborn et al. [6] developed a different DSS, this time for a Swedish HHRP. In order to daily plan workers scheduling and patients' visits, they developed a heuristic based on the matching and set partitioning problems, where previously designed schedules were allocated to workers assuring that all patients were visit exactly once.

Since then, a very interesting amount of works have been published. Single- or multi-period problems, single- or multi-objective, and exact, heuristics, or combined solution approaches can already be found in the literature (see [8] for a very recent literature review). Although our problem is intrinsically a multi-period one, at this first step we addressed it as a single-period problem. Moreover, our problem is quite a small one and our main constraints are to assure that all visits to a patient are assigned to only one team ("loyalty" constraint), patients' time-windows are met, and that all teams have a mandatory lunch break at 1 P.M., which takes place at the day care center. Accordingly, we will focus on single-period problems with time-windows and mandatory breaks.

In 2007, Akjiratikari et al. [1] addressed the scheduling problem for home care workers in UK. These authors developed a particle swarm optimization meta-heuristic to design the visiting routes, so that the total distance traveled is minimized while capacity and time-windows constraints are satisfied. In 2011, Bachouch et al. [3] developed a mixed-integer linear model based on the VRPTW. Their model accounts for workers' skills, lunch breaks, working time regulations, and shared visits to patients. In their work, all patients are visit once, which means no loyalty constraints are needed.

In 2013, Hiermann et al. [9] studied the HHRP in a urban context considering that nurses could use different transportation modes for traveling between visits. They proposed and compared different meta-heuristic approaches and integrated them into a two-stage approach. This work was part of a larger project related with inter-modal transportation in Vienna. Also in Austria, Rest and Hirsh [14] tackle the HHRP as a time-dependent vehicle routing problem since workers travel by public transportations in an urban environment. These authors propose several methods, based on tabu search, to account for time-dependencies and multi-modality in transportation.

The above works have addressed problems with a considerable number of features that are not present in our particular problem at Lisbon. Therefore, a simpler but effective heuristic was needed to address our HHRP. The well-known savings heuristic has been applied in one of the first works to solve the HHRP [2], and it has recently been embedded in a meta-heuristic approach developed by Juan et al. [11]. Given the promising results published in the latter work and its relative simplicity, we decided to adapt it to our problem. Among the issues that appealed us are the existence of only one parameter to tune and the possibility to provide the decision maker with alternative good solutions.

5.3 Problem Description

This work is motivated by a real case study of a Portuguese catholic parish. This community offers several social services to population that lives nearby: meal delivery, activities of the daily living, adult day care, and transportation. The daily schedule of teams of two caregivers has to be planned so that all patients' requests are met. The request vary from twice a day to two days a week. Three teams of two caregivers perform activities of the daily living (such as bathing, dressing, medication assistance, home cleaning, etc.) in each visit. Each team should depart from the Parish Social Center and return there at the end of the day. At 1 P.M. they also go back to the Parish Social Center to have lunch (lunch-break). One of the teams has to arrive one hour earlier to help on preparing the meals. In short, the routing solution must fulfil the following constraints:

- Each patient must be visited by exactly one team.
- All teams depart from, and return to, the Parish Social Centre.
- Each visit must start within a given time window, previously defined.
- Each visit has a pre-defined duration which varies according to the activities performed.
- The working hours for caregivers vary from 08:00 to 16:00, or from 08:00 to 17:00, according to the day of the week.
- Lunch break: there is a mandatory break at the Parish Social Center of one hour duration, starting at 13:00.
- Among the three teams, one must return to the Parish Social Center at 12:00 to help on meals preparation and delivery.
- A patient with more than one visit scheduled for the day must be visited by the same team throughout all visits.

The first four constraints are the traditional ones for the VRPTW if we look into teams as "vehicles" and patients as "customers". The remaining four constraints are specific of the HHRP. Although in vehicle routing problems a customer might be visited more than once a day, the visits can be assigned to different vehicles. However, in the HHRP we are usually dealing with older people, which makes it convenient to assign the same team of nurses that have visited them earlier in the day.

The problem is defined on a graph $G = (N, A)$, the social centre corresponds to nodes 0 and $n + 1$, being the latter a replica of the former. As variables we defined a binary one, x_{ijk}, $(i, j) \in A$, $k \in K$ that has the value 1 if arc *(i, j)* is crossed by team k and 0 otherwise, as well as, a time variable w_{ik}, $i \in N, k \in K$ specifying service starting time at node i by team k. As objective function we considered the total walking distance, where c_{ij}, $(i, j) \in A$, represents the length of arc *(i, j)* (Eq. 5.1).

$$min \sum_{k \in K} \sum_{(i,j) \in A} c_{ij} x_{ijk} \tag{5.1}$$

5.4 Solving Approach

Our solving methodology is based on the MIRHA approach proposed by Juan et al. [11], which combines a classical greedy heuristic with a biased-randomization process and a local search.

The MIRHA Framework

The MIRHA framework is a two phase multi-start method: first, a biased-randomization of a classical heuristic generates an initial solution; then, this initial solution is iteratively improved by using a local search procedure. Being a generic framework, the choices concerning the classical heuristic and the local search strategy depend on the problem under study. In the case of the vehicle routing problem, authors propose the integration of the classical savings heuristic with Monte Carlo simulation as the approach to generate the initial solution [10]. For the local search phase, a divide-and-conquer strategy takes the solution apart allowing for smaller sub-solutions to be improved. One of the advantages of this approach, when compared with other meta-heuristics, is its simplicity and the few number of parameters that require a tuning process.

In many ways, MIRHA is similar to the GRASP meta-heuristic framework [7]. The construction of the solution is based on the evaluation of specific elements and their expected influence on the final solution. Both procedures make use of lists. However, while GRASP limits the number of candidates in the list to be considered and assumes all candidate elements to have the same probability of being selected (uniformly distributed), MIRHA does not limit the number of candidates in the list and it assigns a higher probability to those elements that are more promising (Fig. 5.1).

The savings heuristic starts by building an initial solution where each customer is visited in separated routes, thus having one vehicle for each customer. Then, routes are iteratively merged so that "nearby" customers can be included in the same route. The merging criteria is based on the savings concept: visiting two customers in the same route is "cheaper" than visiting each one directly from the depot (depot–customer–depot). One major disadvantage of the savings heuristic is its greediness, i.e., it always merges the routes connected by the edge at the top of the list of candidates.

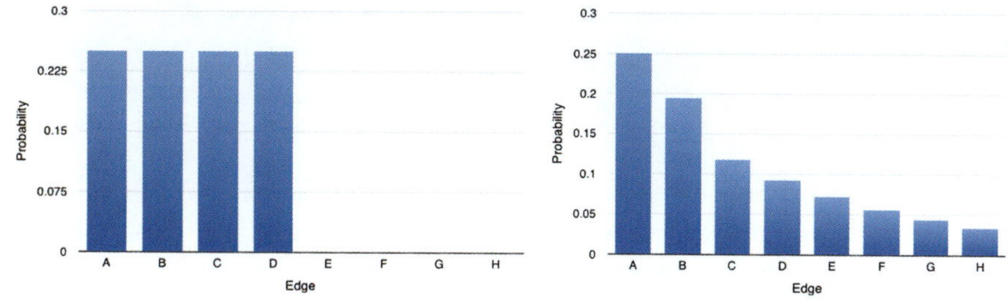

Fig. 5.1 Uniform (left) and Biased (right) randomized selection differences

Based on the savings concept, our algorithm assigns a probability to each item on the savings list, reflecting its quality. Therefore, as one goes down the list, the corresponding probability of being selected also decreases. Some experiments suggest the use of a Geometric probability distribution with parameter α, $0.05 \leq \alpha \leq 0.25$ (also randomly determined by the heuristic). The merging of routes is iteratively carried out until the savings list is empty. To further improve the solution, the heuristic is embedded into a multi-start procedure with a memory mechanism. This last feature stores "the best solution found"; this is to mean, it stores the order in which the nodes were visited in each route and the corresponding distance. When, in a new iteration, a solution contains a route with the same set of nodes as the ones stored in cache, the two solutions are compared (the new and cache one) and the best order will be the one used in the final solution. If the new order is the best one, the cache is updated. This approach is named as the cache procedure and has been successfully applied in [10, 11].

As mentioned above, the HHRP can be viewed as a VRPTW with some additional constraints. Therefore, we have adapted the previously described approach to fit our problem: no capacity constraints, time windows restrictions, and a fix number of routes.

The Adapted Procedure

When analysing patients' time windows, several cases show up: only morning visits, only afternoon visits, more than one visit (at least one in the morning and one in the afternoon), or no time window (for those patients that can be visited at any time during the day). So, taking advantage of these time windows, the MIHRA approach was adapted to fit the HHRP problem as shown in Fig. 5.2.

Firstly a morning solution is created by applying an efficient routing algorithm and assuring this time windows are met. At this first step, only the patients who have to be visited in the morning are considered. Then, the morning solution is used

```
Algorithm Heuristic for the HCP
while (elapsed time < time limit) do
morningsol <- build RandCWSsolution
     if morningsol.number_of_routes == number_of_teams
          afternoonsolTemplate = morningsol
          remove non shared from afternoonsolTemplate
          add afternoon patients
          adjust afternoon solution
          newSol = merge morning and afternoon solutions
          add  all day patients to newSol
     try improvement of newSol with cache memory
if newSol < bestSol
     newSol = bestSol
return bestSol
```

Fig. 5.2 Algorithm: pseudo-code for the proposed solving approach as a template for the afternoon solution, assuring that patients needing more than one visit will be assigned to the same team. The patients needing only one visit are removed from the route, since they have already been visited. The next step inserts patients needing only to be visited during the afternoon. They are added to the route with the minimum inserting time and assuring the time windows. To assure feasibility concerning these time windows, a node is only inserted into a route if the time difference between the two existing nodes is large enough to accommodate the new one. If nodes have very tight time windows and one node cannot be inserted in any of the existing routes, a new one is created. Lastly, those patients who have no constraints regarding the visiting period are inserted in one route again following a minimum insertion criteria and assuring the solution feasibility.

At this point, all patients have been assigned to a team. The final step performs a local improvement considering each route as a travelling salesman problem with time windows and taking advantage of a cache memory, which saves the best results from previous iterations to improve, whenever possible, the current solution.

The major differences between the original MIRHA approach and the one proposed for the HHRP are: *(i)* the morning solution is only accepted if the number of routes is the same as the number of teams; *(ii)* the α parameter of the Geometric distribution is not randomly determined; and *(iii)* time windows are imposed on nodes. Notice that, since teams have a mandatory lunch break, morning and afternoon routes could have been designed independently. However, in that case we could not guarantee that the loyalty constraints were satisfied.

Setting Running Times

In order to determine the running time, some tests were performed. The α value was set to a fixed value of 0.15 since, according to Juan et al. [10], good solutions were achieved for $\alpha \in [0.05, 0.2]$. This α value was then optimized (section below). Three instances were run for each time value. The average distance of each time limit is shown in Fig. 5.3. Given these results, the time limit was set to 500 s.

Setting the Value of α

As mentioned above, the Geometric distribution parameter, α, is fixed instead of being chosen randomly as in previous works. This parameter defines the Geometric

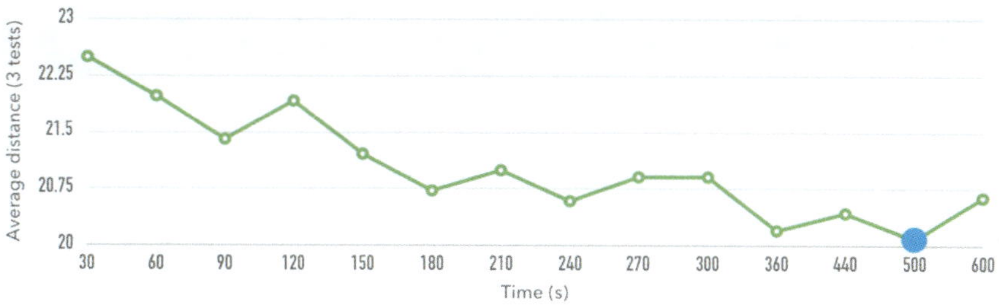

Fig. 5.3 Average distance for different iteration times

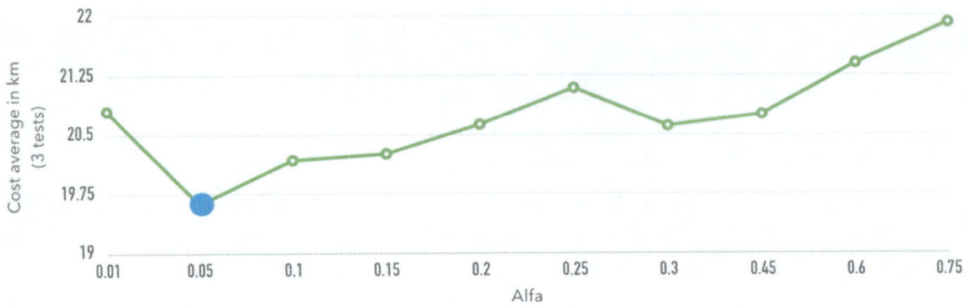

Fig. 5.4 Average distance for different α values

distribution that is used to calculate the probability of selection of each candidate in the savings list. Juan et al. [10] have found near-optimal results with values of α between 0.05 and 0.25. To assess the influence of the parameter α on the performance of our algorithm, we tested 10 different values and limited the runs to 500 s. The average results of three runs are shown in Fig. 5.4. These results allow us to conclude that the values referred in the work of Juan et al. [10] are the α values that provide better objective function values, therefore we set α to 0.05.

5.5 Results

The aforementioned algorithm was coded in Java and run on a personal computer with the OS X 10.11.6, an Intel Core i5 at 2.3 GHz, and 16 GB memory.

Table 5.1 shows the main characteristics of our HHRP instance together with some results. There are between 21–23 patients to visit each day of the week (# nodes) where some of them need to be visited more than once (# multiple visits). Thus, in total, 29–32 visits have to be scheduled and assigned to the three teams. It also presents the total walking and free time (both in minutes). The total walking time

Table 5.1 HHRP instance data by week day. Walking time (objective function) and free times are in minutes

Instance	# Nodes	# Multiple visits	Walking time	Free time total (on street/at centre)
Monday	21	11	195.9	315 (202/113)
Tuesday	22	8	228.9	469 (178/291)
Wednesday	21	9	224.7	306 (151/155)
Thursday	23	5	259.4	352 (125/227)
Friday	21	11	206.2	255 (70/185)

varies from 3 to 4.5 h, an average of 1–1.5 h per team. From meetings we had with the social worker in charge of this service, we know she thought they were working at their full capacity. However, the free time column shows that there is capacity to accommodate more visits. The total free time varies from 4 to about 8 h, representing the free time between visits about 42% of the total.

Figure 5.5 illustrates the routes the teams could perform on Monday morning and afternoon. The node colors indicate when the visits will take place: one morning or afternoon visit (black), visit any time of the day (orange) and multiple visits (green). The morning tours are larger than the afternoon tours since these two periods have different durations: mornings correspond to a 5-h period, while the afternoons have 3 or 4 h, depending on the day. Therefore, most patients with a full day time windows are mostly assigned to the morning visits.

When analysing routes among teams, one sees that team #2 (the red team) has the smallest area to cover and that its morning route has a "subtour". In fact, the "subtour" is caused by two morning visits that have to be made to patient 215, one early in the morning and a second before lunch time. Another aspect are the two "crossings" in team #3 morning route and team #1 afternoon route. This latter crossing can be avoided, as all patients have the same time window (not shown). Lastly, the routes are not balanced in terms of walking distance since, in the heuristic, no mechanism was considered to take this aspect into consideration.

Table 5.2 shows in detail the scheduling plan for team #1 (the yellow team). The first column shows the patient ID and the number of the visit (for instance, patient 267 has the first visit right after 8 a.m., and the second visit in the afternoon). This team has almost no free time since the difference between finishing the work at one patient and starting the work at the next one is spend on walking between both houses.

The HHCP has been also formulated as a MILP model. However, after 5 h, CPLEX was unable to provide solutions with a low gap with respect to the optimal solution. After those 5 h, the gap offered by Cplex was still over 10%.

Fig. 5.5 Monday morning and afternoon visits per team

Table 5.2 Monday schedule for team #1. All values in minutes

Patient ID	Time window	Arrival time	Visit duration
Care Center			
267 (1)	[0, 240]	3	20
299	[0, 480]	28	20
316 (1)	[0, 180]	51	30
280	[0, 180]	82	45
264	[0, 480]	137	20
249	[0, 480]	159	20
255 (1)	[0, 240]	185	20
300	[0, 240]	210	20
Lunch	[300, 300]	300	60
255 (2)	[360, 480]	365	20
267 (2)	[360, 480]	391	20
316 (2)	[360, 480]	419	25
132	[0, 480]	449	20
Care Center		479	

5.6 Final Remarks and Future Work

This work presents a biased-randomized heuristic approach to solve a home health-care routing problem in the city of Lisbon. During the construction phase, our algorithm combines the classical savings heuristics with a biased-randomization procedure. In a second stage, routes are compared with the best route found at the time, which is stored in the cache, to try to improve the overall solution. These stages are embedded in a multi-start framework. Our algorithm accounts for time windows, mandatory lunch breaks, and loyalty between caregivers and patients, which are particular features of the studied problem.

The results show the applicability and adequacy of the approach in solving real-life problems. Finally, it is important to highlight that this algorithm is the first step to create a more sophisticated routing decision support tool for a home care center. The proposed procedure can easily provide more than one (good) schedule, allowing the planner to actively choose what she considers to be the best plan according to her utility function and other preferences that cannot be easily integrated into a mathematical model.

The next steps to take are: *(i)* the development of a local optimization procedure to improve the solution quality even further; and *(ii)* the design of medium and large size instances to test the heuristic in those scenarios. We also aim at extending the solution approach to a 5-day plan, since loyalty has to be assured during all the week.

Acknowledgements This work was partially supported by the Fundação para a Ciência e a Tecnologia (Portuguese Foundation for Science and Technology) through the project UID/MAT/00297/2013 (Centro de Matemática e Aplicações).

References

1. C. Akjiratikarl, P. Yenradee, P.R. Drake, PSO-based algorithm for home care worker scheduling in the UK. Comput. Ind. Eng. **53**(4), 559–583 (2007). ISSN 0360-8352. https://doi.org/10.1016/j.cie.2007.06.002, http://www.sciencedirect.com/science/article/pii/S0360835207000927
2. V.S. Begur, D.M. Miller, J.R. Weaver, An integrated spatial DSS for scheduling and routing home-health-care nurses. Interfaces **27**(4), 35–48 (1997)
3. R.B. Bachouch, A. Guinet, S. Hajri-Gabouj, A decision-making tool for home health care nurses' planning. Supply Chain. Forum **12**(1), 14–20 (2011)
4. S. Bertels, T. Fahle, A hybrid setup for a hybrid scenario: Combining heuristics for the home health care problem. Comput. Oper. Res. **33**(10), 2866–2890 (2006). https://doi.org/10.1016/j.cor.2005.01.015
5. European Commission, Health and long-term care in the European Union. Spec. Eurobarometer **283**, 25–27 (2007)
6. P. Eveborn, P. Flisberg, M. Ronnqvist, Laps care an operational system for staff planning of home care. Eur. J. Oper. Res. **171**(3), 962–976 (2006). ISSN 0377-2217
7. T.A. Feo, M.G.C Resende, Greedy randomized adaptive search procedures. J. Glob. Optim. **6**(2), 109–133 (1995)
8. C. Fikar, P. Hirsch, Home health care routing and scheduling: a review. Comput. Oper. Res. **77**, 86–95 (2017). https://doi.org/10.1016/j.cor.2016.07.019. ISSN 0305-0548
9. G. Hiermann, M. Prandtstetter, A. Rendl, J. Puchinger, G.R. Raidl, Metaheuristics for solving a multimodal home-healthcare scheduling problem. Cent. Eur. J. Oper. Res. **23**(1), 89–113 (2015)
10. A.A. Juan, J. Faulin, J. Jorba, D. Riera, D. Masip, B. Barrios, On the use of Monte Carlo simulation, cache and splitting techniques to improve the Clarke and Wright savings heuristics. J. Oper. Res. Soc. **62**(6), 1085–1097 (2011). https://doi.org/10.1057/jors.2010.29
11. A.A. Juan, J. Faulin, A. Ferrer, H.R. Lourenço, B. Barrios, MIRHA: multi-start biased randomization of heuristics with adaptive local search for solving non-smooth routing problems. Top **21**(1), 109–132 (2013)
12. H.S. Kaye, M.P. LaPlante, C. Harrington, Do noninstitutional long-term care services reduce medicaid spending? Health Aff. **28**(1), 262–272 (2009). https://doi.org/10.1377/hlthaff.28.1.262
13. R. Lahyani, M. Khemakhem, F. Semet, Rich vehicle routing problems: from a taxonomy to a definition. Eur. J. Oper. Res. **241**(1), 1–14 (2015). https://doi.org/10.1016/j.ejor.2014.07.048
14. K.D. Rest, P. Hirsch, Daily scheduling of home health care services using time-dependent public transport. Flex. Serv. Manuf. J. 1–31 (2015). https://doi.org/10.1007/s10696-015-9227-1
15. K.D. Rest, A. Trautsamwieser, P. Hirsch, Trends and risks in home health care. J. Humanit. Logist. Supply Chain Manag. **2**(1), 34–53 (2012). https://doi.org/10.1108/20426741211225993
16. R. Tarricone, A.D. Tsouros, *Home Care in Europe: The Solid Facts* (World Health Organization, Geneva, 2009)

Chapter 6
Planning Health Workforce Training in the Detection and Prevention of Excessive Alcohol Consumption: An Optimization-Based Approach

Joana Faria, Teresa Cardoso-Grilo and Cristina Gomes

Abstract The adequate training of health workforce in the field of excessive alcohol consumption is essential to provide health professionals with the necessary tools for an adequate provision of care, thus leading to a decrease in alcohol consumption. Proper planning of such training is thus essential, but literature in this area is still scarce. This paper proposes an optimization model based on mathematical programming for supporting the planning of health workforce training in the field of excessive alcohol consumption in National Health Service-based countries – the $WFTM^{alcohol}$. The model aims at informing on (i) how many health professionals (physicians and nurses) should be trained per year and health unit, and (ii) which training packages should be available per year. The model allows exploring the impact of considering different objectives relevant in this sector, including the minimization of costs and the maximization of multiple performance indicators. Acknowledging that several sources of uncertainty may affect planning decisions, a sensitivity analysis on key parameters of the model is performed. To illustrate the applicability of the model, a case study based on the *Oeste Sul* ACES in Lisbon is analyzed. Results confirm that there is a shortage of trained professionals in this field in Portugal.

Keywords Health workforce training · Excessive consumption of alcohol
Optimization · Mathematical programming models · Performance indicators

6.1 Introduction

Alcohol consumption is the third reason of sickness and death in the world [6]. In 2014, Portugal was the 11th country with higher alcohol consumption in Europe. The alcohol is responsible for many diseases like mental disorders and cardiovascular

J. Faria (✉) · T. Cardoso-Grilo
Centre for Management Studies of Instituto Superior Técnico (CEG -IST),
Instituto Universitário de Lisboa (ISCTE-IUL), Lisbon, Portugal
e-mail: joana.c.a.faria@gmail.com

C. Gomes
Direção-Geral da Saúde, Lisbon, Portugal

A. I. F. Vaz et al. (eds.), *Operational Research*, Springer Proceedings
in Mathematics & Statistics 223, https://doi.org/10.1007/978-3-319-71583-4_6

diseases, and a higher consumption leads to an increased risk of achieving these conditions. The alcohol consumption represents large costs, both tangible (in order of 123 billion euros) and intangible costs (270 billion euros) [7].

The alcohol consumption is defined by the daily consumption of alcohol in grams [2]: the low risk consumption is defined by a consumption lower than 20 and 40 g in woman and men, respectively; the hazardous consumption is defined by a consumption in between 20 and 40 g in woman and in between 40 and 60 g in men; the harmful consumption is the consumption higher than 40 and 60 g in woman and men, respectively; and dependent consumption is the consumption where the alcohol consumption is faced as priority.

So as to reduce the impact of this consumption in today's society, preventive strategies and an earlier detection play a key role [6]. This prevention and detection is not only essential to improve the quality of life of population, but also to control the costs associated with the provision of care for people with varying levels of alcohol consumption. And in National Health Service (NHS)-based countries, this cost control is even more relevant. It is for this reason that programs focused on the prevention and early detection of alcohol consumptions should be seen as a priority in any NHS-based countries, and the same applies for the adequate training of the health workforce that provides these services.

The prevention and early detection of alcohol consumption in NHS-based countries is typically ensured by health professionals belonging to primary health care centers (PHCC) [13]. PHCC are typically located near urban centers and workplaces, representing the first point of contact with the populations [20]. Thus, the screening should be performed by these professionals, mostly because they can combine the diagnosis of excessive alcohol consumption with the treatment of other diseases. As examples of screening methods, one can have biochemical tests or simple questionnaires. Nevertheless, health professionals do not typically use these methodologies because they do not feel comfortable with this type of intervention. Some studies, like BISTAIRS and PHEPA projects, have proven that the main difficulties felt by doctors include (i) the lack of time and incentives, and (ii) the fear of conflict with patients and frustrating consultations [2, 6, 7]. In fact, according to the mentioned studies, health workforce devoted to these activities are not adequately trained.

It is thus clear that an adequate training of health workforce in the prevention and early detection of alcohol consumption is essential to improve the care provided for the populations, as well as to improve health professionals' knowledge and motivation – it is recognized that a system operating with health professionals with the adequate training in these areas have potential to increase the early detection rates, as well as to increase by 50% the prevention of alcohol consumption [2, 6].

Nevertheless, although recognized as relevant, a lack of research exists to support an adequate planning of the health workforce in areas related to the prevention and early detection of alcohol consumption. Still, when considering other areas, including the healthcare sector in general, mathematical programming models appear as the preferred approach to support the training of professionals (e.g., Wishon et al. [19] and Horn et al. [8], outside the health care sector). Accordingly, mathematical programming models appear as a potential tool to be used to support the planning of workforce in the area of prevention and early detection of alcohol consumption.

Within this context, this paper develops an optimization model based on mathematical programming to support the planning of health workforce training in the field of excessive alcohol consumption in NHS-based countries – herein after mentioned as $WFTM^{alcohol}$ model. The proposed model aims at informing on how many health professionals (both physicians and nurses) should be trained per year and per health unit, as well as which training packages should be made available per year. The model allows exploring the impact of considering different objectives relevant in this sector, such as the minimization of costs and the maximization of multiple performance indicators (both mandatory and optional indicators, with optional indicators being assigned with a lower priority). Acknowledging that several sources of uncertainty may affect planning decisions, a sensitivity analysis on key parameters of the model is also performed.

This paper contributes to the literature in the area by: (i) proposing methods that can be used to support the training of health workforce in areas related to the prevention and early detection of alcohol consumption, which represents an area not studied in existing literature; (ii) exploring the impact of accounting for different policy objectives relevant in this sector, such as the minimization of costs and the maximization of performance indicators; and (iii) proposing a generic approach that can be easily adapted to be applied to other areas in the health care sector.

This paper is organized as follows. Section 6.2 presents a literature review on key studies in the area. Section 6.3 presents some background information. Section 6.4 presents the mathematical details of the proposed model, followed by the presentation of the case study in Sect. 6.5. Final conclusions and lines of further research are presented in Sect. 6.6.

6.2 Literature Review

The review herein presented aims at identifying methodologies used to solve problems related to the training of workforce in general, and health workforce in particular.

Workforce planning in general has been treated extensively in the literature since 1950 [11, 14]. In fact, many different workforce planning perspectives exist, but this paper will be only focused on workforce training. The methodologies most widely used to support the planning of workforce training are: (i) simulation models; (ii) mathematical programming models; (iii) dynamic systems; and (iv) portfolio prioritization models. Nevertheless, when considering the healthcare sector in particular, just few examples can be identified. As a recent example, Ballard et al. [3] presents a simulation model to support the training of the workforce engaged in cardiothoracic transplantation. The proposed simulation model accounts for the uncertainty surrounding the number of transplants and the number of professionals with the skills to do the procedure.

Within the before mentioned approaches, mathematical programming models represent the approach most widely used to support the training of workforce, particularly, linear programming models [14, 18]. This type of approach has been extensively used outside the healthcare sector - e.g., Wishon et al. [19] developed a mathematical programming model to plan the training of workforce in the agricultural sector; and Horn et al. [8] developed a similar model to plan the training in the military sector. But when considering the healthcare sector in particular, there is still little research in this area.

Within the healthcare sector, Lavieri and Puterman [10] proposed a single-objective linear programming model to support the hierarchical planning of nurses, in which the main aim was to determine the optimal number of nurses to train, recruit and promote. This model is applied considering a 20 years' time horizon, and the main goal is to achieve different specialization levels. User-friendly interfaces were also developed so as to enable the use of the proposed model. The authors argue that linear programming models represent the most adequate approach due to its transparency, because it is easy to obtain the optimal solution and also because it is easy to modify and realize a scenario analysis.

Hu et al. [9] have also proposed a mathematical programming model to plan the training, promotion and hiring process of nurses, while aiming at minimizing total costs. An additional model was developed by Senese et al. [15], who developed a linear programming model to support the optimal assignment of medical specialization grants for physicians, while minimizing the gap between supply and demand of physicians.

Table 6.1 summarizes the studies presented before. Based on this table it can be concluded that no study exists accounting for the specificities inherent to the sector of prevention and early detection of alcohol consumption. Furthermore, it can also be concluded that existing studies are mainly focused on objectives related to cost minimization or gap minimization, with no study being also focused in maximizing performance indicators;and most of it are focused in only one profession – nurses or doctors –, and not in both.

Accordingly, it can be concluded that there is space to develop research devoted to the development of planning models based on mathematical programming so as to

Table 6.1 Key features analyzed in existing mathematical programming planning models devoted to the workforce training in the health care sector

	Objectives			Training of nurses	Training of physicians	Alcohol consumption
	Gap	Cost	Others			
Lavieri and Puterman [10]		✓		✓		
Hu et al. [9]	✓	✓		✓		
Senese et al. [15]					✓	
WFTM^{alcohol}		✓	✓	✓	✓	✓

plan the training of both nurses and physicians working in the prevention and early detection of alcohol consumption – the $WFTM^{alcohol}$.

6.3 Excessive Alcohol Consumption Background

The present paper develops the $WFTM^{alcohol}$ model to support the planning of workforce training in the field of excessive alcohol consumption, with background information being presented in this section.

Prevention and Early Detection of Alcohol Consumption

This study aims at proposing a planning model for the health workforce training in the area of prevention and early detection of alcohol consumption in NHS-based countries, and the Portuguese NHS system will be considered as reference. Accordingly, this study considers that:

i. These services are provided in primary health care units;
ii. These services are provided by several health professionals, including general practitioners, nurses specialized in mental health care and non-specialized nurses;
iii. Different performance indicators can be contracted and used to evaluate the performance of each unit in the area of prevention and early detection of alcohol consumption, and different units may have contracted different indicators. These indicators may include:
    a. Mandatory contracted performance indicators;
    b. Optional contracted performance indicators;
    c. Non-contracted performance indicators.
iv. Varying levels of training related to the prevention and early detection of excessive alcohol consumption can be found within professionals in the same unit;
v. A training package including multiple training courses is available, with each specific type of training course having its own objectives;
vi. Each training course made available to a group of health professionals in a given unit may have impact in different performance indicators.

Planning Objectives

Depending on the planning context, different objectives may need to be pursued when training health workforce in the area of prevention and early detection of alcohol consumption. Particularly, cost minimization plays a key role within the current context of severe budget cuts, but the maximization of different performance indicators may also be required. Specifically, three key performance indicators will be considered (these are the ones currently used in Portugal) [12]:

i. Performance indicator A: Percentage of patients aged above 14, with alcohol consumption record (one record, minimum);

ii. Performance indicator B: Percentage of patients aged above 14 with excessive alcohol consumption (the ones with low risk, hazardous or harmful consumption), and with a consultation in the last three years;
iii. Performance indicator C: Percentage of patients aged above 14 with chronic alcohol abuse (the ones with dependent consumption).

Training Package

So as to improve the care provided in the area of prevention and early detection of alcohol consumption within the scope of primary health care units, a training package comprising multiple training courses should be available. Particularly, a comprehensive training package should comprise (i) training courses on the prevention of alcohol related problems, (ii) training courses on the prevention of alcohol related problems specially driven to particular groups, such as pregnant women, young people at risk or older people, and (iii) training courses on screening tools in the area of prevention and early detection of alcohol consumption. Health professionals in each unit may receive a single training course (hereafter mentioned as individual training courses), or a combination of training courses (for instance, a course comprising topics both on prevention and screening tools), depending on the individual objectives of each unit and on other factors, such as the availability of trainers and the type of competences already available in each unit.

Different training courses will have varying levels of impact in the previously mentioned performance indicators – for instance, if a group of professionals belonging to a given health unit receive training on screening tools, it is expected that the competences of this group of professionals on detecting alcohol consumption related problems increases, thus improving the performance indicator A; also, if another group of professionals receive training on the prevention of alcohol related problems, it is expected a improvement in the prevention of these problems, thus improving performance indicators B and C.

6.4 Mathematical Formulation of the Model

The model proposed in this paper – the $WFTM^{alcohol}$ model – aims at supporting the planning of health workforce training on prevention and early detection of alcohol consumption in any country with a National Health System. Relying on the previous context, this planning should follow different objectives, such as the cost minimization or the maximization of multiple performance indicators. Also, it is essential to consider several constraints inherent to this context.

The notation used for the model formulation is presented below, together with the mathematical formulation of the objectives and key constraints of the model.

6.4.1 Notation

Indices and Sets

$p \in P$	Health professionals
$f \in F$	Training courses (individual training courses and combination of individual training courses)
$of \in OF$	Individual training courses
$u \in U$	Health units
$t \in T$	Time periods (in years)
$i \in I$	Performance indicators

Parameters

D_f	The duration of each training course f, in hours
Lo_f	Maximum capacity of health professionals in each individual training course f
Nf_f	Number of trainers for each training course f
Cf_f	Cost per hour associated with each trainer for each training course f, in €
$impact_{i,f}$	Impact in each performance indicator i obtained as a result of each training course f
$target_{i,u}$	The value associated with the target of each performance indicator i in each unit u
$SQ_{i,u}$	The initial value of each performance indicator i in each unit u
$Npu_{p,u}$	Number of health professionals p in each unit u
$Npuf_{p,u,f}$	Number of health professionals p in each unit u with the training course f
$N_{of,f}$	Combination of individual training courses
HF_t	Available work hours in each year t
pi_i	Performance indicator weights, being different if it is mandatory or optional
B_t	Available budget per year t
Nht	Annual number of training hours for each trainer

Variables

$x_{p,u,f,t}$	Number of professionals p in a unit u to be trained with a given training course f at t
$k_{t,f}$	Number of training courses f to be performed in a year t
$y_{p,u,of,t}$	Number of professionals p trained in a given unit u with a given individual training course of in a year t
$w_{p,u,of,t}$	Number of professionals p in a unit u without the individual training course of at a given year t
$j_{p,u,f,of,t}$	Number of professionals p in a unit u missing a given training course f, which includes an individual training of, in a given year t
$indicator_{u,i}$	Value of performance indicator i in a unit u at $t = T$
c_t	Number of hours used to train in a given year t
Z_1	Cost minimization variable
Z_2	Performance indicator maximization variable

6.4.2 Objective Functions

Depending on the planning circumstances, the objectives to be considered may differ.

Within the current context of budgetary cuts, minimization of costs plays a key role. In such circumstances, the objective is to minimize the cost associated with the payment of trainers (health professionals training other health professionals within the scope of the prevention and early detection of alcohol consumption). In this case, the objective function follows Eq. 6.1.

$$Min\ Z_1 = \sum_{t \in T} \sum_{f \in F} k_{t,f} \times Cf_f \times Nf_f \times D_f \tag{6.1}$$

On the other hand, if the focus is on the improvement of multiple performance indicators, the objective is to maximize the improvement found across indicators (Eqs. 6.2–6.3). This maximization is affected by the nature of each indicator, which can be mandatory or optative – and a different weight (pi_i) is attributed to each of these indicators, with a higher weight being assigned to performance indicators with a higher priority.

$$Max\ Z_2 = \sum_{u \in U} \sum_{i \in I} pi_i \times |indicator_{i,u} - SQ_{i,u}| \tag{6.2}$$

$$indicator_{i,u} = \sum_{t \in T} \sum_{f \in F} \sum_{p \in P} x_{p,u,f,t} \times impact_{i,f} + SQ_{i,u} \quad \forall u \in U, i \in I \tag{6.3}$$

6.4.3 Constraints

A set of constraints are considered in the model, and are described below.

Maximum Capacity of Health Professionals Per Training course

Equation 6.4 ensures that the number of professionals that are receiving training course f should not exceed the maximum capacity of that training course, since it is not allowed an overcrowded training.

$$\sum_{u \in U} \sum_{p \in P} x_{p,u,f,t} \leq k_{t,f} \times Lo_f \quad \forall t \in T, f \in F \tag{6.4}$$

Maximum Value of Each Performance Indicator Per Unit

Equation 6.5 ensures that each indicator i does not exceed 1. The maximum value of each indicator is 1, which corresponds to 100%.

$$\sum_{t\in T}\sum_{p\in P}\sum_{f\in F} x_{p,u,f,t} \times impact_{i,f} + SQ_{i,u} \leq 1 \quad \forall u \in U, i \in I \tag{6.5}$$

Maximum Number of Health Professionals that can be Trained Per Unit

Equation 6.6 ensures that the number of professionals p to be trained in a unit u, in a given year t and with the training course f, does not exceed the number of health professionals in that unit without that specific training. For example, if at $t = 0$, in a given unit, exists 10 physicians without any training, and if in the same unit at $t = 1$ there are 3 physicians trained with an individual training $of^{\,2}$, then at $t = 2$ there are only 7 physicians that can be trained with a given training that includes the individual training $of^{\,2}$.

$$x_{p,u,f,t} \leq j_{p,u,f,of,h} \quad \forall p \in P, u \in U, of \in OF, t \in T, t > h \tag{6.6}$$

Budget Constraints and Limit of Training Hours

In the case of a budget limit for the health workforce training within the scope of excessive alcohol consumption, it is necessary to define Eq. 6.7. Equation 6.7 ensures that the budget is not exceeded throughout the planning horizon.

$$\sum_{f\in F} k_{t,f} \times Cf_f \times Nf_f \times D_f \leq B_t \quad \forall t \in T \tag{6.7}$$

In addition, it is also necessary to take into account the maximum number of training hours that are available – Eq. 6.8 prevents that the annual hours that can be used for training are not exceeded.

$$HF_t \times Nht \geq \sum_{f\in F} k_{t,f} \times D_f \quad \forall t \in T \tag{6.8}$$

Targets for Performance Indicators

If specific targets need to be achieved for each performance indicator, Eq. 6.9 is required.

$$\sum_{t\in T}\sum_{p\in P}\sum_{f\in F} x_{p,u,f,t} \times impact_{i,f} + SQ_{i,u} \geq target_{i,u} + SQ_{i,u} \quad \forall u \in U, i \in I \tag{6.9}$$

Number of Health Professionals Receiving Training

Equation 6.10 sets the number of professionals p, in a unit u, trained with an individual training of in a given year t.

$$y_{p,u,of,t} = \begin{cases} \sum\limits_{f\in F} N_{of,f} \times (x_{p,u,f,t} + Npuf_{p,u,f}) \; \forall p \in P, u \in U, of \in OF, t \in T, t = 0 \\ y_{p,u,of,t-1} + \sum\limits_{f\in F} N_{of,f} \times x_{p,u,f,t} \quad \forall p \in P, u \in U, of \in OF, t \in T, t > 0 \end{cases} \tag{6.10}$$

Equation 6.11 sets the number of health professionals p that were not trained in a given unit u with a standard training of in a given year t. Each professional can only have one individual training of each type, so it is essential to control how many health professionals do not have each type of training. However, this equation is defined only for individual trainings of, so it is necessary to control it for all the training courses (including combinations of individual trainings), according to Eq. 6.12.

$$w_{p,u,of,t} = Npu_{u,p} - y_{p,u,of,t} \quad \forall p \in P, u \in U, of \in OF, t \in T \tag{6.11}$$

$$j_{p,u,f,of,t} = N_{of,f} \times w_{p,u,of,t} \quad \forall p \in P, u \in U, of \in OF, t \in T \tag{6.12}$$

Number of Hours used for Training

Equation 6.13 sets the number of hours devoted to training purposes in a given year t.

$$c_t = \sum_{f \in F} k_{t,f} \times D_f \quad \forall t \in T \tag{6.13}$$

6.5 Case Study

In this section we apply the $WFTM^{alcohol}$ model to real data from a Portuguese region to illustrate how it can be used to support the planning of workforce training on prevention and early detection of alcohol consumption. Specifically, the model is applied in the context of *Oeste Sul ACES (Agrupamentos de Centros de Saúde)* in Lisbon.

6.5.1 Dataset and Assumptions Used

The details on the dataset used for applying the model is presented in Table 6.2. The model was applied in the context of *Oeste Sul ACES* in Lisbon, Portugal, which comprises 11 health units and two contracted indicators – Performance Indicator A and Performance Indicator C. The first indicator corresponds to screening activities and the second one corresponds to excessive alcohol consumption identification.

To apply the model, it was considered that 8 individual training courses are available, as shown in Table 6.3, and that 120 training courses are derived from the first ones (by combining individual training courses). It should be noted that, although individual training courses f^1, f^2, f^7 and f^8 represent training courses on screening tools, the specific objectives and tools explored within the scope of each course differs between courses; and the same applies to the individual training courses in the area of prevention of alcohol related problems.

Table 6.2 Dataset used in the model application

	Data	Source	Details
Performance indicators (A and C)	Initial value (Status Quo, SQ)	Interview with an expert	Initial value of performance indicators in each unit
	Target		Target imposed for each indicator in each unit
Training courses	Synergies	SICAD [16] and interview with an expert	Relationship between training courses
	Length of time	SICAD [16]	Duration, in hours for each training
	Capacity	Interview with an expert	Capacity of each training
	Number of trainers	Interview with an expert and SICAD [17]	Number of trainers required for each training
	Cost of trainers	Interview with an expert	Cost per trainer per hour
	Impact	Ref. [1] and interview with an expert	Impact of each training in each indicator
Health professionals (both nurses and physicians)		DGS [5]	
	Number of health professionals	National Institute of Statistics (2014) and interview with an expert	Number of health professionals per unit
	Number of health professionals trained	SICAD [17] and interview with an expert	Number of trained health professionals per unit
Budget		SICAD [17] and interview with an expert	Budget available per year
Time available for training	Percentage of hours available for training	SICAD [17] and interview with an expert	Percentage of hours that trainers dispense of their normal working hours to perform training

Also, several assumptions were made in order to assign impacts do different training courses –e.g., individual training course f^1 has impact on performance indicator A and C (0,04 and 0,005, respectively), and f^5 only have impact on indicator C (0,0035). More details on the how these impacts were derived are available upon request. Also, it was assumed that this impact does not vary over time.

Table 6.3 Summary on training courses and the associated impact in different performance indicators

Training course	Area of training	Indicator with impact
f^1	Screening tools	Performance indicator A
	Prevention of alcohol related problems	Performance indicator C
f^2	Screening tools	Performance indicator A
f^3	Prevention of alcohol related problems	Performance indicator C
f^4	Prevention of alcohol related problems	Performance indicator C
f^5	Prevention of alcohol related problems	Performance indicator C
f^6	Prevention of alcohol related problems	Performance indicator C
f^7	Screening tools	Performance indicator A
f^8	Prevention of alcohol related problems	Performance indicator C

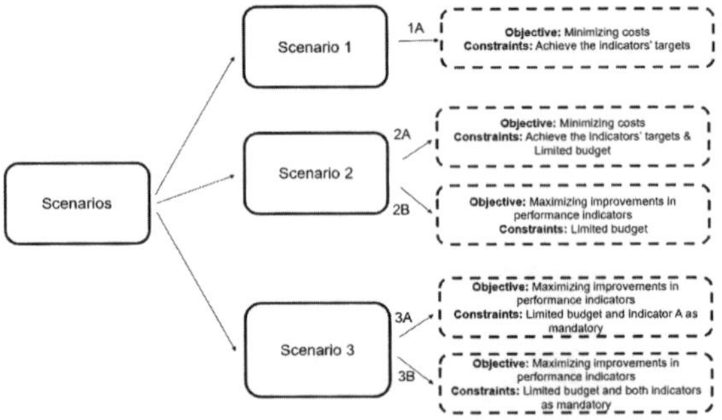

Fig. 6.1 Scenarios' summary

6.5.2 *Scenarios Under Study*

Figure 6.1 summarizes the different scenarios under analysis in this paper.

The first scenario (Scenario 1A) refers to a situation where one wants to plan the workforce training without any restriction in terms of budget, i.e., the question is *"How much would it cost to achieve the pre-defined targets for the contracted performance indicators?"*. The goal of this scenario is thus to analyze how expensive would be the training of professionals so as to achieve the desired indicators, in one year. In this case the objective would be to minimize costs (Eq. 6.1), while performance indicators targets are imposed as constraints (Eq. 6.9).

Two additional scenarios where a budget constraint is imposed were also explored. Two different situations were analyzed:

i. In scenario 2A it is intended to plan the health workforce training when a limited budget exists and when the aim is to achieve the targets imposed for contracted indicators. The key difference between scenario 1A and scenario 2A is related to the time required to achieve these targets – since there is a limited budget per year, it may take more time to achieve those targets, and so the time frame was changed to five years, following indications of experts in the area;

ii. In scenario 2B it is analyzed which are the maximum improvements for different performance indicators, using the available budget. Here the goal is to maximize the improvements in contracted indicators (Eq. 6.2), as an alternative to minimizing costs.

Finally, two additional scenarios were analyzed, in which it is considered that part of the indicators that are currently optional change to mandatory (currently, all the indicators are taken as optional). The following two scenarios were analyzed, with a time frame of five years:

i. In scenario 3A it is proposed to analyze which are the maximum improvements for performance indicators, using the available budget. In this case, indicator A was considered as mandatory (having a weight of 1) and indicator C was optional (having a weight of 0,2);
ii. In scenario 3B it is intended to analyze which are the maximum improvements for performance indicators, using the available budget and when considering both indicators as mandatory. The two performance indicators are assigned with a weight equal to 1.

6.5.3 Results

Results obtained under each scenario are described below. These results were obtained with the General Algebraic Modeling System (GAMS) 23.7 using CPLEX 12.0 on a Two Intel Xeon X5680, 3.33 GHz computer with 12GB RAM.

Scenario 1A

It was found that is possible to achieve the pre-defined targets for the performance indicators in one year, with total costs of around 7 700 €. In such circumstances, a package of training only based on prevention courses (f^1, f^2, f^3, f^6, f^8) should be offered. Still, these costs exceed the current annual available budget, and also the budget currently available for the coming 5 years in Lisbon for areas related to the prevention and early detection of alcohol consumption. These results make it clear the relevance of exploring scenarios where a budget limit is imposed, such as under scenarios 2 and 3.

Scenario 2

Within 5 years, and according to Scenario 2A, the targets imposed for the different performance indicators cannot be achieved using the available budget. For this reason, an additional run of the model was performed, allowing for longer time horizons. Under these new circumstances, it has been found that it is possible to achieve the targets in 8 years, with a cost of around 7 000 € (for the 8 years).

Table 6.4 Key results obtained under the scenarios presented in Fig. 6.1

	Number of professionals to train			Total hours of training	Cost [€]
Scenario	Physicians	Nurses	Total		
Scenario 1A	50	49	99	84	7 700
Scenario 2A	No results were obtained within a 5 years' time frame (due to the limited budget)				
Scenario 2B	112	146	259	70	5 600
Scenario 3A	109	150	259	70	5 600
Scenario 3B	110	149	259	70	5 600

According to Scenario 2B, it can be seen that a total budget of 5 600 € and a total of 70 h of training are required to maximize the two contracted performance indicators. In general terms, it was found that indicator A increases at a great extent in all the units, while indicator C does not exceed the pre-defined target. For example, in one of the units, indicator A increased about 81% when compared to the imposed target, while the highest growth of indicator C was 3,1%.

Scenario 3

In both scenarios 3A and 3B the costs incurred with training is the same found under scenario 2B - 5 600 € and 70 h of training. The training courses that should be ensured are also the same when compared to scenario 2B (combination of individual training courses on prevention – f^1, f^2, f^3 – and on screening tools - f^7). Similarly to scenario 2B, indicator A will always have the highest improvement when compared to improvements in indicator C. It was also found that the value of indicator A is the same for both scenarios 3A and 3B.

Table 6.4 summarizes the results obtained under each of the five scenarios under study. These results make it clear the urgent need for investments in the training of health professionals in the area of prevention and early detection of alcohol consumption, independently of the scenarios under study.

6.5.4 *Sensitivity Analysis*

A set of sensitivity analysis were performed to explore how much sensitive are the previous results to changes in key parameters of the model that are associated with the highest level of uncertainty, namely:

i. The impact caused by each training course in each performance indicator;
ii. The indicator weight assigned to optative indicators.

Scenario 2A was used to explore the impact of increasing the impact in indicators by 20%, and scenario 3A was used to explore the impact in changing the optative indicator's weight from 0,2 to 0,1 and 0,3. These scenarios were selected for this

Table 6.5 Key computational results per scenario

Scenarios	Time [minutes]	Gap	Iterations	Equations	Integer variables	Variables
Scenario 1A	11,45	0	108 652	18 282	2 760	12 078
Scenario 2B	1,8	0	2 303	179 044	13 800	60 212
Scenario 3A	2,2	0	2 649	179 044	13 800	60 212
Scenario 3B	2,1	0	2 247	179 044	13 800	60 212

analysis because these represent the ones that more closely represent the reality in Portugal.

It was found that the initial values used for the impact of each training course per indicator do not allow achieving the pre-defined targets of indicators, but increasing this impact by 20% allows achieving those targets within 5 years. On the other hand, when analyzing the impact of changing the weight assigned to optional indicators, it was found there is no change in the number of trained professionals, cost and hours of training. However, there is only a minimal change in the improvement found for indicator C.

According to these results, it can be concluded that the planning of workforce training is sensitive to changes in the impact of training courses on indicators, as well as on changes in the importance attributed to the optional indicators (with this importance being traduced by the associated weight).

6.5.5 Computational Results

Key computational results obtained when running the model under each scenario are presented in Table 6.5.

6.5.6 Discussion of Results

From Table 6.4 it is clear that if there is not a budget constraint or limited resources (scenario 1A), the costs incurred would be around 7 700 €, with 84 total hours of training in a single year. Nevertheless, in real practice there is a limited budget for training health workforce in the alcohol consumption sector, and the current budget is far from being enough to ensure this level of training in one single year. In general terms, the need for such a high level of training shows that the health workforce currently providing care in the area of prevention and early detection of alcohol consumption in Lisbon are not adequately trained, and so it can be concluded that there is a shortage of trained professionals in this field in Lisbon.

Under scenario 2A it was found that it is impossible to achieve the performance indicators targets within the five years' time frame. This is mainly related with the lower budget that is available per year (on average, 1 294 € per year), when compared to the costs incurred under Scenario 1A (7 700 €).

On the other hand, when the objective is the maximization of the improvement of performance indicators (scenario 2B) there is a significant change in costs, time and training of health professionals, when compared to scenario 2A – this happens because a lower budget is available, limiting the number and type of training courses per year. However, when the objective is to maximize the improvement of the indicators and the only change is on the nature of the indicators (that can be mandatory or optional), there are no significant changes in training time, monetary costs and on the number of professional to be trained – this can be concluded by comparing the results obtained under scenarios 2B, 3A and 3B. It was also found that, whenever there is a limited budget, the training should be the simplest possible (only individual training courses, rather than the combination of different training courses).

6.6 Conclusions and Further Work

The alcohol consumption area had been a major challenge in several European countries, including in Portugal. Some studies have shown that there is a gap in the workforce training, as well as no incentives for health professionals working in this field. Accordingly, there is clearly the need for an adequate planning of health workforce training in the area of prevention and early detection of alcohol consumption. Still, there is a lack of methods devoted to this issue.

Within this setting, this study proposes the development of the $WFTM^{alcohol}$ model. This is an optimization model based on linear programming developed to plan the training of health workforce in the alcohol consumption area in NHS-based countries. This model aims at informing on how many health professionals (both physicians and nurses) should be trained per year and health unit, and which training packages should be made available per year. Multiple objectives relevant in this sector are considered, including the minimization of costs and the maximization of performance indicators (both mandatory and optional indicators).

The key contributes of the present paper are as follows:

1. It is focused on the training of health professionals in the field of excessive alcohol consumption, a health care area not widely studied in the health planning literature;
2. It accounts for the impact of different planning objectives, including the minimization of costs and the maximization of performance indicators;
3. It proposes a generic approach that can be easily adapted and applied to other health care areas.

So as to illustrate the usefulness of the proposed model, it was applied to a Portuguese case study, namely, to *Oeste Sul ACES* in Lisbon, Portugal. The main results confirm that there is a lack of training in excessive alcohol consumption area in Lisbon – in fact, results show that in the unit with the lowest performance for the contracted indicators, the health professionals need to have an average of 2 trainings in distinct areas. By performing a sensitivity analysis, it was confirmed that workforce training is sensitive to changes in the impact of training courses on indicators, as well as on changes in the importance attributed to the optional indicators.

In terms of further research, different lines of research are worth to be pursued. First, it should be explored how to prioritize different indicators. In particular, multiple criteria methods should be employed for that purpose. Specifically, the MACBETH (Measuring Attractiveness by a Categorical Based Evaluation Technique) methodology could be used [4]. Secondly, having recognized the impact of uncertainty in planning decisions, it is also considered as relevant the development of a stochastic model so as to allow for a detailed analysis of uncertainty. Also, having recognized as relevant the analysis of different policies objectives when planning the health workforce training, multi-objective models that allow exploring the joint impact of multiple objectives, such as minimizing costs together with the maximization of performance indicators (and also other policy objectives that may be considered as relevant), should also be pursued. In addition, since sensitivity analysis has confirmed that planning results are sensitive to the impact in performance indicators, there is clearly the need to develop further research so as to build accurate estimates of this impact. Particularly, there is need to estimate this impact over the years while accounting for the type of topics covered in the training courses, for the number of professionals receiving training per course, for the target population to be served, among other factors.

References

1. P. Anderson, A. Gual, J. Colom, Clinical guidelines for the detection and brief interventions [*Recomendações Clínicas para a Deteção e Intervenções Breves*] (2005)
2. P. Anderson, E. Kaner, S. Wutzke, M. Funk, N. Heather, M. Wensing, R. Grol, A. Gual, L. Pas, Attitudes and managing alcohol problems in general practice: an interaction analysis based on findings from a who collaborative study. Alcohol Alcohol. **39**(4), 351–356 (2004). https://doi.org/10.1093/alcalc/agh072
3. T.N.S. Ballard, T.R. Grenda, A.M. Cohn, M.S. Daskin, F. Jacob Seagull, R.M. Reddy, Innovative scheduling solutions for graduate medical education. J. Grad. Med. Educ. **7**(2), 169–170 (2015). https://doi.org/10.4300/jgme-d-14-00581.1
4. Bana Consulting: M-MACBETH (2016), http://www.m-macbeth.com/en/m-home.html
5. DGS, Primary health care centers and hospital units - resources and production 2012 [*Centros de saúde e unidades hospitalares - Recursos e Produção 2012*] (2015)
6. J. Dias, Training in prevention and brief interventions in the area of alcohol consumption change attitudes of health professionals? [*A Formação em rastreio e intervenções breves do álcool muda as atitudes dos profissionais de saúde ocupacional?*] (2015)
7. C. Gomes, General and family medicine and the alcohol consumption approach [*A medicina Geral e Familiar e a Abordagem do Consumo de álcool*] (2010)
8. M.E.T. Horn, T. Elgindy, A. Gomez-Iglesias, Strategic workforce planning for the Australian defence force. J. Op. Res. Soc. **67**(4), 664–675 (2015). https://doi.org/10.1057/jors.2015.85

9. W. Hu, M.S. Lavieri, A. Toriello, X. Liu, Strategic health workforce planning. IIE Trans. **48**(12), 1127–1138 (2016). https://doi.org/10.1080/0740817X.2016.1204488
10. M.S. Lavieri, M.L. Puterman, Optimizing nursing human resource planning in British Columbia. Health Care Manag. Sci. **12**(2), 119–128 (2009). https://doi.org/10.1007/s10729-008-9097-0
11. M.A. Lopes, A.S. Almeida, B. Almada-Lobo, Handling healthcare workforce planning with care: where do we stand? Hum. Resour. Health **13**(1), 13–38 (2015). https://doi.org/10.1186/s12960-015-0028-0
12. Ministry of Health. Health Strategies, V HNP books - health indicators and targets [*Estratégias para a Saúde, V Cadernos do PNS - Indicadores e Metas para a Saúde*] (2011)
13. C. Nunes, C. Correia, C. Ribeiro, C. Santos, L. Marquês, M. Barbosa, M.L Pereira, P. Pardal, T. Seixas, V. Ramos, Primary health care in local units [*Os Cuidados de Saúde Primários nas Unidades Locais de Saúde*] (2012)
14. W.L. Price, A. Martel, K.A. Lewis, A review of mathematical models in human resource planning. Omega **8**(6), 639–645 (1980), http://www.sciencedirect.com/science/article/pii/0305048380900055, https://doi.org/10.1016/0305-0483(80)90005-5
15. F. Senese, P. Tubertini, A. Mazzocchetti, A. Lodi, C. Ruozi, R. Grilli, Forecasting future needs and optimal allocation of medical residency positions: the Emilia-Romagna region case study. Hum. Resour. Health **13**(1), 7 (2015), http://www.ncbi.nlm.nih.gov/pmc/articles/PMC4328064/, https://doi.org/10.1186/1478-4491-13-7
16. SICAD: Training plan in additive behaviors and dependencies [*Plano de Formação em Comportamentos Aditivos e Dependências*] (2015)
17. SICAD: SICAD training courses [Formações SICAD] (2016), http://www.sicad.pt/PT/Formacao/SitePages/FormacaoSICAD.aspx
18. B.D. Stewart, D.B. Webster, S. Ahmad, J.O. Matson, Mathematical models for developing a flexible workforce. Int. J. Prod. Econ. **36**(3), 243–254 (1994), http://www.sciencedirect.com/science/article/pii/0925527394000336, https://doi.org/10.1016/0925-5273(94)00033-6
19. C. Wishon, J.R. Villalobos, N. Mason, H. Flores, G. Lujan, Use of MIP for planning temporary immigrant farm labor force. Int. J. Prod. Econ. **170**, 25–33 (2015), http://www.sciencedirect.com/science/article/pii/S092552731500331X, https://doi.org/10.1016/j.ijpe.2015.09.004
20. World Health Organization: Alma ata declaration (1978)

Chapter 7
Downstream Petroleum Supply Chains' Design and Planning - Contributions and Roadmap

Leão José Fernandes, Susana Relvas and Ana Paula Barbosa-Póvoa

Abstract Petroleum Supply Chains (PSC) networks are complex organizations, strongly affected by competition, environmental regulation and market uncertainty. To improve profits and reduce costs and risks, companies may use mathematical programming for strategic, tactical and operational planning. The current paper identifies the research opportunities and presents our contributions with respect to the strategic and tactical planning of multiple entity, echelon, product, and transportation PSCs, under the context of crude costs, product prices, and customer demand uncertainties. In order to address these gaps, four mixed integer linear programming (MILP) models were developed, namely the individualistic, collaborative, multi-objective stochastic, and robust optimization MILPs. A detailed pricing structure and a piecewise linearization function determine the collaborative economy of scale multi-entity costs, tariffs and prices per route, location and product. A stochastic programming MILP integrates an augmented ϵ-constraint algorithm to simultaneously maximize the expected net present value (ENPV) and minimize risk represented through selected measures. The robust optimization MILP optimizes the worst-case profits considering the crude costs, product prices, and customer demand uncertainties. Test results are presented for the Portuguese downstream PSC.

Keywords Petroleum supply chain · Strategic and tactical planning · Uncertainty and risk management · Stochastic MILP · Robust optimization

---

L. J. Fernandes (✉)
CLC Companhia Logítica de Combustíveis, EN 366, Km 18, 2050-125
Aveiras de Cima, Portugal
e-mail: leao.fernandes@clc.pt

L. J. Fernandes · S. Relvas · A. P. Barbosa-Póvoa
CEG-IST, Instituto Superior Técnico, Universidade de Lisboa, 1049-001 Lisboa, Portugal
e-mail: susana.relvas@tecnico.ulisboa.pt

A. I. F. Vaz et al. (eds.), *Operational Research*, Springer Proceedings
in Mathematics & Statistics 223, https://doi.org/10.1007/978-3-319-71583-4_7

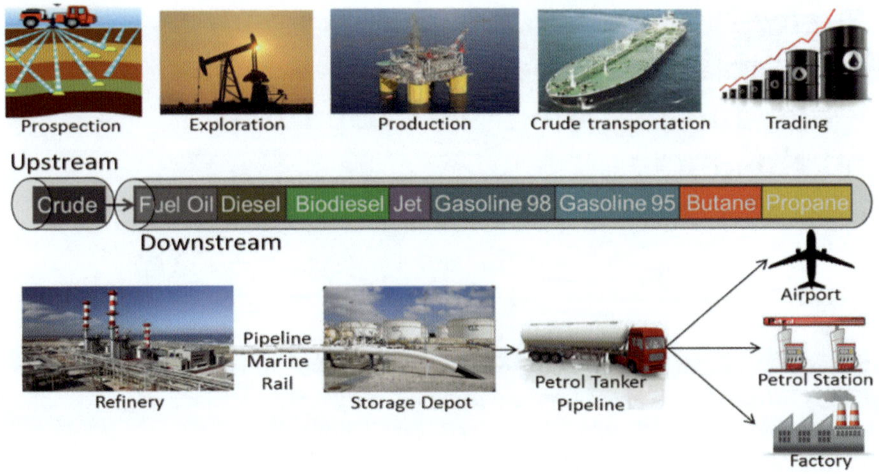

Fig. 7.1 The downstream petroleum supply chain

7.1 Introduction

PSCs are a highly competitive sector that influences country economies, and hence must deliver oil products to markets, timely, cost efficiently, with minimum resources and the lowest risk. The PSC is a sequence of activities that source, transform and deliver petroleum products. This involves huge volumes of product, and therefore implies substantial investment and complex infrastructures to ensure profitable and stable operation in spite of uncertainty. Figure 7.1 presents the PSC that may be broken down into two supply chains: Upstream, comprising of the activities that deal with the crude oil; and Downstream, encompassing the activities related to petroleum products. The Upstream consists of the prospection, exploration, production, marine transportation and trading of the crude oil, whereas the Downstream comprises of the refining, primary and second transportation, storage terminals, and retail commercialization of the petroleum products.

The downstream PSC begins at the refineries, which buy crude oil, produce petroleum products, and either sell the product to petroleum companies for national consumption or export them at wholesale market prices, thereby collecting a refining margin. Refineries support the cost, insurance and freight (CIF) charges in case of exportation. The petroleum companies then transport the product up to a storage depot of convenience using the primary distribution which involves the refinery to depot and inter-depot transportation, and is generally undertaken by pipeline, maritime or railway. Companies may import products, where they pay the wholesale prices and support the CIF charges up to their depots. The products are held at storage depots on behalf of the companies, until their transportation using the secondary distribution to retail clients. The latter comprise of petrol stations or direct clients in the transportation, heating, residential, industrial, commercial, electrical utilities, factories, airports and hospitals wherefrom is recovered a retail profit margin.

7.2 State of the Art

The hike in crude oil prices in 2009 created opportunities to develop pre-salt crude oil and shale oil production, but the industry went into turmoil following the recent oil price meltdown and the costs, prices and demand volatility. Recent developments show that companies and academia have refocused the PSC, with the Downstream PSC Strategic and Tactical Planning in the limelight. We now briefly analyze the PSC literature reviews, and the bibliographical references on refinery, pipeline, and downstream PSC operational and strategic planning, in view of identifying research opportunities.

Literature reviews focus the PSC as in [21] with respect to upstream, midstream, and downstream PSC planning; [31] address the oil refinery scheduling and planning; [30] study crude oil supply chain management; and [22] the strategic, tactical, and operational downstream PSC planning. Reference [30] identify many unaddressed areas: vertical integration of strategic and tactical decisions; horizontal integration of all PSC activities; nonlinear modeling for the formulation of refinery operations; developing efficient multi objective solution techniques for modelling international taxation, transfer prices and environmental impacts. Multi-stage stochastic modelling to study uncertainty problems has been identified as a key area for investigation. Reference [22] identify the need to develop robust optimization decision-support tools to integrate decision levels, PSC characteristics, and simultaneously model uncertainty, risk, resilience, sustainability and collaboration between entities.

Refinery operational planning is modelled in [26] using a mixed integer non-linear program (MINLP) to determine stream flow rates, properties, operational variables, inventory and facility assignments, considering aggregate nodes of production units, tanks and pipelines, interconnecting refineries, terminals and pipelines. Decomposition methods allow increased time periods and scenarios. Reference [20] proposed a supply network and market planning MILP to determine the refinery production rates, and transportation routes from distribution centers to markets. The maximized profits are the differences between revenue and depot and transportation fixed and variable costs. Reference [1] later proposed a two-stage stochastic model for operational planning of crude oil production, processing and distribution to maximize resource utilization and profits. Crude production plans are the first-stage decisions while refinery and petrochemicals production and shipment, storage, lost demand and backlog quantities are the second-stage decisions. Reference [19] integrates the crude oil unloading, refinery production planning, and product distribution models using penalties for unsatisfied demand, optimizing the overall PSC profits.

Multiproduct pipeline scheduling is modelled in [7] using a MILP for transportation between an oil refinery and depots, adopting a continuous time and volume representation. This reduces the model size and processing time, thereby permitting longer time horizons and increased slug pumping flexibility compared to the discrete representations. Reference [28] proposes a MILP to integrate petroleum products' pipeline scheduling with depot inventory management. Reference [6] develop a hierarchical approach to integrate a MILP model using a set of heuristic modules to model pipeline network scheduling. Later, [29] propose an alternative and holistic modelling concept to obtain faster solutions within reduced time frames through a single efficient model so as to avoid decomposition approaches.

Downstream PSC Operational planning was first modelled in [10], using a two-stage stochastic scenario analysis and a deterministic equivalent model. Later, [9] developed deterministic and stochastic models for the logistics tactical and operational planning to minimize overall costs. Reference [24] proposed a MILP for the operational planning of a capacitated transportation network of existing refineries and depots, to satisfy customer demand. Reference [25] extends the MILP to a two-stage stochastic model for capacity expansion to minimize transportation and investment costs. Total inventory and transportation costs are minimized, determining inventory shortages and excesses. More recently, [18] propose a tactical planning model to minimize the production, transportation and inventory planning, hubs and depots costs. Vendors' contributions are maximized with distribution planning and decentralized sales.

Downstream PSC Strategic planning is studied in [27] using a two-stage stochastic MILP initially using Lagrangian and later the Benders (L-Shaped) decomposition methods. Investment and logistics costs are minimized while the portfolio and timings of network capacity expansion projects and inventory decisions are determined for varying demand. They identify that few works have analyzed the presence of uncertainty in oil supply chains when dealing with important characteristics of the PSC. Later, [17] develop a deterministic MILP decision support system (DSS) that maximizes the NPV while determining an investment portfolio and installation times of vessel and transportation capacity expansions. Development of the DSS for strategic planning is seen as an opportunity for the real case Brazilian PSCs.

As seen, most authors identify the need to develop models for the integrated planning of the downstream PSC in order to capture the entire range of the PSC activities, the strategic, tactical and operational decision levels, and multiple entities, besides considering the costs, prices and demand uncertainties. This is an interesting area of research, where our work aims to fill some of the gaps, by developing models for the collaborative planning of the downstream PSCs by considering uncertainty, as will be presented hereafter (for further details on the models and cases solved see [15]).

7.3 Problem Description

To pursue the objectives proposed, we present a general representation that characterizes the integrated downstream PSC strategic, tactical, and operational planning problem under uncertainty as shown in Fig. 7.2, followed by the problem description.

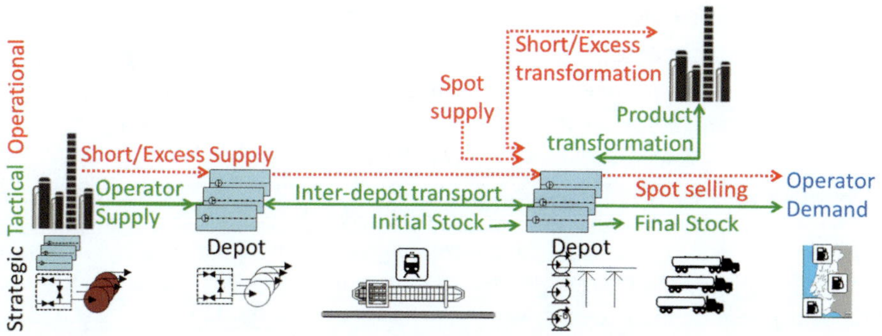

Fig. 7.2 Downstream PSC strategic, tactical and operational planning

Given a set of petroleum companies that produce or import petroleum products at a known set of refineries, and commercialize them to a set of customer regions;

- Determine the optimal current, grassroots, retrofit PSC network designs for the multi-echelon, product, and multi-modal transportation PSC network;
- Strategic decisions of installation, operation, expansion, or closure of storage depots, resource capacities, and multi-modal routes;
- Strategic-tactical decisions of entity investment participations and the economy of scale costs and tariffs per location, route, entity and product;
- Tactical decisions of multi-stage, multi-period transfer volumes and inventories for crude oil and derived products at the production, storage, primary and secondary transportation, importation, exportation, and retail stages;
- Operational decisions of under/over supply, transformation and the spot supply/selling volumes for deficit and excess product demand;

So as to maximize the total network profits, comprising of the oil company direct profits and their consolidated revenues from participated PSC member infrastructures: refineries, depots, transporters, and retailers; and simultaneously reduce the risk of not attaining such profits under the context of uncertainty;

Subject to the production, depot, transportation, resource capacity, inventory, demand restrictions, and crude costs, product prices, and demand uncertainties;

Considering uni-entity and multi-entity networks; pipeline transportation as product flow per time period restricted to the pipeline physical capacities; ship, truck, and wagon transportation as discrete product flow blocks, where the total transported volume for the allowed products and between origin-destination pairs, is restricted to the modal fleet capacity per time period; entities contract product supply volumes per depot and period where imbalances are penalized; logistic operators contract product transformation volumes per refinery, period and technology, where imbalances are penalized; product exchanges permit higher quality products to satisfy lower quality product demand, which is penalized; spot supply satisfy unbalanced demand, incurring spot supply costs; and spot sales deplete the excess product, incurring spot sales costs.

7.4 Research Contributions

The research developed four downstream PSC strategic and tactical planning MILPs and five main works; for details please consult [11–16].

A first work studied the individualistic behavior of the PSC. A MILP was developed that maximizes the total PSC profits for the strategic and tactical planning of downstream PSC networks, by modelling a bi-level transportation network flow formulation where the primary and secondary distribution are subject to restrictions based on the transported product. Strategic decisions of installation, closure or expansion of depot locations, resource capacities, transportation modes, routes, and network affectations; and the tactical decisions of crude oil and petroleum products' importation, exportation, production, storage and transportation volumes are determined. Uni-entity and multi-entity PSC network designs are generated for the current network, grassroots and retrofit networks. Considering the Portuguese PSC, the MILP generates better network designs with higher total PSC profits, when compared the current network to the grassroots and retrofit designs, for the uni-entity and multi-entity networks under individualistic operation. Higher logistic costs are observed in the multi-entity when compared to the uni-entity networks. This suggests that the single entity network improves overall profits and could result in lower tariffs. When reduced the fixed and variable costs, it is observed the installation of additional depots, a reduction of unmet demand, and an increase in the product transfer volumes and total profits.

The second work studied a dynamic collaborative multi-entity downstream PSC. A strategic and tactical planning MILP was developed that maximizes the total PSC profits by incorporating the partially owned logistic operator infrastructure profits. Besides the earlier decisions and multi-stage dynamic inventories, the piecewise linearization is used to determine the economy of scale costs and tariffs. Efficient grassroots and retrofit designs are generated for the uni-entity and multi-entity collaborative networks for the Portuguese PSC, and observed that their total PSC profits are higher than those of the current network. The collaborative MILP generates multi-entity designs whose overall profits improve significantly and approximate those of the more efficient uni-entity designs. The multi-stage inventories optimize the procurement and exportation costs, and benefit from the crude oil and petroleum product price variations. Product families and transportation mode families are used to consistently model the products and/or transportation mode dependent resources. Collaborative planning offsets the limitations of the current entity participation percentages in existing installations. The piecewise linearization function determines the economies of scale entity tariffs and prices, and allow the collaborative and flexible utilization of competitor installation capacity, resulting in considerable savings in costs and tariffs and better profits for the PSC members. The MILP generates network designs and determines the product ownership costs per location as shown in Fig. 7.3, for the products refined at SINES. This is the cost of product at a specific location, inclusive of ex-refinery prices, primary and secondary transportation, depot, and retail costs. Considering product GO (Diesel), its Ex-refinery cost in €/m3 is

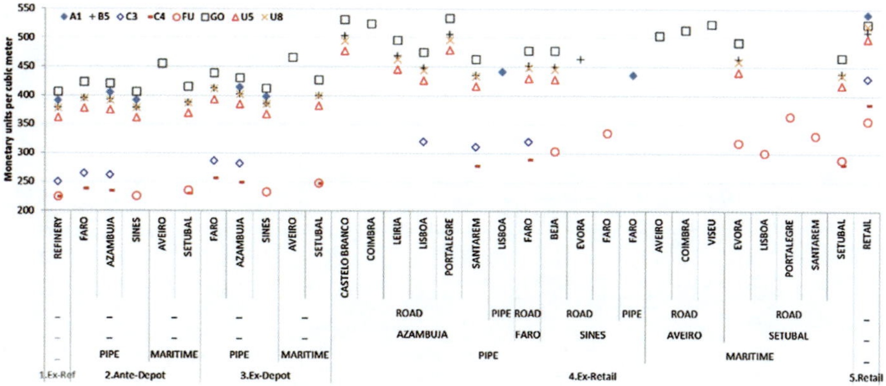

Fig. 7.3 Downstream PSC strategic, tactical and operational planning

406.029, its Ante-Depot cost at AZAMBUJA (includes pipeline transportation) is 420.566 and Ex-Depot cost is 430.767, its ex-retail cost on transportation by road to a near location LISBOA is 475.465, and to a remote location PORTALEGRE is 534.331, which incidentally is higher than the average retail price of 524.524.

To manage demand uncertainty, is developed a stochastic multi-objective planning MILP, which consists in the third work developed. The first-stage strategic decisions are scenario independent, and determine the installation and operation of depot locations, capacities, transportation modes, routes, and economies of scale entity tariffs. The second-stage tactical decisions are scenario dependent, and determine the multistage crude oil and petroleum products inventories and transfer volumes. The stochastic MILP maximizes the expected net present value (ENPV) for the uncertain demand. By considering three 5-year periods of optimistic, realistic and pessimist demand scenarios, are solved for the real-case Portuguese PSC single and multi-entity current, grassroots and retrofit networks designs. The methodology proposed in [5], is used to determine the stochastic solution measures of expected value of perfect information (EVPI), value of stochastic solution (VSS), expected value of the wait-and-see problem (WS), expected recourse problem solution value (RP) and the expectation of the expected value problem (EEV). The stochastic model is enhanced to a multi-objective stochastic MILP by integrating the augmented ε-constraint algorithm proposed in [23], to simultaneously maximize the ENPV and minimize one of the implemented risk measures: variance, variability index, downside risk, value-at-risk (VaR) and conditional CVaR. Improved network designs, achieving higher NPVs are obtained in the presence of demand uncertainty, with the network adjustments made at the tactical level for each scenario. The scenario NPVs cumulative probability distribution are plotted for the risk measures of variance, variability index, downside risk, and CVaR. Considering two conflicting objective functions, the ε-constraint algorithm optimizes first one objective function, divides its solution space into equidistant grid points, introduces a constraint to fix the first objective function at each grid point, and obtains the Pareto-optimal efficient

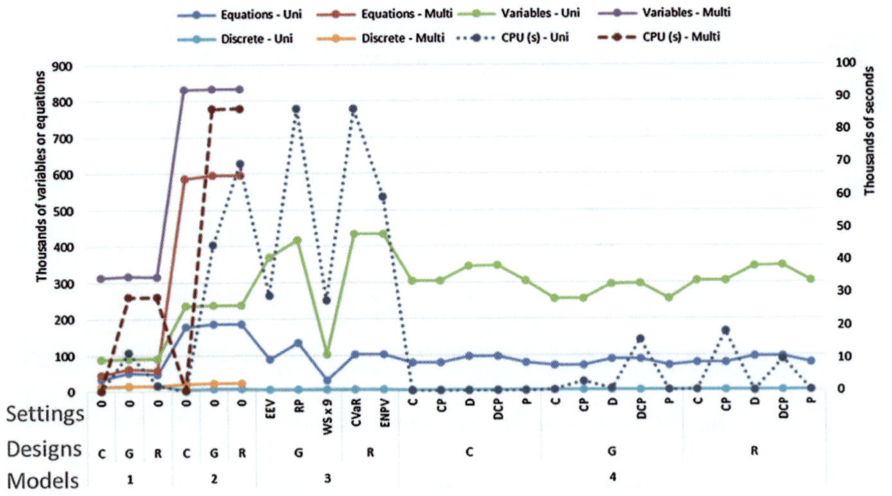

Fig. 7.4 MILP performance statistics

solutions that optimize both the objective functions. The decision maker (DM) may prefer a risk measure to another, based on his risk attitude. If risk averse, the DM may choose the variability index or downside risk measures, whereas if the DM is risk-seeking, the CVaR will be preferred.

The stochastic MILP showed complexity issues and hence, three robust optimization (RO) formulations were developed for modelling the crude costs, product prices, and customer demand uncertainties, for collaborative strategic and tactical planning. This consists in the fourth work developed, and follows the methodology used in [2, 4]. The RO MILPs use polyhedral uncertainty sets to model cost and price uncertainties in the objective function and demand uncertainty in the constraints. This results in computational tractability, as the robust trade-off solutions are obtained considerably faster than with the stochastic MILPs, and hence RO is observed to be adequate for modelling uncertainty for the PSC strategic and tactical planning problem. The uncertainty sets define the range of possible uncertainty, and the budget of uncertainty defines the range of protected uncertainty. The DM may define an acceptable probability for the feasibility of the solutions, based on which is determined a budget of uncertainty. The MILP may be executed specifically for this pre-determined budget of uncertainty, or the model may be executed for various budgets of uncertainty. We take the later approach and present several solutions ranging from the deterministic solution down to the worst-case solutions. Using this approach, the DM may now choose the desired solution from among various available solutions.

The models were developed using GAMS and CPLEX MIP solver, and resolved for 1% gap or 24 hours runtime. Figure 7.4 shows the performance statistics for the four models (x-axis):1,2,3,4 considering the designs: Current, Grassroots and Retrofit and the following uncertainty settings 0:deterministic, EEV, RP, WS, CVaR, ENPV, C:costs, CP:costs,prices, D:demand, and DCP:demand,costs,prices. The number of variables and equations are plotted against the primary y-axis and the CPU time against the secondary y-axis. It is observed that the problem dimension and complexity increase drastically, from the uni-entity to multi-entity networks and from the individualistic to the collaborative to the stochastic MILPs. Interestingly, the RO MILP shows execution times similar to the deterministic equivalent. The development of these MILPs open new frontiers for optimization of the PSC design and planning as presented hereafter.

7.5 Theoretical Implications

The current work takes a complex real world problem of downstream petroleum supply chain design and planning, and progressively applies modelling approaches by increasing the dimension of the problem, first by considering the design problem without inventories, followed by incorporation of inventories and entity collaboration, incorporation of uncertain demand, and finally considering demand, costs and prices uncertainty. As the problem size increases in dimension, the methodology employed increases in complexity, beginning with the bi-level transportation network MILP, followed by the piecewise linearized collaborative MILP for cost, tariff and price differentiation, then the stochastic multi-objective MILP and finally the robust optimization MILP. Besides, the complexity of the model increases from current, to grassroots to retrofit.

The stochastic MILP, however, showed numerical difficulties with only 9 scenarios and only demand uncertainty, let alone simultaneous uncertainty. The stochastic MILP was extended to a multi-objective stochastic MILP by implementing an ε-constraint algorithm, wherein is maximized the net present value and simultaneously minimized a risk measure to obtain Pareto optimal solutions, however it is observed a further increase in computational effort and the need to develop decomposition strategies. On the other hand, the RO MILP showed considerable performance gains when compared to the stochastic multi-objective MILP, even with the increased complexity of simultaneous demand, costs and prices uncertainties. While the stochastic MILP provides only one solution, the multi-objective stochastic MILP provides many Pareto optimal solutions, the RO MILP generates deterministically, many solutions by varying the levels of uncertainty from the optimal to the worst-case. The decision-maker may then make an informed choice among many available solutions based on his risk attitude.

Hence, the current research shows that RO is more suited for the very large scale PSC planning problems, supported by the following advantages: uncertainties may be incorporated without much information on their distributions; demands, costs, prices, and time period uncertainties can be simultaneously treated; large scale supply chain problems can be modelled with numerical tractability; qualitatively similar optimal solutions are obtained as in stochastic programming however the later has serious complexity issues; many solutions are obtained quickly; and finally has the advantage to bring together managers and researchers on a common ground.

7.6 Future Research Perspectives

The downstream PSCs design and planning MILPs developed should be considered as a building block and further research can explore many research directions.

The models were validated for the current, retrofit and grassroots designs, with real large-scale instances based on the Portuguese district-level PSC network. This has practical application as such work could be used to further study the Portuguese PSC by considering the municipality-level product consumption, or could be used to study the integration of the PSC networks in Portugal and Spain. The performance of the PSC could be studied with respect to the impact of differences in the Excise and VAT taxing systems between two neighbor countries. New insights could be arrived as to the effects and benefits of the fiscal harmonization of these taxes between the two countries, especially in the boundary areas, using other economic, sustainability and social indicators.

Although collaborative MILPs were developed, a different approach would be to apply perspectives that are not collaborative, and develop competitive models. For example, in Portugal, there is clearly a leader who dominates many parts of the PSC while the others are followers. Here, the model could develop a competitive or counter strategy based on known or projected strategic or tactical information of a competitor. Such a decision of an entity could directly affect the performance or future decision of the competitor who could launch a counter-strategy as in game theory. An entity could also make investments or hold inventories or sell products at higher prices so as to benefit from a stock out or in order to increase competition.

The MILPs could be extended for operational planning, by modeling decisions for supply shortages and excesses, transformation, and spot supplies and selling of petroleum products. The strategic grassroots and retrofit designs, and the tactical entity tariffs, product flows and inventories could be used to align a contract agreement between the logistic operator and the petroleum companies. This setup can be thought of as minimum volume contracts where each operator is obliged to supply, unload and ensure the inventory volumes to comply within the established contract bounds, and pay a penalty if exceeded or defaulted from these volumes. By fixing the strategic decisions, the models could be run again by using the tactical decisions as the upper and lower bounds for the network volumes, initial and final inventories per location, entity, product, and time period. The operational planning decisions could then be the mismatches (default or excess) between the earlier tactical decisions (contract volumes) and the realized volumes, based on the uncertain demand. The operational planning model could minimize the sum of the shortages and excesses in supply, transformation, spot supply and spot selling volumes which could be appropriately penalized. Similar ideas were proposed in [10], however, an integrated model is not yet available.

The collaborative planning MILPs could also be extended to model resilience by recurring to multi-objective optimization. Reference [8] implemented for a production SC, eleven resilience indicators spread over three categories from the literature. The first being the network design indicators, specifically: the node complexity or number of nodes in the network, the flow complexity or number of forward and reverse flows in the network, the node density or the ratio between the number of total flows in the network and the number of potential flows, and the node criticality or the total number of critical nodes in the network. The second are network centralization indicators such as the out-degree and in-degree centrality based on the actual number of arcs, and the out-degree and in-degree centrality based on the actual amounts of flow circulating in these arcs; and finally the operational indicators such as the ENPV, customer service level, and investment amount.

Another possible extension could be the modelling of sustainability using economic, environmental and social dimensions. Reference [3] salient the need to identify appropriate measures and to follow a systemic and holistic approach that fosters long-term engagements. Here, multi-objective approaches have been suggested to establish trade-offs and build sustainable SCs that are agile, adaptive, and socially balanced. Specifically, society and legislation are putting pressure to resolve impounding issues such as energy consumption, natural resources scarcity, global warming, and climate changes which are so important when designing and planning the downstream PSCs.

7.7 Conclusions

Research concerning the design and planning of multi-entity downstream PSCs under crude costs, product prices, and customer demand uncertainties has been summarized. Four PSC planning MILPs have been presented, specifically the individualistic, collaborative, multi-objective stochastic, and the robust optimization MILPs. Besides determining the strategic and tactical decisions, the MILPs determine individualistic as well as collaborative economy of scale multi-entity costs, tariffs and prices per route, location and product. The augmented ϵ-constraint stochastic MILP simultaneously maximizes the ENPV and minimizes risk measures. The robust optimization MILPs optimize the worst-case profits when considered uncertainties. Test results are presented for the Portuguese downstream PSC. MILP improvement opportunities are identified for operational planning, Iberian-context PSC integration and fiscal harmonization, competitive planning with game theory, resilience modeling for network design, centralization and operational indicators, and sustainability modeling to integrate the economic, environmental and social dimensions.

Acknowledgements The authors thank Fundação para Ciência e Tecnologia (FCT) and Companhia Logística de Combustíveis (CLC) for supporting this research.

References

1. W.B.E. Al-Othman, H.M.S. Lababidi, I.M. Alatiqi, K. Al-Shayji, Supply chain optimization of petroleum organization under uncertainty in market demands and prices. Eur. J. Oper. Res. **3**(189), 822–840 (2008)
2. D. Alem, R. Morabito, Production planning in furniture settings via robust optimization. Comput. Oper. Res. **2**(39), 139–150 (2012)
3. A.P. Barbosa-Póvoa, Process supply chains management - where are we? where to go next? Process Energy Syst. Eng. **23**(2), 1–13 (2014)
4. D. Bertsimas, M. Sim, Robust discrete optimization and network flows. Math. Program. **98**, 43–71 (2003)
5. J.R. Birge, F. Louveaux, *Introduction to Stochastic Programming*, Springer Series in operations Research and Financial Engineering (2011)
6. S.N. Boschetto, L. Magatao, W.M. Brondani, F.J. Neves, L.V.R. Arruda, A.P.F.D. Barbosa-Póvoa, S. Relvas, An operational scheduling model to product distribution through a pipeline network. Ind. Eng. Chem. Res. **49**, 5661–5682 (2010)
7. D.C. Cafaro, J. Cerdá, Optimal scheduling of multiproduct pipeline systems using a nondiscrete MILP formulation. Comput. Chem. Eng. **28**, 2053–2068 (2004)
8. S.R. Cardoso, A.P. Barbosa-Póvoa, S. Relvas, A.Q. Novais, Resilience metrics in the assessment of complex supply chains performance operating under demand uncertainty. Omega **56**, 53–73 (2015)
9. M.A. Dempster, N.H. Pedron, E. Medova, J. Scott, A. Sembos, Planning logistics operations in the oil industry. J. Oper. Res. Soc. **51**, 1271–1288 (2000)
10. L.F. Escudero, F. Quintana, J. Salmerón, Coro, a modeling and an algorithmic framework for oil supply, transformation and distribution optimization under uncertainty. Eur. J. Oper. Res. **3**(114), 638–656 (1999)
11. L.J. Fernandes, S. Relvas, A.P. Barbosa-Póvoa, Strategic network design of downstream petroleum supply chains: single versus multi-entity participation. Chem. Eng. Res. Des. **91**, 1557–1587 (2013)
12. L.J. Fernandes, S. Relvas, A.P. Barbosa-Póvoa, Collaborative design and tactical planning of downstream petroleum supply chains. Ind. Eng. Chem. Res. **44**(53), 17155–17181 (2014)
13. L.J. Fernandes, S. Relvas, A.P. Barbosa-Póvoa, Downstream petroleum supply chain planning under uncertainty, in *Proceedings of 12th International Symposium on Process Systems Engineering and 25th European Symposium on Computer Aided Process Engineering*, vol. 37, pp. 1889–1894 (2015)
14. L.J. Fernandes, S. Relvas, A.P. Barbosa-Póvoa, Petroleum supply chain network design and tactical planning with demand uncertainty, *Operations Research and Big Data*, Springer Series in Studies in Big Data. vol. 15, pp. 59–66 (2015)
15. L.J. Fernandes, S. Relvas, D. Alem, A.P. Barbosa-Póvoa, Robust optimization for petroleum supply chain collaborative design and planning, in *Proceedings of 26th European Symposium on Computer Aided Process Engineering*, vol. 38, pp. 1563–1568 (2016)
16. L.J. Fernandes, S. Relvas, A.P. Barbosa-Póvoa, Downstream petroleum supply chain strategic and tactical planning under uncertainty. Doctoral dissertation, CEG-IST Centre for Management Studies. Instituto Superior Técnico. Universidade de Lisboa, pp. 1–243 (2016)
17. L. Fiorencio, F. Oliveira, P. Nunes, S. Hamacher, Investment planning in the petroleum downstream infrastructure. Int. Trans. Oper. Res. **2**(22), 339–362 (2014)
18. M. Guajardo, M. Kylinger, M. Ronnqvist, Speciality oils supply chain optimization from a decoupled to an integrated planning approach. Eur. J. Oper. Res. **229**, 540–551 (2013)
19. P. Guyonnet, F.H. Grant, M.J. Bagajewicz, Integrated model for refinery planning, oil procuring, and product distribution. Ind. Eng. Chem. Res. **48**, 463–482 (2009)
20. Y. Kim, C. Yun, S.B. Park, S. Park, L.T. Fan, An integrated model of supply network and production planning for multiple fuel products of multi-site refineries. Comput. Chem. Eng. **11**(32), 2529–2535 (2008)

21. A. Leiras, G. Ribas, S. Hamacher, A. Elkamel, Literature review of oil refineries planning under uncertainty. Int. J. Oil Gas Coal Technol. **2**(4), 156–173 (2011)
22. C. Lima, S. Relvas, A.P. Barbosa-Póvoa, Downstream oil supply chain management: a critical review and future directions. Comput. Chem. Eng. **92**, 78–92 (2016)
23. G. Mavrotas, K. Florios, An improved version of the augmented ϵ-constraint method (augmecon2) for finding the exact pareto set in multi-objective integer programming problems. Appl. Math. Comput. Arch. **18**(219), 9652–9669 (2013)
24. S.A. MirHassani, An operational planning model for petroleum products logistics under uncertainty. Appl. Math. Comput. **196**, 744–751 (2008)
25. S.A. MirHassani, R. Noori, Implications of capacity expansion under uncertainty in oil industry. J. Pet. Sci. Eng. **77**, 194–199 (2011)
26. S.M.S. Neiro, J.M. Pinto, A general modeling framework for the operational planning of petroleum supply chains. Comput. Chem. Eng. **28**, 871–896 (2004)
27. F. Oliveira, I.E. Grossmann, S. Hamacher, Accelerating benders stochastic decomposition for the optimization under uncertainty of the petroleum product supply chain. Comput. Oper. Res. **49**, 47–58 (2014)
28. S. Relvas, H.A. Matos, A.P.F.D. Barbosa-Póvoa, J. Fialho, A.S. Pinheiro, Pipeline scheduling and inventory management of a multiproduct distribution oil system. Ind. Eng. Chem. Res. **45**, 7841–7855 (2006)
29. S. Relvas, R. Magatao, A.P. Barbosa-Póvoa, F. Neves, Integrated scheduling and inventory management of an oil products distribution system. Omega **41**, 955–968 (2013)
30. H. Sahebi, S. Nickel, J. Ashayeri, Strategic and tactical mathematical programming models within the crude oil supply chain context - a review. Comput. Chem. Eng. **68**, 56–77 (2014)
31. N.K. Shah, Z. Li, M.G. Ierapetritou, Petroleum refining operations: key issues, advances, and opportunities. Ind. Eng. Chem. Res. **50**, 1161–1170 (2011)

Chapter 8
Efficiency and Capital Structure in Portuguese SMEs

António Fernandes, Clara Bento Vaz and Ana Paula Monte

Abstract This paper aims to analyse the bi-directional relationship between technical efficiency, as a measure of companies' performance, and capital structure, under the agency cost theory as well as the pecking order and trade-off theory, to explain the capital structure decisions. The technical efficiency was estimated by the DEA method and corrected by using a suitable bootstrap to obtain statistical inferences. To test the agency cost hypothesis, asymmetric information hypothesis, risk-efficiency hypothesis and franchise value hypothesis (under pecking order and trade off theories framework), two models were applied using some determinants of capital structure such as size, profitability, tangibility, liquidity as control and explanatory variables through a truncated regression with bootstrapping. From an initial sample of 1024 small and medium sized companies from the interior of Portugal, for the period 2006–2009, a subsample of 210 SMEs from secondary and tertiary sectors was selected. The results suggest that medium sized companies have higher average bias-corrected efficiency than small companies; that short-term leverage is

---

UNIAG—Research unit funded by the FCT - Portuguese Foundation for the Development of Science and Technology, Ministry of Science, Technology and Higher Education. Project No. UID/GES/4752/2016.
NECE—Research Unit in Business Sciences, Beira Interior University, Research unit funded by the FCT - Portuguese Foundation for the Development of Science and Technology, Ministry of Science, Technology and Higher Education. Project No. UID/GES/04630/2013.

A. Fernandes
Applied Management Research Unit (UNIAG), Polytechnic Institute of Bragança, Bragança, Portugal
e-mail: antoniof@ipb.pt

C. B. Vaz (✉)
Centre for Management and Industrial Engineering (CEGI/INESC TEC), Polytechnic Institute of Bragança, Bragança, Portugal
e-mail: clvaz@ipb.pt

A. P. Monte
Applied Management Research Unit (UNIAG) and NECE, Polytechnic Institute of Bragança, Bragança, Portugal
e-mail: apmonte@ipb.pt

A. I. F. Vaz et al. (eds.), *Operational Research*, Springer Proceedings in Mathematics & Statistics 223, https://doi.org/10.1007/978-3-319-71583-4_8

positively related to efficiency and that the companies in the sample follow pecking order theory.

Keywords Data envelopment analysis · Technical efficiency · Capital structure SME · Inland of Portugal

8.1 Introduction

The debate on capital structure of companies has been an issue discussed for long time and there is no consensus on it. Several theories have arisen since the seminal paper of Modigliani and Miller [39]. The most common has been the pecking order theory, hereinafter POT, [40, 42] and the tradeoff theory, hereinafter TOT [16, 38]. POT advocates that firms rank the different sources of capital by giving preference to self-financing, given that information asymmetry increases funding costs [24, 33, 42]. TOT argues that each company has an optimal capital structure that results from the balance between the use of debt (target debt ratio) and tax and other benefits against bankruptcy costs and other costs (such as agency costs) [12, 14, 30].

As Bie and Haan [12] point out, the choice of funding sources is driven by the costs of adverse selection resulting from the asymmetry of information between (more informed) managers and (less informed) investors. The POT is based on the information asymmetric hypothesis. According to this argument, insiders possess more private information about a firm's expectations on returns and potential investment growth, that is, the "true" value of the business [42, 68]. In this sense there is an adverse selection cost that force managers and owners of the companies to prefer to use internal funds, then debt and finally new equity to finance their investment opportunities as the higher risk is perceived and some constraints of external debt financing appear. Smaller and younger companies suffer more from this information asymmetry due to their natural information opacity, as La Rocca et al. [31] pointed out. Therefore, the availability of internally generated funds diminishes the need of external finance and hence it is expected lower debt ratios.

The agency cost hypothesis is another concurrent theory associated with TOT to explain the capital structure decisions. Agency conflicts between managers-ownership and debtors-creditors may explain the behavior and capital structure decisions. Jensen and Meckling [26] defend that low leverage ratios may motivate managers to act on behalf of shareholders' interests, reducing agency costs. High leverage ratios, inducing higher bankruptcy and distress costs, increase the agency costs, limiting the threshold leverage. Also Harris and Raviv [23] as park and jang [45] state that agency costs can be mitigated by use of leverage. The size of the companies can have a positive impact on this relationship, as is stated by several authors [3, 4, 34, 44, 47, 61].

Much literature has tried to validate these theories in different markets and sectors of activity as well as for companies of different sizes [1, 2, 6, 11, 17, 20, 27, 33, 37, 61, 64, 67]. Associated with each theory several determinants have been tested to justify the capital structure and related decisions. Titman and Wessels [63], followed by others such as Frank and Goyal [19] and Guo and Suliman [21] studied factors such as liquidity, profitability, non-debt tax shields, tangibility, growth opportunities, uniqueness of products, among others. Other studies have analyzed, besides companies' characteristics, factors related to markets or macroeconomics characteristics like more market oriented economies or more banked economies [3, 8] or industry [32]. There are also reports of research that have been carried out on SMEs and non-listed or family-owned companies [11, 25, 27, 33, 36]. Some of these studies analyzed factors that influence the capital structure of SMEs, such as asset tangibility; profitability; growth opportunities and level of indebtedness [29]. The empirical results do not always coincide. Most of this research uses panel data analysis or cross-section analysis by ordinary least square (OLS) regressions or generalized method of moments (GMM) to explain the leverage against the above determinants. To measure financial leverage the ratio total debt to total asset is used and sometimes the short-term debt to total assets or long-term debt to total assets is analyzed.

There are some studies on Portuguese market [46, 51–55, 65, 66]. The majority of this research relies on determinants of capital structure or debt ratio using panel data analysis, cross section analysis and linear regressions. These methodologies are applied to SMEs, listed and unlisted companies. None of this research applies technical efficiency as explained or explanatory variables on leverage or capital structure.

In this manner, the main research questions behind this study are (i) will there be an optimal capital structure depending on the efficiency and financial performance of SMEs? (ii) which factors determine the choice of the capital structure of SMEs? To answer these questions a sample of SMEs is used from the interior of Portugal and the technical efficiency is estimated using the DEA method and then two models to explain the bi-directional relationship of efficiency and leverage are used, following a similar research design to [34, 35] with a slightly difference: the bootstrap methods were used. The proposed approach allows more robust efficiency scores to be achieved by using suitable bootstrap methods [59, 60]. These efficiency scores are used to explain the bi-directional relationship of efficiency and leverage, enabling the determination of more robust conclusions, which has an advantage over deterministic approaches used in the literature [34, 35, 50]. In the Portuguese context, as far as we know, this paper is one of the first studies to associate the technical efficiency, as proxy for corporate performance, and leverage (proxy for decisions on companies' capital structure). This methodology may add new contributions to the controversy on leverage theory and SME behavior.

The remainder of the paper is as follows. Section 8.2 presents a concise literature review on performance, efficiency and capital structure. Then, Sect. 8.3 refers to the methodology adopted, describes the data and sample, the DEA method, the research hypothesis formulated on behalf of the previous literature review, the empirical models and definition of variables. Following that, Sect. 8.4 exhibits and discusses the results. It ends with identifiable conclusions, limitations and suggestions for further research.

8.2 Performance, Efficiency and Capital Structure

Most of the research on capital structure and performance relies on unidirectional relationship, independently of the theory used (TOT, POT, or others), measuring performance based on financial ratios such as return on assets (ROA) or return on equity (ROE) or similar [e.g. 13, 29, 33, 43, 46, 62]. However [10] proposed to use profit efficiency as firm performance instead of traditional ratios and looked for the bi-directional relationship between capital structure and performance under the efficiency-risk hypothesis and franchise value hypothesis using a sample of companies from the commercial banking sector. "*Profit efficiency evaluates how close a firm is to earning the profit that a best-practice firm would earn facing its same exogenous conditions. This has the benefit of controlling for firm-specific factors outside the control of management that are not part of agency costs*" [10, p. 1067]. This measure provides a good benchmark for how the firm is expected to perform if agency cost was minimized [10].

Under the efficiency-risk hypothesis, firms that are more efficient may choose higher leverage because higher efficiency reduces the expected costs of bankruptcy and financial distress. On the other hand, under the franchise-value hypothesis, more efficient firms may choose lower leverage to protect the economic rents derived from higher efficiency and the possibility of liquidation [10, 34, 35].

Margaritis and Psillaki [34, 35] used two cross-section models to explain the bi-directional relationship of capital structure, namely leverage and performance, more precisely technical efficiency. Margaritis and Psillaki [34] use the technical efficiency derived from the non-parametric input distance function [56] on a sample of New Zealand SMEs firms. Margaritis and Psillaki [35] measure the technical efficiency through the directional distance function on a sample of French firms from three different manufacturing industries. In model one they relate the technical efficiency obtained through the data envelopment analysis (DEA) model with leverage, measured by the debt ratio, and a group of control variables for firm and market characteristics such as size, tangibility, profitability, growth opportunities, among others.[1] This model intends to test the cost agency hypothesis. Model two relates the debt ratio (total debt to total assets) to the measure of the firm's technical efficiency and a number of factors that have been commonly used in other research to explain the capital structure or leverage. This model is intended to test the risk-efficiency hypothesis and franchise value hypothesis. The control variables used in model two were size, asset structure, profitability, risk and growth (variables related to firm characteristics) as well as market power (as industry characteristics).

[1]Other variables used by Margaritis and Psillaki [34] were risk, measured by the standard deviation of annual earnings before taxes and market power proxied by the concentration index (CI), that represents the market share of the largest four firms in the industry. Margaritis and Psillaki [35] used also the ownership structure. Other determinants used are effective tax paid, measured by the ratio of Tax Paid to Earnings Before Taxes [53] or Non-debt tax shield, measured by the ratio between depreciations and amortizations and total assets [50, 53].

Firm size (measured by the logarithm of the firm's sales) is expected to be positively related to leverage in accordance to TOT [34, 43, 48, 53, 63]. However, the firm's size may be also negatively correlated with leverage since "*size may act as a proxy for the information outside investors have, and that informational asymmetries are lower for large firms which implies that large firms should be in a better position to issue informationally sensitive securities such as equity rather than debt.*" [34, p. 1456]. Also Michaelas et al. [36] and Serrasqueiro et al. [53] pointed out that problems of asymmetric information and moral hazard will be greater for small firms, because of the lack of financial disclosure and their owner-managed nature. Therefore, lenders will be unwilling to lend long-term to small firms on favorable terms (higher long-term debt cost) and therefore, SMEs tend to issue short-term debt to overcome those problems. Consequently, small firms frequently have a higher level of short-term debt than larger firms [53]. Seelanatha [50] used a panel data analysis to estimate the radial technical efficiency scores of Chinese firms from different industries and then considered the impact of a firm's relative efficiency, market share and industry concentration on capital structure decisions.

Asset tangibility (measured by the ratio of fixed tangible assets to total assets) should be positively correlated to leverage, as tangible assets can be used as collateral and mitigate problems of information asymmetry [23, 34, 53]. However some empirical studies have found negative correlation with short-term debt ratios [61].

Profitability (pre-interest and pre-tax operating surplus divided by total assets) should be negatively correlated with leverage in accordance with POT [34, 53], because profitable firms will finance their investments with internal funds and move to external finance only when internal funding is insufficient. Still based on TOT and contracting cost theories we can predict a positive relation between profitability and leverage because the most profitable firms have greater debt capacity, and may take advantage of debt tax-shields [53].

Intangible assets such as future growth opportunities [34, 36, 43, 63] would have a negative correlation with debt since firms with expected growth opportunities would keep low leverage in order to avoid adverse selection and moral hazard costs associated with financing of new investments with new equity capital [34]. This behavior is aligned with TOT. Conversely, according to POT, firms with higher growth opportunities have more needs of funds and when the internal finance is exhausted, firms prefer debt to external equity to finance risky investments and therefore increase leverage. Some empirical studies on SME's capital structure have found positive correlation between growth opportunities and leverage [36]. Growth may be also measured by annual percentage change in earnings [34] or by asset growth as a total asset annual change in percentage [50, 53] or, as suggested by Margaritis and Psillaki [35], by sales growth.

Liquidity (as measured by current ratio: current assets to current liabilities) tends to be positively related to leverage according to TOT [53]. Firms with a lower level of liquidity will face more obstacles in obtaining debt and TOT predicts a positive relationship between liquidity and leverage. Conversely, POT predicts that there is a negative relationship because firms with a high level of liquidity have more internal funds and therefore tend to borrow less [8, 28].

8.3 Methodology

8.3.1 Data and Sample

The database used in this research, from our point of view, is an added value as it contains 1024 companies from whom financial and non-financial data was collected through fiscal documents, known as Simplified Business Information model (Informação Empresarial Simplificada - IES[2]), for the period 2006 to 2009. For the sake of secrecy, the provided information omitted any data that could lead to the identification of the companies and a code was given to the companies in the database. There were only considered firms constituted in the legal form of companies. The basis for this decision was the fact that individual entrepreneurs have a specific IES model. Regarding the activity sector (industry), no restriction was established in the selection phase of the companies. This database is a convenient sample of a population of 60529 companies in 2009 distributed along the interior of mainland of Portugal. In this region were considered the following NUTS III: Douro, Alto Trás-os-Montes, Cova da Beira, Beira Interior Norte, Ave, Dão-Lafães, Beira Interior Sul, and Tâmega. Considering that, the main objective of this research is to analyze the efficient capital structure of companies in the interior region of Portugal, the research variables were defined from the respective accounting documents over the several years. As monetary values are not reported at the same moment in time, they were deflated with the following inflation rates: 2% for 2006, 3% for 2007 and 1% for 2008. See Fernandes [18] for further details about this database.

According to the assumptions of DEA, it is necessary to improve the homogeneity of the companies and remove some outliers from the database. The management schemes of Micro companies can be very different among them while the primary sector has a government support. Thus, to improve the homogeneity of the companies, we only consider the small and medium sized companies from secondary and tertiary sectors following Margaritis and Psillaki [34, 35]; Acaravci [2]. Regarding the outliers, it is essential to exclude from the sample the companies that have different features, such as the ones that are in technical bankruptcy or others that have a negative asset or an unusual other accounting score. In terms of DEA, it is important to mitigate the effect of some extreme observations as these outliers can severely affect the location of the DEA frontier. The extreme observations were removed according to the Andersen and Petersen [7] approach. Therefore, from the initial database of 1024 micro, small and medium sized companies from the interior of Portugal, for the period 2006–2009, a sample of 210 small and the medium sized

[2]IES is a new way for companies to deliver information on-line to public services, through an electronic form, by using a totally dematerialized procedure. This service allows to abiding, at once, the following legal obligations: Deposit of annual accounts in Business Register; Delivery of annual fiscal declaration to Ministry of Finance and Public Administration; Delivery of annual information to National Statistics Institute for statistical purposes; Delivery of information to Portuguese Central Bank. Additional information is available at http://www.ies.gov.pt/.

Table 8.1 Size of companies by sector of activity

Firm size		Activity sector		Total
		Secondary	Tertiary	
Medium	n	44	39	83
	% Size of the company	53.0%	47.0%	100.0%
	% Activity sector	13.3%	12.0%	12.6%
Small	n	287	287	574
	% Size of the company	50.0%	50.0%	100.0%
	% Activity sector	86.7%	88.0%	87.4%
Total	n	331	326	657
	% Size of the company	50.4%	49.6%	100.0%
	% Activity sector	100.0%	100.0%	100.0%

companies (SME) was collected from the secondary and tertiary sectors, involving 657 observations.

Table 8.1 summarizes the distribution of companies (n) in our sample by activity sector and size. The final sample comprising 87.4% of small companies and 12.6% of medium sized companies, involves a similar number of companies from both sectors (50.4 and 49.6% of the companies belong to secondary and tertiary sectors, respectively).

It can be observed from Table 8.2 that, on average, small enterprises have higher profitability (as measured by ROA) than medium sized ones. In terms of Standard Deviation (St. Dev.) it seems that there is not much difference between the two

Table 8.2 Descriptive statistics on profitability and growth, by firm size, for 2006–2009 period

Firm size		ROA[a]	NRT[b]	$\Delta(TNA)$[c]	$\Delta(Turnover)$[d]
Small	Mean	3.53%	−0.57%	6.13%	6.52%
	St.Dev.	9.84%	73.18%	26.09%	52.40%
	n	574	574	574	574
Medium	Mean	3.00%	−3.91%	9.51%	22.67%
	St.Dev.	8.02%	41.07%	51.81%	129.69%
	n	83	83	83	83
Total	Mean	3.46%	−1.00%	8.69%	6.58%
	St.Dev.	9.62%	69.93%	52.12%	26.69%
	n	657	657	657	657

[a]ROA: Return on Assets, measured by the ratio of Earnings Before Taxes (EBT) to TNA, that is a proxy for firm's profitability;
[b]NRT: Net Return on Turnover, measured by the ratio of EBT to firm's Turnover;
[c]$\Delta(TNA)$: is the annual change of TNA, measured by TNA of moment t minus TNA of previous moment $(TNA_t - TNA_{t-1})$;
[d]$\Delta(Turnover)$: is the annual change of Turnover, measured by $(Turnover_t - Turnover_{t-1})$

groups. Attending to the Net Return on Turnover (NRT), small enterprises have also higher profitability, although they present higher variability. In terms of growth (as measured by annual change of Total Net Assets (TNA) or Turnover), it is noted that medium sized enterprises showed higher growth rates than small firms but have higher variability (St. Dev. is higher for medium sized enterprises). This may be related to several factors, namely: more professional management, clearer separation between managers and holders of capital, greater supervision and internal control in the company.

8.3.2 DEA Methodology

The technical efficiency for each company is evaluated from the DEA model, introduced by Charnes et al. [15]. DEA is a non-parametric approach to assess the relative efficiency of a homogeneous set of Decision Making Units (DMUs) in producing multiple outputs from multiple inputs. DEA is used to assess the technical efficiency of the companies in minimizing the resources for a given level of achieved revenues. The technical efficiency reflects the economic perspective of each company in managing the resources for a given level of revenues. DEA allows the identification of the best practices DMUs (the benchmarks) and their linear combination defines the frontier technology that envelops all DMUs observed in the production possibility set (PPS). For the inefficient DMUs located inside the PPS, the magnitude of the inefficiency is derived by the distance to the frontier and a single summary measure of efficiency is estimated.

Consider a set of n DMUs $j\,(j = 1, \ldots, n)$, each consuming m resources (inputs) x_{ij} $(x_{1j}, \ldots, x_{mj})$ to produce s results (outputs) y_{rj} $(y_{1j}, \ldots, y_{sj})$. As the scale size affects the productivity of a company, it is necessary to estimate the pure technical efficiency (hereafter technical efficiency) in reference to the variable returns to scale (VRS) observed frontier [9]. Thus, for an input minimizing perspective the relative efficiency of the assessed DMU$_o$ can be estimated using the linear programming model (8.1):

$$\min \Big\{ \hat{\theta}_{j_o} | \hat{\theta}\, x_{ij_o} \geq \sum_{j=1}^{n} \lambda_j\, x_{ij}, \quad i = 1, \ldots, m \tag{8.1}$$

$$y_{rj_o} \leq \sum_{j=1}^{n} \lambda_j\, y_{rj}, \quad r = 1, \ldots, s$$

$$\sum_{j=1}^{n} \lambda_j = 1$$

$$\lambda_j \geq 0; \quad \forall_{j,i,r} \Big\}$$

The optimum solution of model (8.1), $\hat{\theta}^*_{j_o}$, corresponds to the minimum factor by which the inputs levels can be reduced giving the current level of revenues, corresponding to the relative efficiency of the assessed DMU_o. DEA enables us to identify the efficient DMUs, which the efficiency measure is equal to 1, being the benchmarks. The remaining DMUs are the inefficient units which the efficiency measure is lower than 1, indicating the existence of inefficiencies in managing the resources for a given level of revenues.

To correct the DEA efficiency estimates for bias, the bootstrapping method is used, according to Simar and Wilson [59], which is suitable for use with DEA efficiency estimates, ranging from 0 to 1. Simar and Wilson [59] proposed the smoothed bootstrap method suitable to DEA to estimate the original densities of the non-parametric efficiency scores using kernel smoothing methods combined with a reflection method [58] by mimicking the data generating process (DGP). This procedure was implemented using the statistical software R including the FEAR library, developed by Wilson [69]. Thus, for each DMU there is derived the bias and the bias-corrected efficiency, $\hat{\hat{\theta}}$ as defined in (8.2). These scores are used to assess the company's performance and to analyze the bi-directional relationship.

$$\hat{\hat{\theta}} = \hat{\theta}^*_o - Bias \tag{8.2}$$

To explore the determinants that can be associated with good efficiency levels of the companies, we use the bootstrap-truncated regression formulated according to the double bootstrap method (algorithm #2) proposed by Simar and Wilson [60], in which efficiency scores are bootstrapped in the first stage, as explained before, and then the second step is performed based on the bootstrap truncated regression. This approach is used to investigate the determinant variables on efficiency levels of the companies. Considering the company $j(j = 1, \ldots, n)$ in the time period t $(t = 1, \ldots, m)$, the impact of the regressors, defined by variables z_{jt}, on efficiency score θ_{jt}, is assessed by the model (8.3):

$$\theta_{jt} = \beta_o + z_{jt}\beta + \varepsilon_{jt} \tag{8.3}$$

where β_o is the intercept, β corresponds to the vector of regression coefficients to be estimated and ε_{jt} is the error term with a $N(0, \sigma^2_\varepsilon)$ distribution with a truncation at $(1 - \beta_o - Z_{jt}\beta)$. Note that θ_{jt} corresponds to the efficiency of company j, in year t, estimated by using model (8.1) and corrected by bootstrapping as defined in (8.2).

8.3.3 Hypothesis, Empirical Models and Variables

Having in mind that the goal of this study is to investigate the bi-directional relationship of companies' performance in terms of efficiency and the leverage (or capital structure decision), taking into account the two main theories, TOT and POT as well as the agency theory, the research hypotheses to be tested based on previous literature review are as following:

H_1 (agency cost hypothesis): Performance is improved as higher leverage is expected to lower agency costs and reduce inefficiency. That is, efficiency is positively related to leverage.

H_2 (efficiency risk hypothesis): Firms that are more efficient choose higher leverage ratios because higher efficiency is expected to lower the costs of bankruptcy and financial distress. That is, the leverage is positively related to efficiency.

H_{2a} (franchise-value hypothesis): Firms that are more efficient tend to hold extra equity capital and therefore, all else being equal, choose lower leverage ratios to protect their future income or franchise value. Thus, the leverage is negatively related to efficiency.

Two equations cross-sections models will be used in order to test the proposed hypotheses mentioned above. The research hypothesis H_1 is explored through the truncation regression model for the firm performance model given by equation model (8.4):

$$\hat{\hat{\theta}}_{jt} = \beta_o + Leverage_{jt}\beta_1 + z_{1jt}\beta_2 + \varepsilon_{jt} \tag{8.4}$$

where $\hat{\hat{\theta}}_{jt}$ is the bias-corrected efficiency determined from bootstrapping [59]. The leverage is the short-term debt ratio. The z_{1jt} control variables include the size, the debt cost, the asset structure, the coverage of non-current assets, the ROA and the current ratio.

The research hypotheses H_2 and H_{2a} are investigated through the OLS regression model for the firm leverage given by equation model (8.5):

$$Leverage_{jt} = \alpha_o + \hat{\hat{\theta}}_{jt}\alpha_1 + z_{2jt}\alpha_2 + \upsilon_{jt} \tag{8.5}$$

The z_{2jt} is a vector of control variables that include the bias-corrected efficiency, the size, the debt cost, the asset structure, the coverage of non-current assets, the ROA and the current ratio. The υ_{jt} is a stochastic error term. The statistical inference of OLS regression is performed by bootstrapping, using 2000 replicates.

The control variables used in both models (8.4) and (8.5) are the following:

- Firm size (Size): this variable is measured by ln(sales). If the effect of this variable is positive, it may indicate that large companies use better technology, are more diversified and better managed. A negative effect can be observed in situations where there will be loss of control resulting from inefficient hierarchical structures [5].
- Debt cost (%): this variable corresponds to the cost of external financing sources of capital. This was obtained by dividing the interest expenses by total liabilities. The company is supposed to replace debt by equity, or equity by debt, until the firm's value is maximized [41].

- Short-term (debt ratio) (%): indicates the weight of current liabilities in total assets. The size of the company may influence its financing sources. Small firms, compared to large firms, bear higher costs of issuing long-term debt [63]. The authors consider that small firms may be more leveraged than large firms and may prefer to take short-term loans (through bank loans) instead of issuing long-term debt due to lower fixed costs associated with this alternative.
- Asset structure (%): indicates the proportion of non-current assets that are found in total assets. This is a proxy for tangibility. The existence of asymmetric information and agency costs may induce creditors to require collateralised guarantees [5, 22, 49]. If the company has high investments in land, equipment and other tangible assets, it will usually face lower financing costs compared to a company that is mainly based on intangible assets [5].
- Coverage of non-current asset (%): corresponds to permanent capital (equity + non-current liabilities) to non-current assets. If the value is greater than 1, it is concluded that the company is following the rule of minimum financial equilibrium. Hackbarth [22] considers that the debt-to-firm value ratio captures the degree of leverage. "*The ex ante value of equity prior to the leverage decision differs from the ex post value of equity, i.e., at the time when the debt is already in place. In particular, the ex post value of equity is the value of the perpetual entitlement to the firm's cash flows net of its promised debt service. The ex ante value of equity equals total firm value at* $t = 0$; *i.e., the sum of the ex post value of equity and the issuance value of debt*" [22, p. 397].
- Current Ratio: this ratio is obtained by the quotient of current assets to current liabilities. The indicator is directly related to the working capital being responsible for the changes in the company's debt as a result of deviations from the budget [57].
- ROA (%): corresponds to the ratio of earnings before tax to total net assets. This is one of the profitability ratios commonly used.

8.4 Results

Firstly, the DEA model (8.1) is used to estimate the technical efficiency of Portuguese small and medium sized companies from secondary and tertiary sectors regarding the 2006 and 2009 period. Secondly, we estimate the DEA efficiency of SMEs corrected for the bias [59]. These scores are used to analyze the bi-directional relationship for SMEs from secondary and tertiary sectors, following similar research as Margaritis and Psillaki [34, 35].

8.4.1 *Efficiency*

In this stage, the DEA model (8.1) is used to estimate the technical efficiency for each company in minimizing the current resources required to achieve the observed total revenues. In terms of inputs we identify all the necessary resources to achieve the total revenues. The distance from each company to the efficient production technology is the result of inefficiencies regarding contracting costs, different principle agent objectives, managerial slack or oversight [34, 50]. Thus, the efficient frontier will identify the best practice companies in managing the resources for a given level of revenues. Taking into account this perspective, the DEA model is constructed using a single output (total revenues) and three inputs (capital, operational costs and labor) technology. The total revenues correspond to the total sales and provided services for each company. The capital is measured by the company's total net assets. The operational costs include the cost of goods sold, other operating expenses and the depreciation expenses, excluding staff costs. The labor input is measured by the total number of full-time equivalent employees and working proprietors. The data concerning the inputs and output are summarized in Table 8.3.

Table 8.3 Summary statistics of 210 small and medium sized companies between 2006 and 2009

Variable	Mean	St.Dev.	Min	Max
Capital	2086938	3216828	33828	24500000
Operational costs	1994528	3882617	15382	37800000
Labor	25.61	28.20	9.00	216.00
Total revenues	2788849	4920226	6390	38800000
Debt cost	0.03	0.03	0.00	0.31
Asset structure	0.85	0.61	0.02	4.25
Coverage of non-current assets	0.89	1.33	0.00	20.75
ROA	0.03	0.10	−1.21	0.33
Debt ratio	0.56	0.21	0.04	1.00
Current ratio	0.09	3.04	−33.52	13.12

The technical efficiency of a company in a given year is estimated by comparison to the best practices observed during the period analysed, ranging from 2006 to 2009. These efficiency estimates provide insights into potential improvements by taking into account statistical inference derived through the bootstrapping framework. The correction of the DEA efficiency estimates for bias has been performed by using 2000 bootstrap samples. Table 8.4 summarizes results for the technical efficiency, bias-corrected efficiency, standard error and bias.

Bias-corrected efficiencies reveal that magnitude of the corrected efficiencies are slightly lower than the original efficiencies, although this variation is very small. The bias-corrected efficiency estimates are preferred to the original efficiencies, since they

Table 8.4 Results of original and bootstrapped average efficiency estimates

Year	2006	2007	2008	2009
Original eff. Score $\hat{\theta}$	0.69	0.67	0.69	0.66
Bias-corrected eff. $\hat{\hat{\theta}}$	0.67	0.65	0.68	0.65
Bias	0.019	0.018	0.016	0.014
St.Dev.	0.019	0.018	0.016	0.016

Table 8.5 Average bias-corrected efficiency of companies by sector and size

		Size		
		Small	Medium	Mean by sector
Sector	Secondary	0.70 (287)	0.35 (44)	0.65 (331)
	Tertiary	0.71 (287)	0.37 (39)	0.67 (326)
Mean by size		0.70 (574)	0.36 (83)	

represent a more precise estimate of the true efficiency. Globally, the average bias-corrected efficiency varies between 0.67 (in 2006) and 0.65 (in 2009), indicating that each company should reduce their current operational costs, capital and labor by 33% (in 2006) and 35% (in 2009), on average, achieving the current level of revenues. A slight decrease in 2009 was observed, which could be explained by the financial crisis which erupted in 2008. These scores are used to analyze the bi-directional relationship.

Comparing small and medium sized companies from secondary and tertiary sectors (see the number of involved units in brackets), in Table 8.5, the medium sized companies have the lowest average technical efficiency (0.36), while the small companies have the highest score (0.70). Note that the average technical efficiency is similar in both sectors.

In the next sections, the bias-corrected efficiency of companies are used to explore the bi-directional relationship following the similar research as Margaritis and Psillaki [34, 35].

8.4.2 *Efficiency Determinants Model*

After the identification of the most efficient DMUs, it is intended to analyse the control variables (summarized in Table 8.3) that most contribute to this efficiency using the Eq. (8.4), testing the agency cost hypothesis (H_1). Table 8.6 summarizes the results from the panel data truncated model in terms of coefficients, standard errors and p-values. The total number of observations was 657. The truncated regression

Table 8.6 Truncated regression analysis results

Variable	Coefficient	Bootstrap Std. Err.	p-value
Debt ratio	0.2915	0.1117	0.009[a]
Size	−0.1794	0.0249	0.000[a]
Debt cost	0.1095	0.7054	0.877
Asset structure	−0.1361	0.0409	0.001[a]
Current ratio	0.016	0.0082	0.051[c]
Coverage of non-current assets	−0.0387	0.0217	0.075[c]
ROA	0.7711	0.2433	0.002[a]
Intercept	3.3004	0.3878	0.000[a]

[a]Indicates significance at the 1% level
[b]Indicates significance at the 5% level
[c]Indicates significance at the 10% level

model is statistically significant ($Wald - Chi^2$ test with p-value <0.001), with a pseudo-R^2 equal to 0.25.

The efficiency has a positive and statistically significant relation with short-term debt ratio (leverage). Although we used the short-term leverage, it does not reject the hypothesis H_1, that is the higher the short-term debt, the higher the company's efficiency. This can be explained by higher short-term credit facilities when compared to long-term credit for SME [63]. The firm size has a negative and statistically significant coefficient ($\beta_{Size} = -0.1794$, p-value <0.001), which means that the smaller the company, the higher its efficiency. This result is reinforced by the analysis of the asset structure effect, the current ratio and the coverage of non-current assets. A negative and statistically significant relation of tangibility (asset structure) with efficiency is observed. Although the literature indicates a positive impact of tangibility on efficiency, this is not observed in this sample. A possible reason for this is the need of SMEs to give credit and hold inventories above adequate level. In short, companies in this sample became more efficient with the increase of current assets as well as current liabilities. This observation is strengthened by the result of the coverage of the non-current assets effect on efficiency, which is negative.

Even though we have not used all the same variables in this model as Margaritis and Psillaki [34, 35], we also validate the agency cost hypothesis. We also found similar results on size and tangibility.

8.4.3 Leverage Determinants Models

Table 8.7 summarizes the results from the OLS regression model (8.5), which intends to test the efficiency as a determinant of (short-term) leverage, including other variables commonly presented as capital structure determinants according to POT or TOT. The total number of observations was 657. The OLS regression model is statis-

Table 8.7 OLS regression analysis results

Variable	Coefficient	Bootstrap Std. Err.	p-value
Bias-corrected efficiency	0.0881	0.0366	0.019[b]
Debt cost	−0.8454	0.2475	0.001[a]
Asset structure	−0.0948	0.0258	0.000[a]
Coverage of non-current assets	−0.0521	0.0272	0.057[c]
ROA	−0.458	0.111	0.000[a]
Current ratio	−0.0008	0.0048	0.864
Size	−0.0284	0.0066	0.000[a]
Intercept	1.0691	0.1123	0.000[a]

[a]Indicates significance at the 1% level
[b]Indicates significance at the 5% level
[c]Indicates significance at the 10% level

tically significant ($Wald - Chi^2$ test with p-value <0.001), with a pseudo-R^2 equal to 0.23.

The obtained results are in line with those presented in model one. It is observed that there is a positive and statistically significant effect of efficiency on short-term leverage. This validates H_2 (efficiency risk hypothesis) and rejects the franchise-value hypothesis (H_{2a}). The lower the firm size, the higher is the short-term debt. Note that the variable asset structure (proxy for tangible assets) is negative and statistically significant. This result is reinforced by coverage of non-current assets (also negative) results. This shows that companies follow the rule of minimum financial equilibrium according to POT theory. Analysing the profitability, the ROA is negative and statistically significant related to short-term debt.

8.5 Conclusions and Suggestions for Further Research

This paper reviews some aspects of the empirical literature on capital structure. Although our results are not conclusive, they serve to document empirical regularities that are consistent with existing theory. Some authors, such as Titman and Wessels [63] tried to test several models, including all the hypotheses jointly in the empirical tests. Shyam-Sunder and Myers [57] considered theories as conflicting hypotheses and examined their relative explanatory power. Based on a database of 1024 companies, for the period during the years 2006–2009, a sample of 210 companies was selected. Through the DEA method, we find companies that present an efficient capital structure. Then, we attempt to explain the variables that influence the efficiency of the capital structure of the company, as well as to explain how short-term debt is influenced by efficiency as well as by other variables.

The methodology proposed enables us to estimate more robust technical efficiency scores than Margaritis and Psillaki [34, 35] by using bootstrap methods [59, 60] suitable for the DEA method. These efficiency scores are used to explain the bi-directional relationship of efficiency and leverage, enabling us to determine more robust conclusions.

The main empirical results obtained in this study can be summarized as follows:

- From the efficiency analysis along the time horizon in this research, we observed a slight reduction in the companies' efficiency in 2009. This behavior may be related to the economic and financial crisis that started in 2008.
- The efficiency of companies rises with the increase of short-term accounts (current assets and current liabilities). Given the size of the companies in question, they are able to finance themselves at lower cost using short-term external capital, since they do not have the confidence and guarantees that the banking sector requires to finance them with long-term capital [5].
- More efficient firms choose higher leverage ratio because higher efficiency is expected to lower the costs of bankruptcy and financial distress. That is, the leverage is positively related to efficiency (H_{2a}) following the POT theory, and the efficiency hypothesis.
- The fact that variable size is negative indicates that firms have lost control resulting from an inefficient hierarchical structure in the management of the company.
- Short-term debt ratio was negatively related to company size, probably reflecting the relatively high transaction costs that small firms face when issuing long-term financial instruments. This evidence also supports some of the implications of Titman and Wessels [63].

In general, the companies under study favor the short-term rather than the long-term. Given its size, as well as its environment, current assets present values above those that would be considered normal values. Clients require more credit, just as there is a need for inventories to be higher to avoid the risk of stock out as a result of the long distances traveled. The non-current asset, which would be expected to make the companies more efficient, on the contrary, presents a negative signal. One possible explanation may be that the management of these types of companies is not a professional management, and that is wasting resources. These facts lead to an increase in current ratio (liquidity) as well as an increase in short-term liabilities.

One possible limitation of this research is the time horizon of data. As in this research we focus only on the short-term leverage it would be interesting to test the bi-directional efficiency effect on capital structure using total leverage (total debt ratio) and the long-term leverage (long-term debt ratio). Another suggestion is to test if there is a significant difference if the efficiency is estimated by NACE (Statistical Classification of Economic Activities in the European Community). An important concern that arises is related to estimation methods. Considering a bi-directional relationship, this means that efficiency and leverage are endogenous variables, and thus, other estimation methods can be explored in future developments, as suggested by an anonymous reviewer.

References

1. J. Abor. Determinants of the capital structure of Ghanaian firms, 176. African Economic Research Consortium, Nairobi, 2008
2. S.K. Acaravci, The determinants of capital structure evidence from the Turkish manufacturing sector. Int. J. Econ. Financ. **5**-Capital-Structure, 158–171 (2015), http://www.scopus.com/inward/record.url?eid=2-s2.0-84923040625&partnerID=tZOtx3y1
3. M.A. Acedo-Ramirez, F.J. Ruiz-Cabestre, Determinants of capital structure: United Kingdom versus Continental European countries. J. Int. Financ. Manag. Account. **25**(3), 237–270 (2014)
4. M.A. Acedo-Ramírez, J.C. Ayala-Calvo, J.E. Rodríguez-Osés, Capital structure of small companies in the Spanish footwear sector: relevant factors. SERIEs **2**(2), 155–173 (2012)
5. T. Akhtar, A. Zahir, M.A. Tareq, F. Rabbi, Capital structure and firm efficiency: a case of Pakistan. Indian J. Financ. **10**(2), 50–66 (2016)
6. P. Alves, E.B. Couto, P.M. Francisco, Board of directors' composition and capital structure. Res. Int. Bus. Financ. **35**, 1–32 (2015). https://doi.org/10.1016/j.ribaf.2015.03.005
7. P. Andersen, N.C. Petersen, A procedure for ranking efficient units in data envelopment analysis. Manag. Sci. **39**(10), 1261–1264 (1993)
8. A. Antoniou, Y. Guney, K. Paudyal, M.J. Alderson, B.L. Betker, T.W. Anderson, J.L. Zimmerman, The determinants of capital structure: capital market-oriented versus bank-oriented institutions. J. Financ. Quant. Anal. **43**(1), 59 (2008). https://doi.org/10.1017/S0022109000002751
9. R.D. Banker, A. Charnes, W.W. Cooper, Some models for estimating technical and scale inefficiencies in data envelopment analysis. Manag. Sci. **30**(9), 1078–1092 (1984)
10. A.N. Berger, E.B. Di Patti, Capital structure and firm performance: a new approach to testing agency theory and an application to the banking industry. J. Bank. Financ. **30**(4), 1065–1102 (2006). https://doi.org/10.1016/j.jbankfin.2005.05.015
11. C. Bhaird, B. Lucey, Determinants of capital structure in irish SMEs. Small Bus. Econ. **35**(3), 357–375 (2010). https://doi.org/10.1007/s11187-008-9162-6
12. T.D.E. Bie, L.D.E. Haan, Market timing and capital structure: evidence for Dutch firms. De Economist **155**(2), 183–206 (2007)
13. L. Booth, V. Aivazian, A. Demirguc-kunt, A. Demirgüç-Kunt, V. Maksimovic, Capital structures in developing countries. J. Financ. **56**(1), 87–130 (2001). https://doi.org/10.1111/0022-1082.00320
14. E.C. Charalambakis, D. Psychoyios, What do we know about capital structure? revisiting the impact of debt ratios on some firm specific factors. Appl. Financ. Econ. **22**, 1727–1742 (2012)
15. A. Charnes, W.W. Cooper, E. Rhodes, Measuring the efficiency of decision making units. Eur. J. Oper. Res. **2**(6), 429–444 (1978)
16. H. DeAngelo, R.W. Masulis, Optimal capital structure under corporate and personal taxation. J. Financ. Econ. **8**(1), 3–29 (1980). https://doi.org/10.1016/0304-405X(80)90019-7
17. M. Dong, I. Loncarski, Jt Horst, C. Veld, What drives security issuance decisions: market timing, pecking order, or both? Financ. Manag. **41**(3), 637–663 (2012). https://doi.org/10.1111/j.1755-053X.2012.01213.x
18. A.B. Fernandes, A Importância dos Benefícios Fiscais para as Empresas do Interior de Portugal. Ph.D. in Management, Universidade da Beira Interior, 2013
19. M.Z. Frank, V.K. Goyal, Capital structure decisions: which factors are reliably important? Financ. Manag. **38**(1), 1–37 (2009)
20. J.R. Graham, C.R. Harvey, The theory and practice of corporate finance: evidence from the field. J. Financ. Econ. **60**(2–3), 187–243 (2001). https://doi.org/10.1016/S0304-405X(01)00044-7
21. E. Guo, O. Suliman, Corporate operating characteristics and capital structure: causality testing in heterogeneous panel data. Appl. Financ. Econ. **20**(March), 1901–1922 (2010)
22. D. Hackbarth, Determinants of corporate borrowing: a behavioral perspective. J. Corp. Financ. **15**(4), 389–411 (2009). https://doi.org/10.1016/j.jcorpfin.2009.02.001
23. M. Harris, A. Raviv, The theory of capital structure. J. Financ. **46**(1), 297–355 (1991), http://search.ebscohost.com/login.aspx?direct=true&db=bth&AN=4653270&site=ehost-live&scope=site

24. J. Helwege, N. Liang, Is there a pecking order? evidence from a panel of IPO firms. J. Financ. Econ. **40**(3), 429–458 (1996). https://doi.org/10.1016/0304-405X(95)00851-5
25. S.Z. Islam, S. Khandaker, Firm leverage decisions: does industry matter? N. Am. J. Econ. Financ. **31**, 94–107 (2015). https://doi.org/10.1016/j.najef.2014.10.005
26. M.C. Jensen, W.H. Meckling, Theory of the firm: managerial behavior, agency costs and ownership structure. J. Financ. Econ. **3**(4), 305–360 (1976). https://doi.org/10.1016/0304-405X(76)90026-X
27. K. Jõeveer, Firm, country and macroeconomic determinants of capital structure: evidence from transition economies. J. Comp. Econ. **41** (2013). https://doi.org/10.1016/j.jce.2012.05.001
28. A.D. Jong, R. Kabir, T.T. Nguyen, Capital structure around the world: the roles of firm- and country-specific determinants. J. Bank. Financ. **32**(9), 1954–1969 (2008)
29. M. Kebewar, S.M.N.A. Shah, The effect of debt on corporate profitability: evidence from French service sector. Bruss. Econ. Rev. **56**(1), 43–60 (2013)
30. A. Kraus, R.H. Litzenberger, A state-preference model of optimal financial leverage. J. Financ. **28**(4), 911–922 (1973). https://doi.org/10.1111/j.1540-6261.1973.tb01415.x
31. M. La Rocca, T. La Rocca, A. Cariola, Capital structure decisions during a firm's life cycle. Small Bus. Econ. **37**(1), 107–130 (2011). https://doi.org/10.1007/s11187-009-9229-z
32. M.T. Leary, M.R. Roberts, Do peer firms affect corporate financial policy? J. Financ. **69**(1), 139–178 (2014)
33. J. López-Gracia, F. Sogorb-Mira, Testing trade-off and pecking order theories financing SMEs. Small Bus. Econ. **31**, 117–136 (2008). https://doi.org/10.1007/s11187-007-9088-4
34. D. Margaritis, M. Psillaki, Capital structure and firm efficiency. OP-J. Bus. Financ. Account. **34**(9), 1447–1469 (2007). https://doi.org/10.1111/j.1468-5957.2007.02056.x
35. D. Margaritis, M. Psillaki, Capital structure, equity ownership and firm performance. J. Bank. Financ. **34**(3), 621–632 (2010). https://doi.org/10.1016/j.jbankfin.2009.08.023
36. N. Michaelas, F. Chittenden, P. Poutziouris, Financial policy and capital structure choice in UK SMEs: empirical evidence from company panel data. J. Bus. Econ. **12**(1992), 113–130 (1999). https://doi.org/10.1023/A:1008010724051
37. A.D. Miguel, J. Pindado, Determinants of capital structure: new evidence from Spanish panel data. J. Corp. Financ. **7**(1), 77–99 (2001). https://doi.org/10.1016/S0929-1199(00)00020-1
38. M.H. Miller, Debt and taxes. J. Financ. **32**(2), 261–275 (1977). https://doi.org/10.1111/j.1540-6261.1977.tb03267.x
39. F. Modigliani, M.H. Miller, The cost of capital, corporation finance and the theory of investment. Am. Econ. Rev. **XLVIII**(3) (1958)
40. S.C. Myers, Determinants of corporate borrowing. J. Financ. Econ. **5**(2), 147–175 (1977). https://doi.org/10.1016/0304-405X(77)90015-0
41. S.C. Myers, The capital structure puzzle. J. Financ. **39**(3), 574–592 (1984). https://doi.org/10.1111/j.1755-053X.2010.01090.x
42. S.C. Myers, N.S. Majluf, Corporate financing and investment decisions when firms have information that investors do not have. J. Financ. Econ. **13**(2), 187–221 (1984). https://doi.org/10.1016/0304-405X(84)90023-0
43. A. Ozkan, Determinants of capital structure and adjustment to long run target: evidence from UK company panel data. J. Bus. Financ. Account. **28**(1–2), 175–198 (2001). https://doi.org/10.1111/1468-5957.00370
44. Ö. Öztekin, Capital structure decisions around the world: which factors are reliably important? J. Financ. Quant. Anal. **50**(3), 301–323 (2015). https://doi.org/10.1017/S0022109014000660
45. K. Park, S. Jang, Capital structure, free cash flow, diversification and firm performance: a holistic analysis. Int. J. Hosp. Manag. **33**, 51–63 (2013). https://doi.org/10.1016/j.ijhm.2013.01.007
46. P. Proença, R.M.S. Laureano, L.M.S. Laureano, Determinants of capital structure and the 2008 financial crisis: evidence from Portuguese SMEs. Science **150**, 182–191 (2014)
47. M. Psillaki, N. Daskalakis, Are the determinants of capital structure country or firm specific. Small Bus. Econ. **33**(3), 319–333 (2009). https://doi.org/10.1007/s11187-008-9103-4

48. R.G. Rajan, L. Zingales, What do we know about capital structure? some evidence from international data. J. Financ. **50**(5), 1421–1460 (1995)
49. J.H. Scott, Bankruptcy, secured debt, and optimal capital structure: comment. J. Financ. **32**(1), 1–19 (1977). https://doi.org/10.1111/j.1540-6261.1979.tb02085.x
50. S.L. Seelanatha, The determinants of capital structure: evidence from China. Econ. Manag. Financ. Mark. **5**(4), 106–126 (2010)
51. Z. Serrasqueiro, A. Caetano, Trade-off theory versus pecking order theory: capital structure decisions in a peripheral region of Portugal. J. Bus. Econ. Manag. **16**(2), 445–466 (2014). https://doi.org/10.3846/16111699.2012.744344
52. Z. Serrasqueiro, P.M. Nunes, J. Leitão, M.J.R. Armada, Are there non-linearities between SME growth and its determinants? a quantile approach. Ind. Corp. Chang. **19**(4), 1071–1108 (2010). https://doi.org/10.1093/icc/dtp053
53. Z. Serrasqueiro, F. Matias, L. Salsa, Determinants of capital structure: new evidence from Portuguese small firms. Dos Algarves Multidiscip. E-J. **28**, 13–28 (2016). http://www.dosalgarves.com/rev/N28/2rev28.pdf
54. Z.S. Serrasqueiro, P.J.M. Nunes, Performance and size: empirical evidence fand rom Portuguese SMEs. Small Bus. Econ. **31**, 195–217 (2008). https://doi.org/10.1007/s11187-007-9092-8
55. Z.S. Serrasqueiro, M.R. Armada, P.M. Nunes, Pecking order theory versus trade-off theory: are service SMEs'capital structure decisions different? Serv. Bus. **5**(4), 381–409 (2011). https://doi.org/10.1007/s11628-011-0119-5
56. R.W. Shephard, *Theory of Cost and Production Functions* (Princeton University Press, Princeton, 1970)
57. L. Shyam-Sunder, S.C. Myers, Testing static tradeoff against pecking order models of capital structure. J. Financ. Econ. **51**, 219–244 (1999). https://doi.org/10.1016/S0304-405X(98)00051-8
58. B.W. Silverman, *Density Estimation for Statistics and Data Analysis* (Chapman and Hall, London, 1986)
59. L. Simar, P.W. Wilson, Sensitivity analysis of efficiency scores: how to bootstrap in nonparametric frontier models. Manag. Sci. **44**(1), 49–61 (1998)
60. L. Simar, P.W. Wilson, Estimation and inference in two-stage, semi-parametric models of production processes. J. Econ. **136**, 31–64 (2007)
61. F. Sogorb-Mira, How SME uniqueness affects capital structure: evidence from a 1994–1998 Spanish data panel. Small Bus. Econ. **25**(5), 447–457 (2005). https://doi.org/10.1007/s11187-004-6486-8
62. R. Stulz, Managerial discretion and optimal financing policies. J. Financ. Econ. **26**(1), 3–27 (1990). https://doi.org/10.1016/0304-405X(90)90011-N
63. S. Titman, R. Wessels, The determinants of capital structure choice. J. Financ. **43**(1), 1–19 (1988). https://doi.org/10.1111/j.1540-6261.1988.tb02585.x
64. D. Vasiliou, N. Daskalakis, Institutional characteristics and capital structure: a cross-national comparison. Global Financ. J. **19**(3), 286–306 (2009). https://doi.org/10.1016/j.gfj.2008.09.002
65. E. Vieira, Determinantes da estrutura de capital das empresas portuguesas cotadas. Revista Portuguesa E Brasileira de Gestão, pp. 38–51 (2013)
66. E.S. Vieira, A.J. Novo, A estrutura de capital das PME: Evidência no mercado português. Estudos Do ISCA **2**(2), 1–19 (2010). https://doi.org/10.1234/ei.v0i2.578
67. S.C. Vogt, The role of internal financial sources in firm financing and investment decisions. Rev. Financ. Econ. **4**(1), 1–24 (1994). https://doi.org/10.1016/1058-3300(94)90002-7
68. R. Watson, N. Wilson, Small and medium size enterprise financing: a note on some of the empirical implications of a pecking order. J. Bus. Financ. Account. **3**, 1–23 (2002). https://doi.org/10.1111/1468-5957.00443
69. P.W. Wilson, Fear: a software package for frontier efficiency analysis with R. Socio-Econ. Plan. Sci. **42**(December), 247–254 (2008)

Chapter 9
Evaluating Suppliers in the Olive Oil Sector Using AHP

Dalila B. M. M. Fontes, Teresa Pereira and Elisabete Dias

Abstract This work proposes a multi-criteria decision making approach to help assessing and selecting suppliers in the olive oil sector. Olive oil is a protected agricultural product, by region and origin certificate. Therefore to select a supplier, it is of utter importance to inspect and test (taste, colour, smell, density, among others) the olive oil in addition to the supplying company. The identification of possible suppliers was done in two stages: firstly, the region of origin from which to choose possible suppliers was identified and then potential suppliers were evaluated on a set of characteristics for which minimum threshold values were set. From this study, which is not part of the research reported here, we were able to identify the suppliers of interest. Due to the several characteristics and characteristic dimensions used to choose a supplier we resort to the Analytic Hierarchy Process to rank them, this way allowing for a better choice. The rank obtained is robust as the top ranked supplier remains the same for any reasonable change in the criteria weighs and in the evaluation of the suppliers on each criterion. The involved company found the results of value, as well as the lessons learned by addressing the supplier evaluation problem using a more systematic approach.

Keywords Multi-criteria decision making · AHP · Olive oil sector

---

D. B. M. M. Fontes (✉)
LIAAD/INESC TEC, Faculdade de Economia da Universidade do Porto,
Rua Dr. Roberto Frias, 4200-464 Porto, Portugal
e-mail: fontes@fep.up.pt

T. Pereira
ISEP – School of Engineering, Polytechnic Institute of Porto and CIDEM, Porto, Portugal
e-mail: mtp@isep.ipp.pt

E. Dias
Faculdade de Economia da Universidade do Porto, Rua Dr. Roberto Frias, 4200-464 Porto, Portugal
e-mail: elisabeted@gmail.com

A. I. F. Vaz et al. (eds.), *Operational Research*, Springer Proceedings
in Mathematics & Statistics 223, https://doi.org/10.1007/978-3-319-71583-4_9

9.1 Introduction

This work proposes a multi-criteria decision making (MCDM) methodology to evaluate and select suppliers for an olive oil distribution company.

Supplier selection is a strategic decision that significantly influences the firms competitive advantages [15, 20]. The choice of suppliers is a key aspect in the management of enterprises, since the costs of the raw materials is a major burden on their final products. Most companies spend a large part of their earnings on purchases [7]. Therefore, a reliable selection process needs to be established to ensure a proper selection of suppliers, since it significantly reduces purchasing costs, increases production rate, and assurance of the final product quality, thus improving competitiveness and business performance. Such a process must consider several quantitative and qualitative criteria, which are to be defined and evaluated by the company's management [8]. The supplier selection process is not only beneficial for the company looking for a supplier, but also for the companies that established themselves as potential suppliers. The former can choose better; while the latter by becoming aware of the essential and preferential customer requirements can be more focused, thus improving the efficiency and effectiveness of their activity [9]. However, selecting the right supplier is a very difficult task [12].

The Supplier evaluation and selection problem has been extensively studied and many research methodologies have been proposed over the years. Several reviews have been published, see e.g., [4, 5, 8, 19, 20]. The comprehensive review by [20] examines 221 papers published in leading journals from 1990 to 2015. The authors divided and classified the literature in six streams and their main finding are summarized below:

- Stream-1 — reports research on problem approaches, being clear that more than one criterion should be considered. Most of the approaches proposed fall in the MCDM arena.
- Stream-2 — investigates the criteria to be used, however this stream has not played a major role since 1990.
- Stream-3 — reports the inclusion of green and sustainable issues. According to the authors, the first such works have been published only in 2006 and 2009, respectively [10, 11].
- Stream-4 — considers research on strategic issues within supplier selection. This stream seems to be in an early stage of development although it has been considered for a long time.
- Stream-5 — integrates product development with supplier selection. This stream was identified as the one attracting less research.
- Stream-6 — in this stream the quality criterion plays a critical role. Several works dating from the early 2000s onwards focus on operations, i.e., process quality or assembly quality.

The field of MCDM has experience a significantly growth and at the same time it has joined forces with other scientific knowledge domains. A clear and thorough work showing the MCDM birth and evolution is provided by [3]. Reference [8] review the main approaches used to address the supplier selection problem. Among the approaches reviewed are analytic hierarchy process (AHP), analytic network process (ANP), case-based reasoning (CBR), data envelopment analysis (DEA), fuzzy set theory, genetic algorithm (GA), mathematical programming, simple multi-attribute rating technique (SMART), and their hybrids. The authors main findings were: (i) AHP approaches, considering both stand alone and integrated ones, are the most popular; which supports our methodological choice and (ii) from the hundreds of criteria proposed, quality was the most relevant, followed by delivery, price/cost, manufacturing capability, service, management, technology, research and development, finance, flexibility, reputation, relationship, risk, and safety and environment; which also support and confirm our findings.

The remainder of this document is organized as follows. Section 9.2 describes the MCDM approach used in this work, namely the AHP. Section 9.3 presents and details the application of the AHP to a real-world decision-making problem in the olive oil industry. Finally, Sect. 9.4 concludes the paper.

9.2 Multi-criteria Decision Making Approach

Multi-criteria decision making methodologies are particularly appropriated to address supplier evaluation and selection problems since, in addition to involving several dimensions (here translated into criteria) on which to evaluate the suppliers, some of these dimensions are quantitative in nature, while others are qualitative and thus subjective. More often than not, these criteria are conflicting and hence their simultaneous optimization is not possible. Furthermore, the number of possible suppliers, here termed alternatives, is small and each has its own known characteristics. MCDM is known for being able to incorporate Decision Makers (DMs) subjectivity through their involvement in the analysis and evaluation processes [2]. Thus, assisting them in making their own choice by providing a ranking of the alternative solutions using the DMs preferences and evaluations. Therefore, supplier selection belongs to the class of multi-criteria decision making problems in which the company needs to identify the top priorities to select the "best" supplier, based on its needs and suppliers information [1]. Thus, it is better addressed by a two stage procedure, whereby a set of potential interesting solutions is first identified, followed by an analysis along several dimensions of each of the alternative solutions. The main goal of this analysis is to assist the decision maker in the process of identifying the most preferred solution(s) from a set of possible ones (explicitly or implicitly defined). The alternative solutions differ in the extent in which they achieve the objectives, since none of them has the best performance for all objectives; otherwise the choice would be trivial.

The use of MCDM in real-world problems is a non-linear recursive process involving several steps, which vary in number with the specific approach. Nevertheless, it is possible to outline the following critical steps (common to most approaches and studies, although not always corresponding to the same number of steps):

1. Structuring – begins with a contextualization of the problem and then moves onto the definition of the alternatives and criteria.
2. Evaluation – includes the evaluation of each alternative on each of the criterion defined, as well as the relative importance of the criteria.
3. Preparation of recommendations – this step involves obtaining a global score for each alternative making use of the individual scores of each one of them on each criterion and of the criteria weights (both obtained in the previous step). The set of alternatives is then ranked based on the overall scores. The process may also involve a sensitivity analysis of the results to changes in scores or criteria, in order to infer on the robustness of the outcome of the MCDM.

It is important to note that the results yielded by a MCDM process are not prone to generalisations, in the sense that they only apply to the set of alternatives that were evaluated [6].

It follows a brief description of the Analytic Hierarchy Process (AHP) as this is the approach used in this study, although others exist.

9.2.1 AHP – Analytic Hierarchy Process

The origin of the AHP goes back to the 70s and was first proposed by Thomas L. Saaty [16]. This methodology is based on mathematics and psychology and provides a framework for structuring decision problems by allowing for the representation and quantification of all problem elements. The AHP considers a discrete set of possible solutions, which are then analysed and ranked according to their value regarding a set of relevant characteristics, i.e., criteria, previously identified.

The main reasons behind the wide applicability and acceptability of AHP are: (i) easy to understand – its simple and intuitive nature allows for the decision-makers to understand how the recommendations are obtained; (ii) active participation - the decision makers are involved in every step of the decision analysis; (iii) hierarchy – helps in identifying the importance of the factors involved in the problem; (iv) subjectiveness - its ability to deal with subjective assessments (feelings and judgments), which are then converted into numerical values that reflect the decision makers opinion and preferences; (v) pairwise comparison – it is easier to perform and it helps to articulate the relative importance of criteria and to quantify the relative contributions of the alternatives on the criteria; (vi) inconsistency measure – helps to avoid inconsistent judgments, since by identifying them the decision makers can repeatedly work through their inconsistent judgments until they obtain acceptable results. Nevertheless, some drawbacks have also been pointed out [13, 18]: (i) the compensatory effect of trade-offs, and (ii) the time involved in the process of gathering the DM's evaluations. The first is a consequence of the additive aggregation and allows for the degradation of performance on a criterion to be compensated by good performance in other(s); while the second is due to the substantial number of pairwise comparisons when dealing with a large number of criteria (over ten) and/or alternatives. For

Table 9.1 Scale of relative scores [16, 17], even values have intermediate meaning

Score value (a_{ij})	Interpretation
1	i and j are equally important
3	i is slightly more important than j
5	i is more important than j
7	i is strongly more important than j
9	i is absolutely more important than j

Table 9.2 Average random consistency [16]

Size of the matrix (n)	3	4	5	6	7	8
Random consistency (RI)	0.58	0.90	1.12	1.24	1.32	1.41

these reasons, the AHP is sometimes complemented with another MCDM method, see e.g., [14].

AHP basic idea is converting subjective assessments of relative importance into a set of overall scores and weights. The assessments are relative and subjective, since they are provided and reflect the opinion of a specific decision maker and are based on pairwise comparisons.

The pairwise comparisons are performed by asking the decision maker to compare pairs of criteria and to indicate, for each pairwise comparison, their relative contribution to the overall goal using the nine-point intensity scale proposed by [16, 17], see Table 9.1. The comparisons are then represented through a squared matrix $n \times n$, where n is the number of criteria. Each matrix element a_{ij} represents the importance of criterion i relative to criterion j. If $a_{ij} > 1$ then criterion i is more important than criterion j; otherwise criterion j is more important. Moreover, $a_{ij} \times a_{ji} = 1$.

However, care must be taken when using the pairwise comparisons as they may lead to inconsistencies in the judgments. Therefore, a consistency check needs to be performed by computing the Consistency Ratio (CR) as [16, 17]:

$$CR = \frac{CI}{RI}, \quad \text{with} \quad CI = \frac{\lambda_{max} - n}{n - 1}, \tag{9.1}$$

where λ_{max} denotes the maximum eigenvalue of the pairwise comparison matrix, n represents the matrix size (in this case, the number of criteria), and the acronyms CR, CI and RI stand for Consistency Ratio, Consistency Index, and Random Consistency Index, respectively. RI is obtained from a randomly generated pairwise comparison matrix of size n (see Table 9.2). If $CR > 0.10$ the decision maker's judgments reveal an unacceptable degree of inconsistency and, thus, the entries of the pairwise comparison matrix need to be amended before proceeding to the next step of the AHP.

The output of AHP consists of a set of priorities, or weights, one for each criteria and a set of relative performance scores, for each alternative in each criterion. The former, i.e., the vector of priorities or criteria weights, is obtained by finding the first eigenvector of the pairwise comparisons matrix; while the latter can be obtained by separate pairwise comparisons on the set of alternatives in each criterion or by simply rating each alternative for each criterion, by identifying the grade that best describes them. Afterwards, a global score is obtained for each alternative by computing for each the weighted sum of the performance in each criterion.

9.3 Case Study: Evaluating Olive Oil Suppliers

This case study involves a small Portuguese company of olive oil distribution. This company distributes exclusively Portuguese olive oils and its portfolio of suppliers covers all areas of Portugal and features a range of olive oils of different varieties and qualities. Although most of the company business is in the domestic market; its international business is growing and currently it is already present in six European countries. Olive oil is the companys asset, therefore the selection of suppliers is a crucial decision and as such the directors of three departments have been involved, namely: domestic trade, foreign trade, and quality control. Three main reasons have led to the need to select new suppliers: (i) replace the current supplier, (ii) looking for another variety of olive oil and/or increasing the amount to buy, and (iii) finding different categories of olive oil (price and quality).

9.3.1 Problem Context and Definition

Olive oil is a protected agricultural product, by region and origin certificate, having specific denominations.[1] Among which there is Protected Designation of Origin (PDO): identifies products produced, processed, and prepared in a specific geographical area, using the recognised know-how of local producers and ingredients from the specific region. These are products whose characteristics are linked to their geographical origin. They must adhere to a precise set of specifications and may bear the PDO logo. In Portugal there are six different PDO olive oil regions: PDO Trás-os-Montes; PDO Beira Interior; PDO Ribatejo; PDO Alentejo Interior; PDO Norte Alentejano, and PDO Moura. This certification is supported by regional governmental entities for unique characteristics and organoleptic tests.

This case study refers to the Trás-os-Montes PDO and was done with the support of the Inter-professional Association of Trás-os-Montes and Alto Douro olive producers[2] (AIATMAD). Several factors, such as (1) the climate conditions, maturation, latitude and the type of soils, (2) the dominant traditional varieties ("Verdeal",

[1]The European Union agricultural product quality policy can be found in http://ec.europa.eu/agriculture/quality/index_en.htm.

[2]Translation from Associação Inter-profissional do Azeite de Trás-os-Montes e Alto Douro.

"Madural", "Cordovil", "Cobran çosa" and "Negrinha do Freixo"), (3) the olive oil extraction process (hot or cold process, manual or mechanical, etc.), and (4) the chemical and organoleptic characteristics of the olive oil, led to the creation of the PDO. The AIATMAD provided all information regarding the DOP certification and organoleptic and chemical tests, which were used both to eliminate from further consideration some potential suppliers and to evaluate the suppliers of interest.

9.3.2 Decision Makers, Alternatives and Criteria

The goal of this case study is to use a MCDM approach to rank the suppliers of an olive oil distribution company. As usual in any MCDM process, the first step is to identify the decision makers (DMs) and then in collaboration with them identify company objectives. The first meeting allowed for the identification of the decision makers, namely directors of the domestic trade, foreign trade, and quality control departments. In this meeting it was decided that gathering the perspectives of each decision maker was a very important issue, thus they would not work as a single DM (consensus). Furthermore, for simplicity and due to time constraints, it was also decided to assign weight to the evaluations performed by each director to reflect the knowledge differentiation. A weight of 40% was assigned to the director of the quality control department and to each of the remaining two department directors a weight of 30% was assigned. It is considered that the director of the quality control department should have a higher weighting given his expertise in the production phase and his regular interaction with the suppliers.

A series of meetings followed and their purpose evolved with the project. Initially, the main purpose was to define the characteristics the company seeks for in its suppliers. Then based on these findings and on literature as set of possible criteria was brought to the discussion, during which it was shorted to a 17 items list. From this list, each DMs chose and ranked the top 5 criteria. A weight was computed for each possible criteria using these rankings (1-worst and 5-best) and the DMs' weights, which was used to select the 5 criteria to evaluate the suppliers on, see Table 9.3. The characteristics identified were also used to define the set of potential suppliers. Candidate suppliers have to satisfy a number of threshold values for the company to consider them as potential suppliers; otherwise would be disregarded.

This way, only a restricted set of very interesting suppliers needs to be considered, which leads to a more efficient and less time/effort consuming process. For this specific study three suppliers (S1, S2, and S3) are considered. In the following meetings, the DMs were asked to assess the relative importance of the previously identified criteria and to appraise each of the alternative suppliers in regards to their individual contribution to the five criteria selected, namely:

- Product quality (qualitative) – olive oil quality is usually inferred from its chemical (acidity ph, peroxides index, and ultraviolet absorption) and organoleptic (sensory analysis: smell, colour, and taste) characteristics.

Table 9.3 List of 17 potential criteria, from company objectives and literature review

Criteria	DM1 quality	DM2 domestic	DM3 foreign	Weighted sum
Product quality	5	4	5	4.7
Product variety				0
Brand value				0
Product availability				0
Product cost	2	5	4	3.5
Delivery lead time				0
Technical capacity		2	3	1.5
Payment flexibility	1	3		1.3
Financial capacity				0
Supplier location				0
Deadlines fullfilment	4		1	1.9
Flexibility				0
Behavior and honesty		1		0.3
Safety				0
Refund policies				0
Attitude (criticism and complaints)				0
Product reliability	3		2	1.8

- Product cost (quantitative) – olive oil net cost per litre; each supplier quotes his/her own price in euros per litre.
- Suppliers technical capacity (qualitative) – perceived technical knowledge of the production process and in place monitoring of the entire olive oil production path (from the olives, harvesting and handling, to the olive oil production, packaging, and delivery).
- Lead time (quantitative) – percentage of fulfilled deliveries (in time, quantity, and quality).
- Product reliability – small fluctuations of product quality over time. This measured by chemical analysis over the production year, using historical information, since in any delivery the olive oil quality is assessed.

Table 9.4 AHP pairwise comparison matrix for the chosen criteria and the corresponding criteria weights

	PQ	PC	TC	LT	PR	Criteria weight
Product quality (PQ)	1	5	5	5	9	0.547
Product cost (PC)	0.2	1	1	1	7	0.145
Technical capacity (TC)	0.2	1	1	1	5	0.131
Lead time (LT)	0.2	1	1	1	7	0.145
Product reliability (PR)	0.11	0.14	0.2	0.14	1	0.032

9.3.3 *Elicitation of the Criteria Weights*

A meeting was then promoted with the DMs to analyze the results of the aggregation explained above an validate the criteria chosen. The final evaluation of the five criteria, which is reported in Table 9.4, was obtained by consensus having the aggregated values as a starting point. This evaluation was done by performing pairwise comparisons using Saatys nine-point intensity scale (1–9 as described in Table 9.1) and the values obtained are reported in Table 9.4 together with the resulting criteria weights.

As it can be seen from the AHP results, the product quality was deemed the most important criterion ($w_{PC} = 54.7\%$) and product reliability the least important one ($w_{PR} = 3.2\%$). The other three criteria have similar importance ($w_{PC} = w_{LT} = 14.5\%$ and $w_{TC} = 13.1\%$). The considerable weight assigned to the quality index criterion is consistent with the marketing idea that quality is one of the most important product features, being critical to product sales.

To check the consistency of the DMs judgments we computed the maximum eigenvalue of the pairwise comparison matrix as $\lambda_{max} = 5.211$ and the consistency index as $CI = \frac{\lambda_{max}-n}{n-1} = 0.053$. The random consistency index is $RI = 1.12$, see Table 9.2. Resulting in a consistency ratio of $CR = \frac{CI}{RI} = 0.047$. It can be concluded that consistency is ensured, since the consistency ratio obtained is below the 10% threshold ($4.72 < 10$). Hence, it can be said that the DMs have been coherent in their judgments and, thus, the criteria weights obtained can be used in the decision making process.

9.3.4 *Supplier Ranking*

As said before, the output of the analysis is a ranking of the suppliers considered. In order to do so, the suppliers need to be evaluated on the defined criteria. These evaluations were provided by the decision makers. More specifically, each decision maker provided a comparative evaluation for each pair of suppliers on each criterion. These evaluations were then aggregated by using the DMs weights, i.e., for each

Table 9.5 AHP pairwise comparison matrices for the three suppliers and the corresponding supplier scores, regarding each criterion

	S1	S2	S3	Suppliers score
S1	1	1	0.2	0.143
S2	1	1	0.2	0.143
S3	5	5	1	**0.714**

(a) Product quality

	S1	S2	S3	Suppliers score
S1	1	0.2	3	0.193
S2	5	1	7	**0.724**
S3	0.333	0.143	1	0.083

(b) Product cost

	S1	S2	S3	Suppliers score
S1	1	3	0.333	0.260
S2	0.333	1	0.2	0.106
S3	3	5	1	**0.633**

(c) Technical capacity

	S1	S2	S3	Suppliers score
S1	1	0.333	5	0.283
S2	3	1	7	**0.643**
S3	0.2	0.143	1	0.074

(d) Lead time

	S1	S2	S3	Suppliers score
S1	1	5	3	**0.633**
S2	0.2	1	0.333	0.106
S3	0.333	3	1	0.260

(e) Product reliability

supplier and each criterion the three evaluations (one from each DM) were aggregated using a 0.4 weight for the director of the quality control department and a weight of 0.3 for each of the other two directors. The aggregated evaluations were then rounded to the nearest integer and fed to AHP in order to obtain suppliers scores, see Table 9.5.

From the values reported in Table 9.5 it can be observed that the best performing supplier is not the same for all criteria. Supplier 1 has the best performance regarding product reliability criterion, while supplier 2 has the best performance both regarding product cost and lead time. Finally, supplier 3 is the best one in product quality and technical capacity. The consistency ratio for each criterion was verified to be below 10% and thus, the weights obtained can be used in the decision-making process as the DMs have been coherent in their judgments.

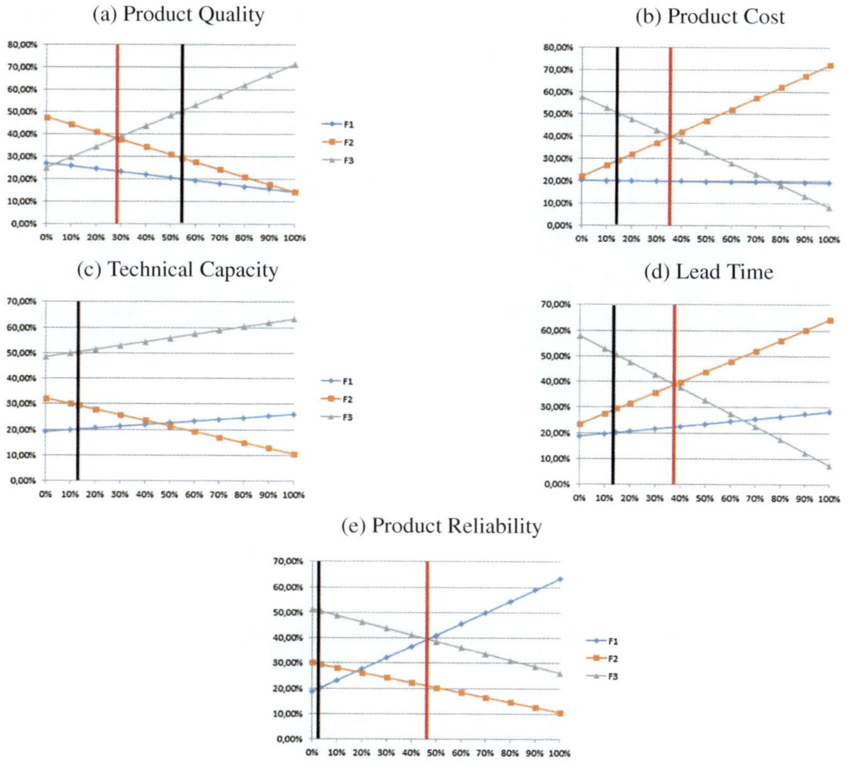

Fig. 9.1 Global impact due to the variation of the criteria weights

The supplier's global scores are given by the weighted sum of the scores obtained for each criterion. The weights are the ones reported in the last column of Table 9.4 and suppliers individual scores are reported in the last column of Table 9.5. The values obtained for global scores are 0.202 for supplier 1, 0.294 for supplier 2, and 0.503 for supplier 3. Therefore, the suppliers are then ranked in the following order: supplier 3, supplier 2, and supplier 1.

9.3.5 Sensitivity Analysis

Since some steps of the MCDM process can be permeated by subjectivity and uncertainty, we validated our results by performing sensitivity analysis in order to determine how the final ranking of alternatives would change under different criteria weighting schemes. The results obtained can be seen in the graphs of Fig. 9.1 for the five criteria used in the analysis.

The vertical black lines in the graphs of Fig. 9.1 show the current value of the weight for each criterion; while the red vertical lines show the value that the weights need to have for the first ranked supplier to change. As it can be seen, the variation needed is of a very large magnitude. For product quality, its weight would have to drop down from 54.7% to under 30%; while for product cost, the weight would have to go up from 14.5% to approximately 35%. For the technical capacity, no variation, regardless of the magnitude, would lead to a change in the top ranked supplier. The lead time would require an increase from 14.5% to almost 40%, and finally the product reliability would need to increase to around 48% from its current 3%. Therefore, it can be conclude that the ranking obtained is very robust.

9.4 Conclusions

This work uses multi-criteria decision making (MCDM) to address the problem of supplier selection of an olive oil distribution company. MCDM is a formal quantitative approach to assist the decision-making process at various levels, including the organization and synthesis of available information, as well as enhanced decision-making with a holistic and structured approach to the problem. The application of multi-criteria analysis method encourages discussion and deeper analysis of the problem within various departments. This reflection throughout the process collects and summarizes data in order to choose the correct criteria to evaluate alternative suppliers in order to rank them. Amongst the many MCDM methods available we chose to use the AHP, since this method is easy to understand an intuitive, thus fostering greater involvement of the DMs, which in turn leads to higher likelihood of its implementation. The suppliers ranking obtained reflects the relative quality of the compromise reached by each alternative in relation to the criteria set by the decision makers. The best-ranked supplier had a score significantly larger than the other two. Furthermore, the top ranked supplier remains the same for any reasonable change in the criteria weighs and in the evaluation of the suppliers on each criterion. In addition, the evaluations performed by the DMs have consistency ratio values well below the threshold value (10%). Therefore, we can conclude that the results obtained are valid and robust.

The study was conducted with a small family business, therefore making any lessons learned and conclusions drawn very specific, thus not prone to generalizations. However, the DMs involved found the logic behind the proposed approach easy to understand. Moreover, the DMs recognized that although time consuming, the approach forced introspection and reflection, which in turn allowed for a better understanding of the problem structure, characteristics, and consequences. The authors acknowledge that this study is limited in ways that suggest opportunities for future research, such as the consideration of group decision and the use of fuzzy AHP due to the vagueness of DMs judgments and uncertainty associated with the product quality and quantity, for example due to weather.

Acknowledgements We acknowledge the financial support of Projects "NORTE-01-0145-FEDER-000020", financed by the North Portugal Regional Operational Programme (NORTE 2020), under the PORTUGAL 2020 Partnership Agreement and PTDC/EEIAUT /2933/2014, financed through the European Regional Development Fund (ERDF) and FEDER /COMPETE2020-POCI/FCT.

References

1. P. Agarwal, M. Sahai, V. Mishra, M. Bag, V. Singh, A review of multi-criteria decision making techniques for supplier evaluation and selection. Int. J. Ind. Eng. Comput. **2**(4), 801–810 (2011)
2. C.A. Bana e Costa, Processo de apoio à decisão: problemáticas, actores e acções. Florianópolis: ENE (Escola de Novos Empreendedores da UFSC) (Santa Catarina, Brazil, 1995)
3. J. Bragge, P. Korhonen, H. Wallenius, J. Wallenius, Scholarly communities of research in multiple criteria decision making: a bibliometric research profiling study. Int. J. Inf. Technol. Decis. Mak. **11**(02), 401–426 (2012)
4. L. De Boer, E. Labro, P. Morlacchi, A review of methods supporting supplier selection. Eur. J. Purch. Supply Manag. **7**(2), 75–89 (2001)
5. Z. Degraeve, E. Labro, F. Roodhooft, An evaluation of vendor selection models from a total cost of ownership perspective. Eur. J. Oper. Res. **125**(1), 34–58 (2000)
6. A.E. Dooley, D.C. Smeaton, G.W. Sheath, S.F. Ledgard, Application of multiple criteria decision analysis in the new zealand agricultural industry. J. Multi-Criteria Decis. Anal. **16**(1–2), 39–53 (2009)
7. K. Goffin, F. Lemke, M. Szwejczewski, An exploratory study of 'close' supplier-manufacturer relationships. J. Oper. Manag. **24**(2), 189–209 (2006)
8. W. Ho, X. Xu, P.K. Dey, Multi-criteria decision making approaches for supplier evaluation and selection: a literature review. Eur. J. Oper. Res. **202**(1), 16–24 (2010)
9. W. Ho, P.K. Dey, M. Lockström, Strategic sourcing: a combined QFD and AHP approach in manufacturing. Supply Chain Manag.: Int. J. **16**(6), 446–461 (2011)
10. P. Humphreys, A. McCloskey, R. McIvor, L. Maguire, C. Glackin, Employing dynamic fuzzy membership functions to assess environmental performance in the supplier selection process. Int. J. Prod. Res. **44**(12), 2379–2419 (2006)
11. A.B.L.S. Jabbour, C.J.C. Jabbour, Are supplier selection criteria going green? case studies of companies in Brazil. Ind. Manag. Data Syst. **109**(4), 477–495 (2009)
12. F.-H.F. Liu, H.L. Hai, The voting analytic hierarchy process method for selecting supplier. Int. J. Prod. Econ. **97**(3), 308–317 (2005)
13. C. Macharis, J. Springael, K. De Brucker, A. Verbeke, PROMETHEE and AHP: the design of operational synergies in multicriteria analysis: strengthening PROMETHEE with ideas of AHP. Eur. J. Oper. Res. **153**(2), 307–317 (2004)
14. M. Oliveira, D.B.M.M. Fontes, T. Pereira, Evaluating vehicle painting plans in an automobile assembly plant using an integrated AHP-PROMETHEE approach. Int. Trans. Oper. Res. (2015)
15. J. Rezaei, T. Nispeling, J. Sarkis, L. Tavasszy, A supplier selection life cycle approach integrating traditional and environmental criteria using the best worst method. J. Clean. Prod. **135**, 577–588 (2016)
16. T.L. Saaty, Axiomatic foundation of the analytic hierarchy process. Manag. Sci. **32**(7), 841–855 (1986)
17. T.L. Saaty, How to make a decision: the analytic hierarchy process. Eur. J. Oper. Res. **48**(1), 9–26 (1990)
18. J. Seydel, Data envelopment analysis for decision support. Ind. Manag. Data Syst. **106**(1), 81–95 (2006)
19. C.A. Weber, J.R. Current, W.C. Benton, Vendor selection criteria and methods. Eur. J. Oper. Res. **50**(1), 2–18 (1991)

20. A. Wetzstein, E. Hartmann, W.C. Benton Jr., N.-O. Hohenstein, A systematic assessment of supplier selection literature–state-of-the-art and future scope. Int. J. Prod. Econ. **182**, 304–323 (2016)

Chapter 10
Optimal Control Strategies for an Advertisement Viral Diffusion

João N. C. Gonçalves, Helena Sofia Rodrigues and M. Teresa T. Monteiro

Abstract The process of diffusing viral marketing campaigns through social networks can be modeled under concepts of mathematical epidemiology. Based on a Susceptible-Infected-Recovered (SIR) epidemiological model, the benefits of optimal control theory on the diffusion of a real viral advertisement are studied. Two optimal control strategies that could help marketers to maximize the spread of information and minimize the costs associated to it in optimal time windows are analyzed and compared. The uniqueness of optimality system is proved. Numerical simulations show that high investment costs in publicity strategies do not imply high overall levels of information diffusion. This paper contributes to the current literature by studying a viral marketing campaign using real numerical data.

Keywords Optimal control theory · Viral marketing · SIR epidemiological model · Information diffusion strategies

10.1 Introduction

Marketing is a valuable tool to orient and increase the performance of a company [15]. However, traditional marketing strategies are having difficulty to meet demanding conditions of consumers [13]. Therefore, to create competitive advantages against other companies, marketing professionals have been trying to design attractive and

---

J. N. C. Gonçalves (✉) · M. T. T. Monteiro
Department of Production and Systems, Algoritmi R&D Center,
University of Minho, Braga, Portugal
e-mail: jncostagoncalves@gmail.com

M. T. T. Monteiro
e-mail: tm@dps.uminho.pt

H. S. Rodrigues
School of Business Studies, Polytechnic Institute of Viana do Castelo, Valença, Portugal
e-mail: sofiarodrigues@esce.ipvc.pt

H. S. Rodrigues
Department of Mathematics, Center for Research and Development in Mathematics
and Applications (CIDMA), University of Aveiro, Aveiro, Portugal

A. I. F. Vaz et al. (eds.), *Operational Research*, Springer Proceedings
in Mathematics & Statistics 223, https://doi.org/10.1007/978-3-319-71583-4_10

viral campaigns, based on strategies that allow to revert the drawbacks of traditional marketing. One of these strategies is known as Viral Marketing (VM), which refers to the process that takes advantage of word-of-mouth to replicate and diffuse a marketing message into a large set of customers [9]. VM aims to reach a large audience with a low cost associated to it, by exploiting network effects which in turn maximize information diffusion. Moreover, VM has proved to be a sustainable marketing strategy by avoiding the necessity of establishing a direct contact with targeted individuals [11]. Nonetheless, some downsides can be highlighted to VM, such as its uncontrollable nature or even the difficulty in controlling the timing and success of the transmission phenomena (see [19] and the references cited therein).

Optimal control theory is an extension of the calculus of variations that seeks to find control strategies for a dynamic system [12], and gives some insights on its usefulness by optimizing marketing strategies that maximize the spread of information without losing much money.

Over time, optimal control problems applied to marketing have been proposed and discussed (see, e.g., [4–6] and the references cited therein). More precisely, these research studies focus on study optimal strategies to promote information diffusion in social networks and environments.

Considering that VM can be modeled by epidemiological models [9, 14, 17], this article studies the dynamics and impact of a real VM advertisement - Dove Real Beauty Sketches, based on real data presented in [18]. Produced in 2013, Dove Real Beauty Sketches is a publicity campaign that focuses on state the definition of beautiful, promoting self-esteem and changing the perception of beauty [2]. However, one of the major difficulties in diffusing a viral message relates to create mechanisms that convince and persuade individuals to spread it in the best time window. Thus, by choosing this campaign as an example of how to build a successful marketing strategy, our aim is to study optimal policies and time intervention strategies that could help marketing professionals to increase information diffusion with low cost in future campaigns. For that, a controlled SIR epidemiological model proposed in [4] is analyzed and applied to Dove's advertisement. Additionally, some numerical simulations related to different investment cost scenarios are performed.

The paper is organized as follows. Section 10.2 presents the mathematical problem according to the marketing campaign real data. In Sect. 10.3, an optimal control problem is formulated and studied. Section 10.4 presents numerical simulations related to the comparison of different marketing strategies and guidelines regarding the best control policies to apply in different investment scenarios. Conclusions are carried out in Sect. 10.5.

10.2 SIR Epidemiological Model and Properties

In this section, the SIR epidemic model without control is formulated and basic properties within a marketing context are established. This model subdivides the population into three mutually–exclusive compartments: susceptible individuals, who correspond to the portion of the target population who can spread the marketing message (S); infected individuals, who correspond to the set of population who encourage the spreading of the message among social networks (I); recovered individuals, who stop diffusing the marketing message (R).

The dynamics, over time t, of the mutually–exclusive compartments can be described by the following system of ordinary differential equations:

$$\begin{cases} \dfrac{dS(t)}{dt} = -\beta \dfrac{S(t)I(t)}{N} \\ \dfrac{dI(t)}{dt} = \beta \dfrac{S(t)I(t)}{N} - \gamma I(t) \\ \dfrac{dR(t)}{dt} = \gamma I(t)\,, \end{cases} \qquad \begin{cases} S(0) = S_0 > 0 \\ I(0) = I_0 > 0 \\ R(0) = 0\,. \end{cases} \tag{10.1}$$

The total population and its rate of change at time t are given by

$$N = S + I + R \Leftrightarrow \frac{dS}{dt} + \frac{dI}{dt} + \frac{dR}{dt} = 0\,, \forall t \in [0, \infty)\,. \tag{10.2}$$

Hence, since $R(t) = N - S(t) - I(t)$, the system (10.1) can be reduced to

$$\begin{cases} \dfrac{dS(t)}{dt} = -\beta \dfrac{S(t)I(t)}{N} \\ \dfrac{dI(t)}{dt} = \beta \dfrac{S(t)I(t)}{N} - \gamma I(t)\,. \end{cases} \tag{10.3}$$

Exploring the dynamics of the system (10.3), when the marketing message is appealing, individuals leave the class S and move to class I at a rate β. Over time, individuals in the class I stop sharing the message and move to class R at a rate γ. It should be emphasized that the variables $S(t)$, $I(t)$ and $R(t)$, parameters β and γ, and initial conditions are non-negative. Thus, it can be shown that the solutions of the SIR system are also non-negative (see, e.g., [3]).

Henceforth, further analyses consider a selected set of parameter values, in part computed based on numerical algorithms described in the literature and in part taken from real numerical data related to Dove's campaign (see Table 10.1). It is also appropriate to reformulate the model (10.3) in terms of the fractions of S, I and R, by considering $s = \dfrac{S}{N}$, $i = \dfrac{I}{N}$, $r = \dfrac{R}{N}$.

Using FMINSEARCH routine (see [7]) from MATLAB optimization toolbox, included in the algorithm proposed in [10], parameters estimation was performed based on the first seven days of Dove's marketing campaign, where the maximum peak of infection is attained. Obtained by the inequality $\dfrac{di(t)}{dt} > 0$ for $s \approx 1$, the basic reproduction number of the normalized SIR model $\left(\mathscr{R}_0 = \dfrac{\beta}{\gamma}\right)$ expresses the number of secondary transmissions made by a single infected individual, within a susceptible set of popu-

Table 10.1 Parameters and initial conditions

Symbol	Description	Estimated value	References
β	Predisposition to share the marketing message	67.5244	[10]
γ	Cessation of diffusion of the marketing message	65.0751	[10]
N	Total population at the campaign launching	10^9	[16]
$S(0)$	Number of susceptible individuals at the campaign launching	10^9–30.000	[18]
$I(0)$	Number of infected individuals at the campaign launching	30.000	[18]
$R(0)$	Number of recovered individuals at the campaign launching	0	[18]

lation [14]. It can be proved that if $\mathscr{R}_0 < 1$ the marketing message is not widespread. However, in this case, $\mathscr{R}_0 > 1$, which confirms that Dove's campaign was a viral epidemic (see, e.g., [3] for more details on the dynamics of $\mathscr{R}_0$ in this model).

10.3 Optimal Control Problem

As proposed in [4], an optimal control problem related to the model analyzed so far is formulated. After the normalization of the model (10.3), two control functions $u_1(t)$ and $u_2(t)$, $\forall t \in [0, 6]$, are added. Hence, considering $r(t) = 1 - s(t) - i(t)$, the resultant state system of ordinary differential equations with optimal control is given by

$$\begin{cases} \dfrac{ds(t)}{dt} = -\big(\beta + u_2(t)\big)s(t)i(t) - u_1(t)s(t) \\[2ex] \dfrac{di(t)}{dt} = \big(\beta + u_2(t)\big)s(t)i(t) + u_1(t)s(t) - \gamma i(t)\,, \end{cases} \qquad \begin{cases} s(0) = \dfrac{S(0)}{N} \\[2ex] i(0) = \dfrac{I(0)}{N}\,. \end{cases} \tag{10.4}$$

The control functions $u_1(t)$ and $u_2(t)$ express the recruitment of susceptible individuals to act as spreaders within the targeted population (e.g., via advertisements in mass media) and the encouragement of infected individuals to continue to spread the marketing message into their social network (e.g., through vouchers, rewards, monetary stimuli), respectively [4].

Let t_f be the considered campaign deadline. The set of admissible control functions is defined as

$$\Omega = \left\{ (u_1(\cdot), u_2(\cdot)) \in (L^2(0, t_f))^2 \mid 0 \leqslant u_1(t) \leqslant u_{1\max} \wedge 0 \leqslant u_2(t) \leqslant u_{2\max},\ \forall t \in [0, t_f] \right\}.$$

For $u_i : i \in \{1, 2\}$, note that the application of the control policy is maximum if $u_i = u_{i\,\max}$, and minimum if $u_i = 0$. The main objective is to find the optimal values u_1^* and u_2^* for the controls u_1 and u_2, in such a way that the state trajectories s and i are the solution of the system (10.4) over $[0, t_f]$ and maximize the objective functional (10.5). As a trade–off, the optimal control problem consists in maximize the spreading of information at the final time and minimize the intervention costs related to the application of the control policies, i.e.,

$$\max_{\Omega} J(u_1(\cdot), u_2(\cdot)) = r(t_f) + i(t_f) + \int_0^{t_f} -\left[Bu_1^2(t) + Cu_2^2(t)\right] dt\,, \qquad (10.5)$$

subject to (10.4), where the non-negative constants B and C represent the weights of the investment cost associated to the control signals u_1 and u_2, respectively. The quadratic structure of the weighted controls is consistent with related literature (see, e.g., [4, 8]). The existence of the optimal controls can be proved using existence results of optimal solutions studied in [1].

Under the Pontryagin's Maximum Principle (PMP) [12] and considering the optimal control problem (10.4) and objective functional (10.5), if $(u_1^*(\cdot), u_2^*(\cdot))$ is a control pair that is optimal for the problem, then there exists a nontrivial Lipschitz continuous mapping, called *adjoint vector*, $\lambda : [0, t_f] \to \mathbb{R}^2$, $\lambda(t) = (\lambda_1(t), \lambda_2(t))$, such that

$$\frac{ds}{dt} = \frac{\partial H}{\partial \lambda_1}, \quad \frac{di}{dt} = \frac{\partial H}{\partial \lambda_2}$$

and

$$\frac{d\lambda_1}{dt} = -\frac{\partial H}{\partial s}, \quad \frac{d\lambda_2}{dt} = -\frac{\partial H}{\partial i}\,,$$

where the function H defined by

$$\begin{aligned} H(s(t), i(t), u_1(t), u_2(t), \lambda_1(t), \lambda_2(t)) &= -(Bu_1^2(t) + Cu_2^2(t)) \\ &+ \lambda_1(t)\left[-(\beta + u_2(t))s(t)i(t) - u_1(t)s(t)\right] \\ &+ \lambda_2(t)\left[(\beta + u_2(t))s(t)i(t) + u_1(t)s(t) - \gamma i(t)\right] \end{aligned}$$

is called the *Hamiltonian*. At time t, let s^* and i^* be the optimal state trajectories. Thus, according to the PMP it follows that

$$\begin{cases} \dfrac{d\lambda_1}{dt} = \lambda_1(t)\left[(\beta + u_2^*(t))i^*(t) + u_1^*(t)\right] - \lambda_2(t)\left[(\beta + u_2^*(t))i^*(t) + u_1^*(t)\right] \\ \dfrac{d\lambda_2}{dt} = \lambda_1(t)\left[(\beta + u_2^*(t))s^*(t)\right] - \lambda_2(t)\left[(\beta + u_2^*(t))s^*(t) - \gamma\right]\,, \end{cases} \qquad (10.6)$$

with transversality conditions $\lambda_1(t_f) = 0$ and $\lambda_2(t_f) = 1$, since $r(t_f)$ and $i(t_f)$ appear as payoffs terms in the objective functional (10.5).

In addition, by setting $\frac{\partial H}{\partial u_1}$ and $\frac{\partial H}{\partial u_2}$ to zero and due to the boundedness of the control functions $u_1(t)$ and $u_2(t)$ on Ω, the optimal controls $u_1^*(t)$ and $u_2^*(t)$ are characterized by:

$$u_1^*(t) = \min\left\{\max\left\{\frac{s^*(t)\Big(\lambda_2(t) - \lambda_1(t)\Big)}{2B}, 0\right\}, u_{1\max}\right\}, \tag{10.7}$$

and

$$u_2^*(t) = \min\left\{\max\left\{\frac{s^*(t)i^*(t)\Big(\lambda_2(t) - \lambda_1(t)\Big)}{2C}, 0\right\}, u_{2\max}\right\}. \tag{10.8}$$

At this point, it is possible to derive the *optimality system*, consisting of the state system (10.4), the adjoint system (10.6) and transversality conditions with the characterizations (10.7) and (10.8).

Theorem 10.1 (Uniqueness of Optimality System) *Given the initial value problem (10.4) and the objective functional (10.5), the optimal solution $(s^*(t), i^*(t))$ with associated optimal control functions $u_1^*(t)$, $u_2^*(t)$ and the adjoint functions $\lambda_1(t)$, $\lambda_2(t)$ are unique for t_f sufficiently small.*

Proof Let $(s, i, \lambda_1, \lambda_2)$ and $(\overline{s}, \overline{i}, \overline{\lambda_1}, \overline{\lambda_2})$ be two solutions of the optimality system.

Consider $s = e^{\phi t}a_1, i = e^{\phi t}a_2, \lambda_1 = e^{-\phi t}b_1, \lambda_2 = e^{-\phi t}b_2$. Analogously, consider $\overline{s} = e^{\phi t}\overline{a_1}, \overline{i} = e^{\phi t}\overline{a_2}, \overline{\lambda_1} = e^{-\phi t}\overline{b_1}, \overline{\lambda_2} = e^{-\phi t}\overline{b_2}$, where ϕ is a constant.

Let

$$u_1(t) = \min\left\{\max\left\{\frac{a_1(b_2 - b_1)}{2B}, 0\right\}, u_{1\max}\right\},$$

$$u_2(t) = \min\left\{\max\left\{\frac{e^{\phi t}a_1a_2(b_2 - b_1)}{2C}, 0\right\}, u_{2\max}\right\},$$

and

$$\overline{u_1}(t) = \min\left\{\max\left\{\frac{\overline{a_1}(\overline{b_2} - \overline{b_1})}{2B}, 0\right\}, u_{1\max}\right\},$$

$$\overline{u_2}(t) = \min\left\{\max\left\{\frac{e^{\phi t}\overline{a_1}\overline{a_2}(\overline{b_2} - \overline{b_1})}{2C}, 0\right\}, u_{2\max}\right\}.$$

Henceforth, for the sake of simplicity, it will be considered $u_1, \overline{u_1}, u_2, \overline{u_2}$ instead of the above characterizations. Using these assumptions, the first state equation of the optimality system becomes

$$e^{\phi t}\frac{da_1}{dt} + \phi e^{\phi t} a_1 = -(\beta + u_2)e^{2\phi t} a_1 a_2 - u_1 e^{\phi t} a_1 \,. \tag{10.9}$$

Now, the equations for s and $\overline{s}$, i and $\overline{i}$, λ_1 and $\overline{\lambda_1}$, λ_2 and $\overline{\lambda_2}$ are subtracted. Then, each of these equations is multiplied by an appropriate difference of functions and integrated from 0 to t_f. Next, the four integral equations are added and some estimates are performed. Below, we illustrate one of these manipulations using the Eq. (10.9).

$$\begin{aligned}
&\frac{1}{2}\left(a_1(t_f) - \overline{a_1}(t_f)\right)^2 + \phi \int_0^{t_f} (a_1 - \overline{a_1})^2 \, dt \\
&= -\int_0^{t_f} e^{\phi t}\Big[(\beta + u_2)a_1 a_2 - (\beta + \overline{u_2})\overline{a_1}\overline{a_2})\Big](a_1 - \overline{a_1}) \, dt - \int_0^{t_f} (u_1 a_1 - \overline{u_1}\overline{a_1})(a_1 - \overline{a_1}) \, dt \\
&= -\int_0^{t_f} e^{\phi t}\Big[\beta\Big((a_1 - \overline{a_1})a_2 + \overline{a_1}(a_2 - \overline{a_2})\Big) + (u_2 - \overline{u_2})a_1 a_2 + \overline{u_2}(a_1 - \overline{a_1})a_2 \\
&\quad + \overline{u_2}\overline{a_1}(a_2 - \overline{a_2})\Big](a_1 - \overline{a_1}) \, dt - \int_0^{t_f} \Big[(u_1 - \overline{u_1})a_1 + \overline{u_1}(a_1 - \overline{a_1})\Big](a_1 - \overline{a_1}) \, dt \\
&\leqslant D \int_0^{t_f} (a_1 - \overline{a_1})^2 + (b_1 - \overline{b_1})^2 + (b_2 - \overline{b_2})^2 \, dt + E e^{\phi t_f} \int_0^{t_f} (a_1 - \overline{a_1})^2 + (a_2 - \overline{a_2})^2 \, dt \\
&\quad + F e^{3\phi t_f} \int_0^{t_f} (b_1 - \overline{b_1})^2 + (b_2 - \overline{b_2})^2 \, dt \,,
\end{aligned}$$

where D, E and F are constants. After estimate all the four equations of the optimality system, and noting that $e^{\phi t_f} \leqslant e^{3\phi t_f}$, all integral equations are combined producing the following inequality:

$$\begin{aligned}
&\frac{1}{2}\Big[\left(a_1(t_f) - \overline{a_1}(t_f)\right)^2 + \left(a_2(t_f) - \overline{a_2}(t_f)\right)^2 + \left(b_1(0) - \overline{b_1}(0)\right)^2 + \left(b_2(0) - \overline{b_2}(0)\right)^2\Big] \\
&\quad + \phi \int_0^{t_f} (a_1 - \overline{a_1})^2 + (a_2 - \overline{a_2})^2 + (b_1 - \overline{b_1})^2 + (b_2 - \overline{b_2})^2 \, dt \\
&\leqslant \widetilde{D} \int_0^{t_f} (a_1 - \overline{a_1})^2 + (a_2 - \overline{a_2})^2 + (b_1 - \overline{b_1})^2 + (b_2 - \overline{b_2})^2 \, dt \\
&\quad + \widetilde{F} e^{3\phi t_f} \int_0^{t_f} (a_1 - \overline{a_1})^2 + (a_2 - \overline{a_2})^2 + (b_1 - \overline{b_1})^2 + (b_2 - \overline{b_2})^2 \, dt \,.
\end{aligned}$$

Rearranging the terms, the result is

$$\begin{aligned}
&\frac{1}{2}\Big[\left(a_1(t_f) - \overline{a_1}(t_f)\right)^2 + \left(a_2(t_f) - \overline{a_2}(t_f)\right)^2 + \left(b_1(0) - \overline{b_1}(0)\right)^2 + \left(b_2(0) - \overline{b_2}(0)\right)^2\Big] \\
&\leqslant \left(\widetilde{D} + \widetilde{F} e^{3\phi t_f} - \phi\right) \int_0^{t_f} (a_1 - \overline{a_1})^2 + (a_2 - \overline{a_2})^2 + (b_1 - \overline{b_1})^2 + (b_2 - \overline{b_2})^2 \, dt \,,
\end{aligned}$$

where $\widetilde{D}$ and $\widetilde{F}$ depend on the coefficients and the bounds of a_1, a_2, b_1, b_2. By choosing $\phi > \widetilde{D} + \widetilde{F}$ and $t_f < \dfrac{1}{3\phi} \log\left(\dfrac{\phi - \widetilde{D}}{\widetilde{F}}\right)$, it therefore follows that

$$
\begin{aligned}
0 &\leqslant \frac{1}{2}\Big[\big(a_1(t_f) - \overline{a_1}(t_f)\big)^2 + \big(a_2(t_f) - \overline{a_2}(t_f)\big)^2 + \big(b_1(0) - \overline{b_1}(0)\big)^2 + \big(b_2(0) - \overline{b_2}(0)\big)^2\Big] \\
&\leqslant \left(\widetilde{D} + \widetilde{F}e^{3\phi t_f} - \phi\right) \int_0^{t_f} (a_1 - \overline{a_1})^2 + (a_2 - \overline{a_2})^2 + (b_1 - \overline{b_1})^2 + (b_2 - \overline{b_2})^2 \, dt \\
&\leqslant 0\,,
\end{aligned}
$$

which implies that

$$
\left(\widetilde{D} + \widetilde{F}e^{3\phi t_f} - \phi\right) \int_0^{t_f} (a_1 - \overline{a_1})^2 + (a_2 - \overline{a_2})^2 + (b_1 - \overline{b_1})^2 + (b_2 - \overline{b_2})^2 \, dt = 0.
$$

Thus, knowing that $\left(\widetilde{D} + \widetilde{F}e^{3\phi t_f} - \phi\right) < 0$, we have $a_1 = \overline{a_1}$, $a_2 = \overline{a_2}$, $b_1 = \overline{b_1}$, $b_2 = \overline{b_2}$, and $(s, i, \lambda_1, \lambda_2) = (\overline{s}, \overline{i}, \overline{\lambda_1}, \overline{\lambda_2})$.

Remark 10.1 Since the state system (10.4) is autonomous, the proof of theorem 10.1 holds for any time t_f.

10.4 Numerical Results and Discussion

In this section, the influence of the optimal control strategies incorporated in the SIR model (10.4) with objective functional (10.5) is studied. The main goal is to provide insights related to when and which control strategies should be applied to maximize the spreading of information and minimize costs.

Using MATLAB software to solve the optimality system, numerical results were obtained using a forward-backward fourth-order Runge–Kutta scheme. More detailed information on this numerical scheme is presented in [8]. Numerical simulations consider a campaign deadline $t_f = 6$, which corresponds to the Dove's campaign data on the first seven days.

In what follows, two approaches are considered. Firstly, in Sect. 10.4.1, simulations of the control weights are performed using both optimal controls u_1^* and u_2^*, in order to assess which control weights induce a higher objective functional. Secondly, in Sect. 10.4.2, these pairs are used to model and compare scenarios related to high and low investment costs in publicity actions.

10.4.1 Simulation of the Control Weights Using u_1^* and u_2^*

Firstly, using both optimal controls u_1^* and u_2^*, the variation of the control weights on both infected individuals ($i(t)$) and control signals is studied over $[0, 6]$. The simulations of the control weights are performed under three strategies described in Table 10.2.

The choice of the values for B and C was based on several experimental simulations that aimed at obtaining the results that best described the reality of each strategy. For the sake of consistency, since at the beginning of a viral epidemic the main goal is to attract susceptible individuals, it is conceivable to start with an intermediate investment cost in detriment of a lower one. For this reason, the above strategies do not consider $B < 1$. The simulations regarding the three strategies are illustrated in Fig. 10.1. Figure 10.1a shows that around the first day, the fraction of infected individuals is higher when the investment costs in fostering people who had already been in contact with the marketing message are low, namely for $B = 1$ and $C = 10^{-3}$.

Regarding the Strategy 2 (Fig. 10.1b), by fixing $B = 1$, as the control weight C increases, infection levels do not vary. Thus, it is plausible to conclude that higher costs in implement further publicity strategies to foster infected individuals do not result in higher levels of spreading.

In what concerns the Strategy 3, Fig. 10.1c reports that as the control weight B increases, the number of people who have contact with the intended message is diminishing all the time and the maximum peak of infection is attained increasingly late.

Transversally, whatever the strategy considered, the smaller the investment costs neither in fostering infected individuals to continue to spread the message, nor in recruit susceptible individuals to act as spreaders, the greater the levels of information spreading.

Overall, Table 10.3 presents the values for the objective functional (10.5) by varying the control weights for each strategy.

Table 10.2 Optimal control marketing strategies

# Strategy	Marketing context
Strategy 1: $B = 1$ and $C \in \{x \mid x = 10^{-i}$ and $i = 1, \ldots, 3\}$	Low investment costs in encouraging infected individuals to continue to spread the marketing message (e.g., exploiting social networks such as Facebook and Twitter)
Strategy 2: $B = 1$ and $C \in \{x \mid x = 10^{i}$ and $i = 1, \ldots, 3\}$	High investment costs in encouraging infected individuals to continue to spread the marketing message (e.g., monetary rewards and stimuli, expensive promotional gifts and international trips)
Strategy 3: $B \in \{x \mid x = 10^{i}$ and $i = 0, \ldots, 3\}$ and $C = 1$	Increasing the investment costs in recruiting susceptible individuals to act as spreaders

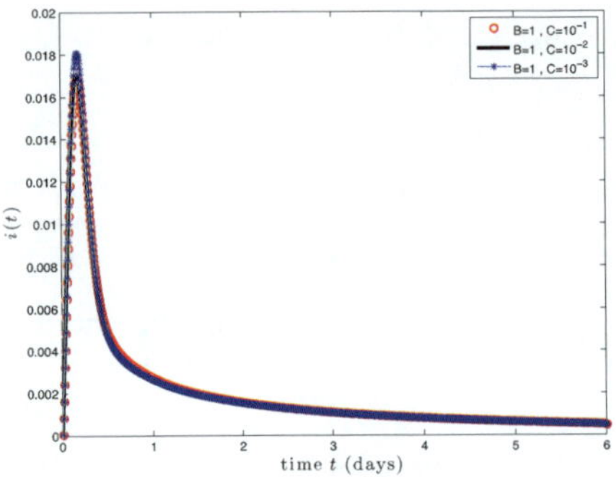

(a) $i(t)$ for Strategy 1

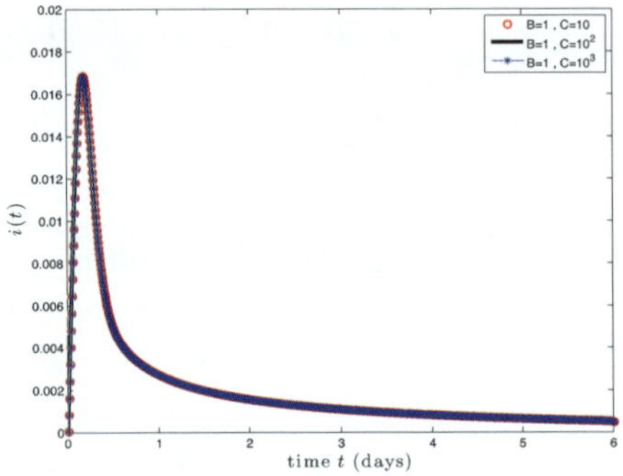

(b) $i(t)$ for Strategy 2

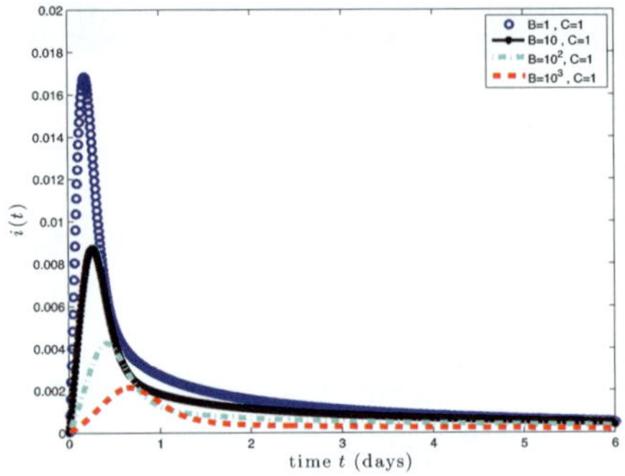

(c) $i(t)$ for Strategy 3

Fig. 10.1 Variation of the control weights on $i(t)$ for the different marketing strategies

Table 10.3 Summary of objective functionals varying control weights (B, C)

Strategy 1		Strategy 2		Strategy 3	
(B, C)	$\approx J(u_1^*, u_2^*)$	(B, C)	$\approx J(u_1^*, u_2^*)$	(B, C)	$\approx J(u_1^*, u_2^*)$
$\mathbf{(1, 10^{-3})}$	**0.699766**	$\mathbf{(1, 10)}$	**0.698692**	$\mathbf{(1, 1)}$	**0.698693**
$(1, 10^{-2})$	0.698815	$(1, 10^2)$	0.698691	$(10, 1)$	0.463743
$(1, 10^{-1})$	0.698704	$(1, 10^3)$	0.698691	$(10^2, 1)$	0.273901
–	–	–	–	$(10^3, 1)$	0.159022

In Table 10.3, recalling the aim of maximize the objective functional (10.5), the pairs of control weights that induce a higher cost functional are highlighted in *bold*, for each strategy. The choice of the highlighted pairs was based on the aim of portray low, high and equal investment cost scenarios, respectively, to study their effects on the diffusion of the marketing message. These three scenarios are simulated in the next section.

10.4.2 Control Experiments for Different Investment Costs Scenarios

In this Section, scenarios based on the control weights highlighted in Table 10.3 are simulated, varying the control functions $u_1(t)$ and $u_2(t)$, $\forall t \in [0, 6]$. If the company has monetary funds to invest in extra publicity actions either on susceptible or already infected individuals, upper control policies are tested. The upper control is defined as the maximum control value for which the cost functional (10.5) is positive, and it represents the maximum application of a control measure. Henceforth, let $u_{1\max}$ and $u_{2\max}$ be the upper controls related to the control functions u_1 and u_2, respectively.

10.4.2.1 Low Investment Costs Scenario Using Control Weights $(B, C) = (1, 10^{-3})$

In Fig. 10.2a it is possible to note that the fraction of infected individuals is significantly higher whenever control is applied. In this regard, the implementation effect of the controls u_1 and u_2 is assessed, separately and combined, over [0, 6], see Fig. 10.3.

Observing Fig. 10.3a, b at the launch of the advertisement, the best policy is to implement the optimal control combination (u_1^*, u_2^*), in order to rapidly attain the maximum peak of infection at the end of the first day of Dove's campaign (see Fig. 10.2a). Then, during the next 2 days, the control u_2 is at the upper bound (see, Fig. 10.3b), suggesting in this time window that the best policy is to apply $(0, u_2^*)$ in such a way as to encourage infected individuals to continue to spread the message. Hence, at the end of $t = 2$, when the levels of recovery begin to increase, the pair $(u_1^*, 0)$ should be implemented in order to minimize the rate of recovered individuals by attracting susceptible individuals to diffuse the intended message, see Fig. 10.3a.

Notice that, in terms of the objective functional, despite of infection levels attain a maximum level using $(u_{1\max}, u_{2\max})$, $J(u_{1\max}, u_{2\max}) \approx 0.002095$, which means that a double investment in both control policies compromises the objective of mini-

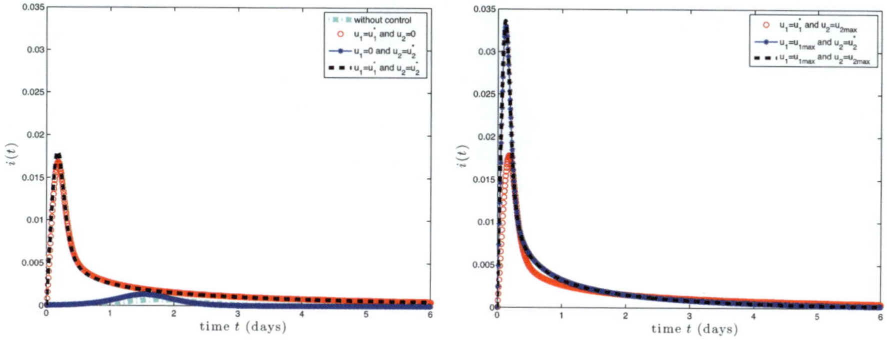

(a) Optimal controls (b) Mixing upper and optimal controls

Fig. 10.2 $i(t)$ with $B = 1$, $C = 10^{-3}$, $u_{1\max} = 0.4$, $u_{2\max} = 1$

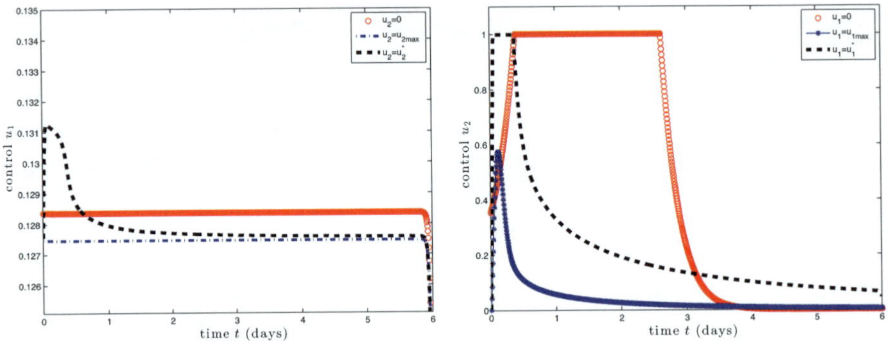

(a) Optimal control u_1

(b) Optimal control u_2

Fig. 10.3 Optimal controls u_1 and u_2 with $B = 1$, $C = 10^{-3}$, $u_{1\,\max} = 0.4$, $u_{2\,\max} = 1$

mize costs. In contrast, based on Table 10.3, $J(u_1^*, u_2^*) \approx 0.699766$, that is, the simultaneous use of the control functions u_1^* and u_2^* fulfills the proposed trade-off. This is underpinned by the fact that, for the control intervention $(u_1^*, u_{2\,\max})$, the cost functional (10.5) is almost the same as the obtained when the optimal controls u_1^* and u_2^* are applied ($J(u_1^*, u_{2\,\max}) \approx 0.695835$). These arguments show the importance of the control u_1 to attain the maximum peak of infection at the beginning of the campaign.

10.4.2.2 High Investment Costs Scenario Using Control Weights $(B, C) = (1, 10)$

In this scenario, analogously to the previous one, the fraction of infected individuals is higher with the implementation of control policies than without it, see Fig. 10.4a.

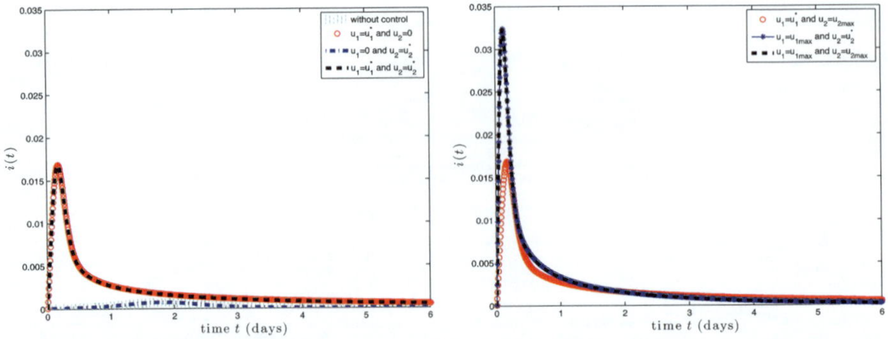

(a) Optimal controls

(b) Mixing upper and optimal controls

Fig. 10.4 $i(t)$ with $B = 1$, $C = 10$, $u_{1\,\max} = 0.4$, $u_{2\,\max} = 0.01$

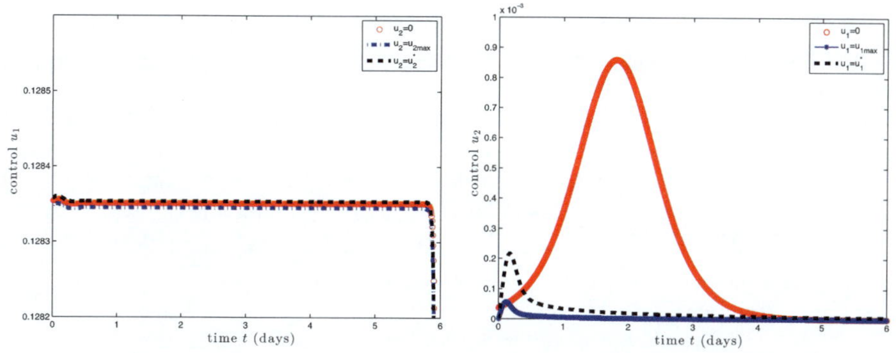

(a) Optimal control u_1 (b) Optimal control u_2

Fig. 10.5 Optimal controls u_1 and u_2 with $B = 1$, $C = 10$, $u_{1\,\max} = 0.4$, $u_{2\,\max} = 0.01$

However, Fig. 10.5 illustrates that the magnitude of the control u_2 is significantly lower than the magnitude of u_1. By linking this finding with Fig. 10.4a, it is possible to infer not only that the optimal control measure u_2^* applied by itself has no direct influence on the information dissemination, but also that by applying $(u_1^*, 0)$, the levels of information diffusion are satisfactory.

When upper control measures are applied, the infection levels increase substantially, see Fig. 10.4b. However, the use of $(u_{1\,\max}, u_{2\,\max})$ results in a residual objective functional $(J(u_{1\,\max}, u_{2\,\max}) \approx 0.001591)$. At this point, upper control policies for u_2 are disadvantageous, inasmuch as the adoption $(u_1^*, u_{2\,\max})$ leads to the same infection levels as the obtained by using both optimal controls and $J(u_1^*, u_2^*) \approx 0.698692$, see Fig. 10.4a, b. Furthermore, $J(u_1^*, 0) \approx 0.698678$, which means that the control u_1^* is a sufficient condition to achieve the proposed trade–off.

Analogously to the previous strategy, it also can be noticed that the control u_1 has much more influence on the information spreading than u_2.

In order to avoid redundancy, the control simulations using $B = C = 1$ were omitted. However, in this case, u_1 has much more influence on the information spreading than u_2. In addition, to meet the objective functional, optimal control strategies overlap the efforts imposed by upper control policies.

10.5 Conclusion

This paper applies optimal control theory to a real viral marketing campaign, by using real numerical data. The uniqueness of optimality system is proved. We show that the spreading of information attains high levels of dissemination at much faster rates when optimal control theory is applied. It is observed that when B is fixed and C decreases, the levels of information diffusion tend to grow up. In contrast, when B is fixed and C increases, the levels of information spreading do not vary and are lower than the values obtained for smaller levels of C.

In order to improve the timing of the information diffusion, we recommend, for each scenario, optimal time windows to apply the control policies u_1 and u_2. Moreover, we conclude that the infection transmission process is maximized by using optimal control policies related to u_1 and u_2. However, it should be noted that due to the chaotic and quasi-unpredictable nature of viral campaigns, the success of a campaign depends on a constant monitoring and controlling on the part of marketing professionals. Regarding the parameters estimation, reaching a high level of fitting accuracy is not a trivial task. At this point, mathematical modeling reveals a fruitful tool to maximize the diffusion of marketing contents and minimize the costs of running a campaign.

To sum up, optimal control theory plays a key role on the effective diffusion of viral marketing campaigns, by providing not only higher levels of infection than the obtained without using it, but also by speeding up the transmission process within the target audience.

Acknowledgements The authors would like to acknowledge the comments and suggestions from the reviewers, which improved the quality of the paper. This work was supported in part by the Portuguese Foundation for Science and Technology (FCT - Fundação para a Ciência e a Tecnologia), through CIDMA - Center for Research and Development in Mathematics and Applications, within project UID/MAT/04106/2013; and through Algoritmi R&D Center, under COMPETE: POCI-01-0145-FEDER-007043 within the Project Scope: UID/CEC/00319/2013.

References

1. W.H. Fleming, R.W. Rishel, *Deterministic and Stochastic Optimal Control*, vol. 1 (Springer Science & Business Media, New York, 2012)
2. E. Gray, Dove's 'real beauty sketches' ad campaign tells women 'you're more beautiful than you think'. The Huffington Post (2013)
3. H.W. Hethcote, The mathematics of infectious diseases. SIAM Rev. **42**(4), 599–653 (2000)
4. K. Kandhway, J. Kuri, How to run a campaign: optimal control of SIS and SIR information epidemics. Appl. Math. Comput. **231**, 79–92 (2014)
5. A. Karnik, P. Dayama, Optimal control of information epidemics, in *2012 Fourth International Conference on Communication Systems and Networks (COMSNETS)* (IEEE, 2012), pp. 1–7
6. R. Kheirandish, A.S. Krishen, P. Kachroo, Application of optimal control theory in marketing: What is the optimal number of choices on a shopping website? Int. J. Comput. Appl. Technol. **34**(3), 207–215 (2009)
7. J.C. Lagarias, J.A. Reeds, M.H. Wright, P.E. Wright, Convergence properties of the nelder-mead simplex method in low dimensions. SIAM J. Optim. **9**(1), 112–147 (1998)
8. S. Lenhart, J.T. Workman, *Optimal Control Applied to Biological Models* (CRC Press, Boca Raton, 2007)
9. J. Leskovec, L.A. Adamic, B.A. Huberman, The dynamics of viral marketing. ACM Trans. Web (TWEB) **1**(1), 1–39 (2007)
10. M. Martcheva, *Introduction to Mathematical Epidemiology*, vol. 61 (Springer, Berlin, 2015)
11. A. Mochalova, A. Nanopoulos, A targeted approach to viral marketing. Electron. Commer. Res. Appl. **13**(4), 283–294 (2014)
12. L.S. Pontryagin, *Mathematical Theory of Optimal Processes* (CRC Press, Boca Raton, 1987)
13. L. Porter, G.J. Golan, From subservient chickens to Brawny men: a comparison of viral advertising to television advertising. J. Interact. Advert. **6**(2), 4–33 (2006)

14. H.S. Rodrigues, M.J. Fonseca, Can information be spread as a virus? viral marketing as epidemiological model. Math. Methods Appl. Sci. **39**(16), 4780–4786 (2016)
15. R.T. Rust, T. Ambler, G.S. Carpenter, V. Kumar, R.K. Srivastava, Measuring marketing productivity: current knowledge and future directions. J. Market. **68**(4), 76–89 (2004)
16. S. Schroeder, Youtube now has one billion monthly users (2016)
17. K. Sohn, J.T. Gardner, J.L. Weaver, Viral marketing-more than a buzzword. J. Appl. Bus. Econ. **14**(1), 21–42 (2013)
18. L. Stampler, How dove's' real beauty sketches' became the most viral video ad of all time. Business Insider (2013)
19. M. Woerndl, S. Papagiannidis, M. Bourlakis, F. Li, Internet-induced marketing techniques: critical factors in viral marketing campaigns. Int. J. Bus. Sci. Appl. Manag. **3**(1), 33–45 (2008)

Chapter 11
The Two-Dimensional Strip Packing Problem: What Matters?

Alvaro Neuenfeldt Júnior, Elsa Silva, A. Miguel Gomes and José Fernando Oliveira

Abstract This paper presents an exploratory approach to study and identify the main characteristics of the two-dimensional strip packing problem (2D-SPP). A large number of variables was defined to represent the main problem characteristics, aggregated in six groups, established through qualitative knowledge about the context of the problem. Coefficient correlation are used as a quantitative measure to validate the assignment of variables to groups. A principal component analysis (PCA) is used to reduce the dimensions of each group, taking advantage of the relations between variables from the same group. Our analysis indicates that the problem can be reduced to 19 characteristics, retaining most part of the total variance. These characteristics can be used to fit regression models to estimate the strip height necessary to position all items inside the strip.

Keywords Strip packing problems · Cutting and packing problems · Principal component analysis · Knowledge discovery

11.1 Introduction

In the 2D-SPP the aim is to pack a set of rectangular items inside a rectangular object with a fixed width, minimizing the height dimension of the object that is infinite. The small items can be rotated, orthogonally positioned without overlapping and completely inside the object, also the This description fits in the definition of cutting

---

A. Neuenfeldt Júnior · E. Silva (✉) · A. Miguel Gomes · J. F. Oliveira
INESC-TEC, Faculty of Engineering, University of Porto, Porto, Portugal
e-mail: emsilva@inesctec.pt

A. Neuenfeldt Júnior
e-mail: alvaro.l.junior@inesctec.pt

A. Miguel Gomes
e-mail: agomes@fe.up.pt

J. F. Oliveira
e-mail: jfo@fe.up.pt

A. I. F. Vaz et al. (eds.), *Operational Research*, Springer Proceedings in Mathematics & Statistics 223, https://doi.org/10.1007/978-3-319-71583-4_11

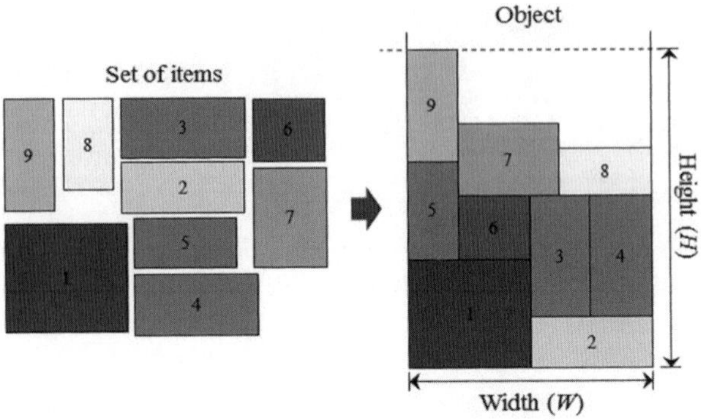

Fig. 11.1 A general view of the rectangular 2D-SPP

and packing problems and indeed the 2D-SPP can be classified as an open dimension problem [23]. An example can be found in Fig. 11.1.

Over the years a considerable number of test problem instances appeared in the literature to test the different heuristics that have been developed to solve the 2D-SPP. However, none of the developed heuristics were able to solve efficiently all the existing test problem instances and 2D-SPP variants.

The test problem instances are generally created with the use of some problem generators which were developed considering specific characteristics, methodologies and input parameters. As a consequence, it is possible to find data instances in the literature with different characteristics and combinations between items and object shape variation [19].

Data mining techniques can be used to facilitate a better understanding of the test problem instances characteristics, ensuring that the main details of the problem is known [21].

In this paper, we conduct an exploratory research to find the most relevant test problem instances characteristics for the two-dimensional rectangular 2D-SPP. An initial set of variables (more than fifty) are used to represent these characteristics, and the PCA was chosen as a technique to explore the relations among variables and to convert them in a smaller number of components. A sample of 1217 test problem instances extracted from the literature is explored.

A similar approach was developed by López–Camacho et al. [16], where a PCA was considered to develop two-dimensional analysis with the aim of better understand the structure of the two-dimensional (and three-dimensional) bin packing problems. This information was used to compare the performance of heuristics with a wide set of characteristics provided by test problem instances found in the literature. López–Camacho et al. [17] also developed a hyper-heuristic approach and compared the computational results with other approaches using the components developed through the PCA.

11.2 Strip Packing Problem Characteristics

Problem generators are one of the most efficient ways to replicate real-world applications of different problems over controlled scenarios, ensuring the reproducibility of the test problem instances. Problem generators can be used as one source of information to study the characteristics of packing problems. A literature review about problem generators for the 2D-SPP is presented below and will serve as a base for the development of explanatory variables.

Wang and Valenzela [22] developed two important factors to generate test problem instances for the rectangular placement problems: the maximum area ratio between all pairs of items; and the aspect ratio, to identify the variation level between the largest and the smallest dimension of each item. As consequence, the maximum aspect ratio identifies the item that has a greater difference in its dimensions.

Regardless the total number of items, larger values for aspect ratio and area ratio indicates a more significant variability in items shape, which allows the generation of more heterogeneous data instances. For example, “nice” instances have items of smaller size and variability, which indicate a homogeneous behavior. In contrast, “path” instances have items with larger size and variability, which characterizes test problem instances with higher heterogeneity.

Bortfeldt and Gehring [6] used the degree of heterogeneity among the items as one of the primordial factors to generate data instances. The degree of heterogeneity measures the ratio between the total number of different items in comparison to the total number of items of an test problem instance. This characteristic is reaffirmed by Bortfeldt [5] as one of the most important aspects to be considered in cutting and packing problems.

In Berkey and Wang [4] is considered the width ratio, that is calculated using the object width and items dimensions. This measure is one of the most important to develop test problem instances for both two-dimensional bin packing problems and 2D-SPP. The influence of width ratio on the quality of solutions for the strip packing was verified mainly in test problem instances with a smaller number of items ($n < 100$). Smaller width ratios indicated a greater probability of obtaining lower object heights.

The 2DCPackGen problem generator proposed by Silva et al. [20] is able to generate data instances for all two-dimensional (and three-dimensional) rectangular cutting and packing problems. In specific, for the 2D-SPP, the number of different item types, the item type demand and size and shape of the items are some of the measures that influences the assortment of items.

Leung et al. [15] combined in a dataset with 16 test problem instances some characteristics of instance *gcut*13 by Beasley [1] and instances *cx* by [9] to obtain items with different maximum and minimum areas using the generator proposed by Wang and Valenzela [22]. The main objective was to evaluate the capacity of some heuristics to solve problems with a strongly heterogeneous items shape variation, where the position of the larger items into the object is a determinant factor to obtain good solutions.

Table 11.1 Test problem instances variables summary

Variable	Definition
Area	
$arearatioextr$	$arearatioextr = area_{max}/area_{min}$
$arearatioperc$	Ratio between percentile 90% and percentile 10% of $area_i$
$arearatioquart$	Ratio between third quartile and first quartile of $area_i$
$areacompnormal$	Compensation between the sum of 50% larger $area_i$ and the sum of 50% smaller $area_i$
$areacompquart$	Compensation between the sum of 25% larger $area_i$ and the sum of 25% smaller $area_i$
$areacompextr$	Compensation between the sum of 10% larger $area_i$ and the sum of 10% smaller $area_i$
$areamean$	Mean value of $area_i$
$areamed$	Median value of $area_i$
$areastdev$	Standard deviation of $area_i$
Perimeter	
$perimratioextr$	$perimratioextr = perimeter_{max}/perimeter_{min}$
$perimratioperc$	Ratio between percentile 90% and percentile 10% of $perimeter_i$
$perimratioquart$	Ratio between third quartile and first quartile of $perimeter_i$
$perimcompnormal$	Compensation between the sum of 50% larger $perimeter_i$ and the sum of 50% smaller $perimeter_i$
$perimcompquart$	Compensation between the sum of 25% larger $perimeter_i$ and the sum of 25% smaller $perimeter_i$
$perimcompextr$	Compensation between the sum of 10% larger $perimeter_i$ and the 10% smaller $perimeter_i$
$perimmean$	Mean value of $perimeter_i$
$perimmed$	Median value of $perimeter_i$
$perimstdev$	Standard deviation of $perimeter_i$
Dimensions	
$vardim$	$vardim = W/[\sum(d1_i + d2_i)/n]$
$dimratioextr$	$dimratioextr = dimension_{max}/dimension_{min}$
$dimratioperc$	Ratio between percentile 90% and percentile 10% of $dimension_i$
$dimratioquart$	Ratio between third quartile and first quartile of $dimension_i$
$dimcompnormal$	Relation between the sum of 50% larger $dimension_i$ and the sum of 50% smaller $dimension_i$
$dimcompquart$	Compensation between the sum of 25% higher $dimension_i$ and the sum of 25% smaller $dimension_i$
$dimcompextr$	Compensation between the sum of 10% higher $dimension_i$ and the sum of 10% smaller $dimension_i$
$dimmean$	Mean value of $dimension_i$
$dimmed$	Median value of $dimension_i$
$dimstdev$	Standard deviation of $dimension_i$

(continued)

Table 11.1 (continued)

Variable	Definition
Width dimensions	
widthdimratioextr	$widthdimratioextr = widthdimension_{max}/widthdimension_{min}$
widthdimratioperc	Ratio between percentile 90% and percentile 10% of $widthdimension_i$
widthdimratioquart	Ratio between third quartile and first quartile of $widthdimension_i$
widthdimcompnormal	Compensation between the sum of 50% larger $widthdimension_i$ and the sum of 50% smaller $widthdimension_i$
widthdimcompquart	Compensation between the sum of 25% larger $widthdimension_i$ and the sum of 25% smaller $widthdimension_i$
widthdimcompextr	Compensation between the sum of 10% larger $widthdimension_i$ and the sum of 10% smaller $widthdimension_i$
widthdimmean	Mean value of $widthdimension_i$
widthdimmed	Median value of $widthdimension_i$
widthdimstdev	Standard deviation of $widthdimension_i$
Proportions	
aspectratio	$aspectratio = (D1/D2)/[\sum(d1_i/d2_i)/n]$
propratioextr	$propratioextr = proportion_{max}/proportion_{min}$
propratioperc	Ratio between percentile 90% and percentile 10% of $proportion_i$
propratioquart	Ratio between third quartile and first quartile of $proportion_i$
propcompnormal	Compensation between the sum of 50% larger $proportion_i$ measures and the sum of 50% smaller $proportion_i$
propcompquart	Compensation between the sum of 25% larger $proportion_i$ measures and the sum of 25% smaller $proportion_i$
propcompextr	Compensation between the sum of 10% larger $proportion_i$ measures and the sum of 10% smaller $proportion_i$
propmean	Mean value of $proportion_i$
propmed	Median value of $proportion_i$
propstdev	Standard deviation of $proportion_i$
Other	
n	Total number of items in the test problem instance
coefficient	$coefficient = [(\Sigma d1_i/n) + (\Sigma d2_i/n)]/2$ Average items dimensions values
heterogeneity	$heterogeneity = nt/n$ Proportion between the quantity of different items (nt) for all n
heterognt	Measure of heterogeneity considering only types of items with more than one item
difcoefficient	For all n, the total number of different items dimensions
objdimratio	Number of times that the object lower-bound is bigger than the object width
itdimratio	Number of times that the items the maximum items dimension is bigger than the minimum items dimensions
maxcoefficient	10% larger items dimensions values
mincoefficient	10% smaller items dimensions values

To help the generation of experimental data to cover more practical applications for the knapsack problem, Hall and Posner [10] used a wide range of factors to identify the test problem instances difficulty. Three characteristics can be explored in the 2D-SPP context: the total number of items (data instances size); the coefficients of all items values; and the heterogeneity (relation between the size and the number of different test problem instances).

All the concepts and parameters about the problem generators previously described are used to develop the exploratory variables to study the characteristics of the rectangular 2D-SPP. These variables were created considering both items and object shape variation, as well as some intrinsic factors of the test problem instances. Table 11.1 describes the 56 variables defined for this study, divided into six groups ($Area$, $Perimeter$, $Dimensions$, $Widthdimensions$, $Proportions$, and $Other$) accordingly to their origin and level of similarity. To simplify the variables calculation five reference parameters for each item i were defined:

- $area_i = A/a_i$: Ratio between the object area (A) and item i area (a_i);
- $perimeter_i = P/p_i$: Ratio between the object perimeter (P) and item i perimeter (p_i);
- $dimension_i = W/[(d1_i + d2_i)/2]$: Average dimension of item i compared to the object width (W). $d1_i$ is the largest item dimension and $d2_i$ is the smallest item dimension;
- $proportion_i = (D1/D2)/(d1_i/d2_i)$: Level of proportion between the object and item dimensions. $D1$ is the largest object dimension and $D2$ is the smallest object dimension;
- $widthdimension_i = W/d1_i$: Size of the largest item dimensions ($d1_i$) compared to the object width (W).

Small letters (i.e. a_i) represent items dimensions, capital letters (i.e. A) are used to object dimensions and extended words are reserved for the definition of the reference parameters (i.e. $area_i$) and for variables (i.e. $arearatio$).

11.3 Test Problem Instances

In this section, the most frequently benchmark data instances used over the years for the 2D-SPP are identified. Table 11.2 describes the main characteristics of these test problem instances, organized by name, number of instances, minimum and maximum number of items, organization and source.

In Hopper [12], the total number of items and the objects width and height similarities were used to generate twenty-one test problem instances divided in seven classes. Different items shape were randomly generated with a maximum ratio between the items dimensions equal to seven. The object shape varies for dimensions ratio between one and three. Hopper and Turton [11] generated all test problem instances with the same characteristics, but the first 35 test problem instances (*NTn*) corre-

Table 11.2 Test problem instances

Dataset	Instances[a]	Items[b]	Organization	Source
C	21	17–197	7 classes ($C1$–$C7$)	Hopper and Turton [11]
NTn	35	17–199	7 classes ($NTn1$–$NTn7$)	Hopper [12]
NTt	35	17–199	7 classes ($NTt1$–$NTt7$)	Hopper [12]
N	13	10–3152	1 class ($N1$–$N13$)	Burke et al. [7]
cx	7	50–15000	1 class (cx)	Ferreira and Oliveira [9]
iy	170	16–32768	11 classes ($i4$–$i15$)	Imahori and Yagiura [13]
cgcut	3	23–623	1 class ($cgcut$)	Christofides and Whitlock [8]
bwmv	300	20–100	6 classes ($C01$–$C06$)	Berkey and Wang [4]
bwmv	200	20–100	4 classes ($C07$–$C10$)	Martello and Vigo [18]
ngcut	12	7–22	1 class ($ngcut$)	Beasley [2]
gcut	13	10–50	1 class ($gcut$)	Beasley [1]
zdf	16	580–75032	1 class (zdf)	Leung and Zhang [14]
AH	360	1000	6 classes ($AH1$–$AH6$)	Bortfeldt [5]
beng	10	20–200	1 class ($beng$)	Bengtsson [3]
nice	36	25–5000	8 classes ($nice1$–$nice5t$)	Wang and Valenzela [22]
path	36	25–5000	8 classes ($path1$–$path5t$)	Wang and Valenzela [22]

[a]Total number of test problem instances
[b]Minimum and maximum number of items

sponds to guillotine patterns, while the next 35 (NTt) corresponds to non-guillotine patterns. The data instances are classified into seven classes, according to the total number of items.

The *N* test problem instances proposed by Burke et al. [7] were generated with constant values for the dimensions of the object. In a second moment, these objects were randomly divided in small items. In Ferreira and Oliveira [9] the main focus was to create very heterogeneous items, with extreme differences between the maximum and minimum items dimensions. Imahori and Yagiura [13] used the generator proposed by Wang and Valenzela [22] to develop the iy test problem instances. The main characteristic is the exponential variation of the total number of items per data instance, varying from 32 to 32768 items.

Christofides and Whitlock [8] prefixed a value for the object area of each $cgcut$ test problem instance, and the item's dimensions were generated according to a uniform distribution. As a consequence, items dimensions are proportional to the object. The ten classes of the $bwmv$ instances developed by Berkey and Wang [4] ($C01$–$C06$) and later updated by Martello and Vigo [18] ($C07$–$C10$) were proposed using some bin packing problem parameters. Each class has a total of 50 test problem instances, with item's dimensions uniformly generated.

In Beasley [2] and Beasley [1], the $ngcut$ and $gcut$ instances were generated based on the object width parameter. Guillotinable cuts were also considered for both x and y coordinates. Leung et al. [15] divided the zdf instances in two different groups. The first one ($zdf1$–$zdf8$) is composed of medium and large number of items, varying from 580 to 2532 items. The remaining ($zdf9$–$zdf16$) are considered extra-large data instances, varying from 5032 to 75032 items.

In Bortfeldt [5], the AH test problems set was developed using two parameters, varying in uniform distributions: items heterogeneity; and the ratio between the object width and items average width. Bengtsson [3] developed ten *beng* instances, based on the industrial cutting processes found in the manufacturers of excavators and mechanical shovels.

As mentioned before, Wang and Valenzela [22] developed one of the most used problem generators for the 2D-SPP. The process is recursive based on the variation of items area, according to some parameters defined by the user, and constant object dimensions ($W = 1000$ and $H = 1000$). The $nice$ instances have a high items shape similarity. In contrast, $path$ instances have extreme shape variations.

The data instances for the 2D-SPP presented have some differences, related to items and object shape variations and intrinsic test problem instances characteristics. The main reason for these effects is the use of different types of problem generators. As a consequence, the total number of test problem instances representing each of the variables listed in the previous session is not uniform.

11.4 Principal Component Analysis

For the exploratory analysis the use of PCA is proposed, with the aim of decrease the 56 variables presented in Table 11.1. The objective is to reduce the problem dimension to a more manageable size, retaining as much as possible the original information, by aggregating similar variables into principal components.

The work was developed in two steps. In a first moment, the consistency of each of the six groups of variables from Table 11.1 was checked, by analyzing the correlation coefficients between pairs of variables of each group. A linear correlation is used as a quantitative measure to validate the assignment of predictors to groups. A value of 0.75 for the correlation coefficient was used. The remaining part of this section specifies the procedures performed in all groups. Due to space constraints, we have chosen to show in detail only the results obtained for groups $Area$ and $Proportions$. However, the conclusions proposed at the end of this study considers the results of all groups.

To exemplify this first step, Tables 11.3 and 11.4 summarize the correlation coefficients between variables in groups $Area$ and $Proportions$, respectively. A total of 14 and 16 coefficient correlations higher than the reference value are found for these groups, and all variables have at least one high correlation coefficient that justifies the group coherence. Variables with low positive or negative correlation coefficients do not represent the same characteristic of the problem. To facilitate the interpretation of the problem and maintain the information provided by the test problem instances in the original format, the input data was not previously normalized. In some situations, the correlation may have been suffered small effects of any outliers or obvious unusual cases.

Table 11.3 Correlation coefficient matrix for group *Area*

	arearatioperc	***arearatioquart***	***areacompnormal***	***areacompquart***	***areacompextr***	***areamean***	***areamed***	***areastdev***
arearatioextr	0.44	0.45	0.38	0.45	0.44	0.89	0.73	0.81
arearatioperc		0.73	0.79	0.89	0.91	0.36	0.14	0.51
arearatioquart			0.82	0.80	0.67	0.38	0.17	0.51
areacompnormal				0.95	0.82	0.32	0.09	0.53
areacompquart					0.94	0.38	0.12	0.61
areacompextr						0.36	0.11	0.57
areamean							0.88	0.88
areamed								0.57

Table 11.4 Correlation coefficient matrix for group *Proportions*

	propratioextr	***propratioperc***	***propratioquart***	***propcompnormal***	***propcompquart***	***propcompextr***	***propmean***	***propmed***	***propstdev***
aspectratio	−0.21	−0.25	−0.28	−0.31	−0.25	−0.25	0.97	0.97	0.80
propratioextr		0.81	0.72	0.78	0.78	0.80	−0.13	−0.13	−0.01
propratioperc			0.92	0.95	0.97	0.97	−0.21	−0.22	−0.11
propratioquart				0.97	0.96	0.95	−0.25	−0.25	−0.15
propcompnormal					0.97	0.96	−0.27	−0.28	−0.15
propcompquart						1.00	−0.22	−0.22	−0.12
propcompextr							−0.21	−0.21	−0.12
propmean								1.00	0.91
propmed									0.90

Table 11.5 Variance explained by each component for all groups

Group	Component	Variance (%)	Cumulative (%)	Group	Component	Variance (%)	Cumulative (%)
Area	*areacomp*	62.48		Perimeter	*perimcomp*	75.36	
	areastats	24.92	87.40		*perimstats*	20.16	95.52
Dimensions	*dimcomp*	77.75		Proportions	*propcomp*	59.76	
	dimstats	18.00	95.75		*propstats*	33.39	93.15
Width dimensions	*widthdimcomp*	79.13					
	widthdimstats	16.39	95.52				

In a second moment, PCA is used individually for each group to reduce the dimensions of the problem. All the PCA are conducted with orthogonal rotation (varimax), and all requirements were reached: the ratio between the sampling size and the number of variables is greater than five to one; a minimum of 0.5 for overall Kaiser–Meyer–Olkin measure of sampling adequacy; and the Bartlett test of sphericity is statistically significant (<0.001).

As a result, two components with eigenvalues greater or equal to one were extracted for each of the first five groups, namely: $areacomp$ and $areastats$ for group $Area$; $perimcomp$ and $perimstats$ for $Perimeter$; $dimcomp$ and $dimstats$ for $Dimensions$; $propcomp$ and $propstats$ for $Proportions$; and $widthcomp$ and $widthstats$ for group $Width\ dimensions$. For the remaining group, $Other$, it was not possible to extract a small number of components, since the variables in this group are not related with each other (small correlation coefficients). As a result, a total of 19 characteristics was obtained for the 2D-SPP, 10 components and the original nine variables from group $Other$.

Table 11.5 presents the percentage of variance explained for the components individually in the first five groups. An average of 93% of the data variation is explained by the components extracted, higher than the variation of 91% obtained if the PCA was used with all the variables simultaneously.

Figure 11.2 represents the variables' projections along the extracted components for groups $Area$ and $Proportions$. In $Area$ there is a clear difference between the variables with high positive factor loadings for each component. High values for variables based on ratios (i.e. $areratioperc$) and compositions (i.e. $areacompextr$) establish the component $areacomp$. In contrast, $arestats$ is based on classical statistical measures, such as mean, median and standard deviation variables. One exceptions is $arearatioextr$, which is a ratio variable but influences more significantly the $arestats$ component.

Group $Proportions$ shows a similar behavior, with component $propcomp$ influenced by all ratio and composition variables, and the $propstats$ with all classical statistical measures and the correlated variable $aspectratio$.

Figure 11.3 presents the distribution of the 1217 test problem instances according to the components of $Area$ and $Proportions$. To a better visualization of the data, the scores for each instance are normalized to a scale between zero and one, according to the maximum and minimum values found for each component.

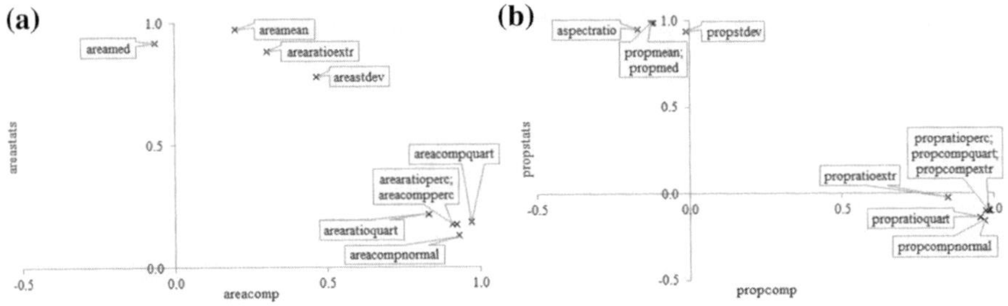

Fig. 11.2 Relations between variables and components for groups *Area* and *Proportions*

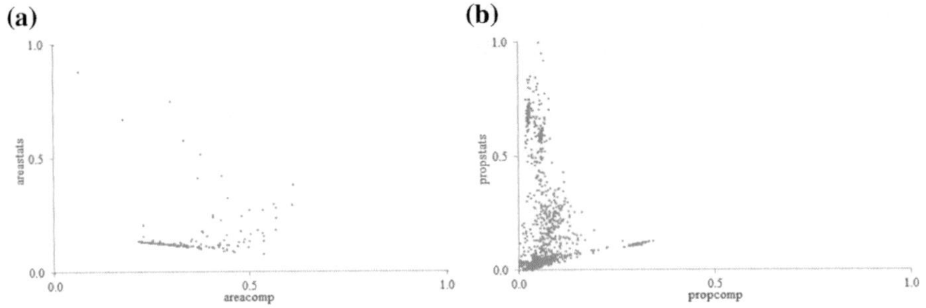

Fig. 11.3 Distribution of the 1217 test problem instances among components of groups *Area* and *Proportions*

In the first graph, the highest values for *areastats* are found for seven test problem instances *zdf* (*zdf*10–*zdf*16), a consequence of the high standard deviation and mean value between the largest and smallest items of each data instance, which can be verified through variables *areastdev* and *areamed*, respectively. Figure 11.2 shows the strong influence between these variables and *areastats*.

Instance *cx*50 has items with a strip shape, which means that the difference between the largest and smallest items dimensions is very high, and the object has a square shape, a fact that reflects the high value of the test problem instance for *areacomp* and *propcomp*. Some similar effects can also be found, with less amplitude, for *cx*100 and some *path* and *iy* instances. In *propstats* a total of four *AH* instances (*AH*14, *AH*36, *AH*50 and *AH*55) have a high dispersion between the average value of items proportion compared to the object. As a consequence, these are test problem instances that have a high degree of heterogeneity.

For both dispersion graphs, almost all test problem instances have similar values. In *Proportions* the test problem instances are located near the center of coordinates graph. In *Area* almost all instances are between 0.2 and 0.4 for *areacomp* and 0.1 and 0.2 for *areastats*. Therefore, these instances have few differences between them, leading to few variations in the variables and affecting the analysis of the characteristics of the problem.

11.5 Conclusions

In this paper, we conduct an exploratory research to identify the most significant characteristics for the two-dimensional rectangular 2D-SPP. Initially, a total of 56 variables were considered, based on parameters and characteristics found in the most used problem generators.

To reduce the complexity the PCA was used to reduce the problem dimensionality. Our analysis suggests that 19 components can explain the problem consistently. A relevant result is the similarity showed by the test problem instance from the literature.

In a second moment, it was verified that the number of test problem instances that represent the possible items and object shape variation must be improved. As a consequence, during the research the need of generation of new test problem instances was evident, to overcome the drawback of some missing characteristics in the existing test problem instances in the literature.

This study helps in the development of more efficient heuristics for solving the 2D-SPP, by providing a more accurate information on the characteristics of the problem. Future work will describe the relation between the components developed (named as features) with a dependent variable, using regression models in order to allow the prediction of the object height to be used according to the test problem instances characteristics. Also, new variables will be studied in order to complement the characterization of the problem.

Acknowledgements The second author was supported by FCT – Fundação para a Ciência e a Tecnologia within the grant SFRH/BPD/98981/2013. The research was partially supported by ERDF European Regional Development Fund through the Operational Programme for Competitiveness and Internationalisation - COMPETE 2020 Programme within project "POCI-01-0145-FEDER-006961", and by National Funds through the Portuguese funding agency, FCT – Fundação para a Ciência e a Tecnologia as part of project "UID/EEA/50014/2013".

References

1. J.E. Beasley, An exact two-dimensional non-guillotine cutting tree search procedure. Oper. Res. **33**(1), 49–64 (1985)
2. J.E. Beasley, Algorithms for unconstrained two-dimensional guillotine cutting. J. Oper. Res. Soc. 297–306 (1985)
3. B.E. Bengtsson, Packing rectangular pieces - a heuristic approach. Comput. J. **25**(3), 353–357 (1982)
4. J.O. Berkey, P.Y. Wang, Two-dimensional finite bin-packing algorithms. J. Oper. Res. Soc. 423–429 (1987)
5. A. Bortfeldt, A genetic algorithm for the two-dimensional strip packing problem with rectangular pieces. Eur. J. Oper. Res. **172**(3), 814–837 (2006)
6. A. Bortfeldt, H. Gehring, A hybrid genetic algorithm for the container loading problem. Eur. J. Oper. Res. **131**(1), 143–161 (2001)
7. E.K. Burke, G. Kendall, G. Whitwell, A new placement heuristic for the orthogonal stock-cutting problem. Oper. Res. **52**(4), 655–671 (2004)

8. N. Christofides, C. Whitlock, An algorithm for two-dimensional cutting problems. Oper. Res. **25**(1), 30–44 (1977)
9. E.P. Ferreira, J.F. Oliveira, A note on Fekete and Schepers' algorithm for the non-guillotinable two-dimensional packing problem (Technical report, FEUP, 2005)
10. N.G. Hall, M.E. Posner, Generating experimental data for computational testing with machine scheduling applications. Oper. Res. **49**(6), 854–865 (2001)
11. E. Hopper, B.C.H. Turton, An empirical investigation of meta-heuristic and heuristic algorithms for a 2D packing problem. Eur. J. Oper. Res. **128**, 34–57 (2001)
12. E. Hopper, Two-dimensional packing utilising evolutionary algorithms and other meta-heuristic methods. PhD thesis, University of Wales, Cardiff, 2000
13. S. Imahori, M. Yagiura, The best-fit heuristic for the rectangular strip packing problem: An efficient implementation and the worst-case approximation ratio. Comput. Oper. Res. **37**(2), 325–333 (2010)
14. S.C.H. Leung, D. Zhang, A fast layer-based heuristic for non-guillotine strip packing. Expert Syst. Appl. **38**(10), 13032–13042 (2011)
15. S.C.H Leung, D. Zhang, K.M. Sim, A two-stage intelligent search algorithm for the two-dimensional strip packing problem. Eur. J. Oper. Res. **215**(1):57–69 (2011)
16. E. López-Camacho, H. Terashima-Marín, G. Ochoa, S.E. Conant-Pablos, Understanding the structure of bin packing problems through principal component analysis. Int. J. Prod. Econ. **145**(2), 488–499 (2013)
17. E. López–Camacho, H. Terashima–Marin, P. Ross, G. Ochoa, A unified hyper-heuristic framework for solving bin packing problems. Expert Syst. Appl. **41**(15), 6876–6889 (2014)
18. S. Martello, D. Vigo, Exact solution of the two-dimensional finite bin packing problem. Manag. Sci. **44**(3), 388–399 (1998)
19. J.F. Oliveira, A.N. Júnior, E. Silva, M.A. Carravilla, A survey on heuristics for the two-dimensional rectangular strip packing problem. Pesqui. Oper. **36**(2):197–226 (2016)
20. E. Silva, J.F. Oliveira, G. Wäscher, 2DCPackGen: a problem generator for two-dimensional rectangular cutting and packing problems. Eur. J. Oper. Res. **237**(3), 846–856 (2014)
21. K. Smith–Miles, L. Lopes, Measuring instance difficulty for combinatorial optimization problems. Comput. Oper. Res. **39**(5), 875–889 (2012)
22. P.Y. Wang, C.L. Valenzela, Data set generation for rectangular placement problems. Eur. J. Oper. Res. **134**(2), 378–391 (2001)
23. G. Wäscher, H. Haußner, H. Schumann, An improved typology of cutting and packing problems. Eur. J. Oper. Res. **183**(3), 1109–1130 (2007)

Chapter 12
Banking Risk as an Epidemiological Model: An Optimal Control Approach

Olena Kostylenko, Helena Sofia Rodrigues and Delfim F. M. Torres

Abstract The process of contagiousness spread modelling is well-known in epidemiology. However, the application of spread modelling to banking market is quite recent. In this work, we present a system of ordinary differential equations, simulating data from the largest European banks. Then, an optimal control problem is formulated in order to study the impact of a possible measure of the Central Bank in the economy. The proposed approach enables qualitative specifications of contagion in banking obtainment and an adequate analysis and prognosis within the financial sector development and macroeconomy as a whole. We show that our model describes well the reality of the largest European banks. Simulations were done using MATLAB and BOCOP optimal control solver, and the main results are taken for three distinct scenarios.

Keywords Banking risk · Contagion spread · Epidemic approach · Optimal control

---

This work is part of first author's Ph.D., which is carried out at University of Aveiro under the Doctoral Program in Applied Mathematics MAP-PDMA, of Universities of Minho, Aveiro and Porto.

O. Kostylenko (✉) · D. F. M. Torres
Center for Research and Development in Mathematics and Applications (CIDMA), Department of Mathematics, University of Aveiro, 3810-193 Aveiro, Portugal
e-mail: o.kostylenko@gmail.com

D. F. M. Torres
e-mail: delfim@ua.pt

H. S. Rodrigues
Center for Research and Development in Mathematics and Applications (CIDMA), School of Business Studies, Polytechnic Institute of Viana do Castelo, 4930-678 Valença, Portugal
e-mail: sofiarodrigues@esce.ipvc.pt

A. I. F. Vaz et al. (eds.), *Operational Research*, Springer Proceedings in Mathematics & Statistics 223, https://doi.org/10.1007/978-3-319-71583-4_12

12.1 Introduction

Mathematical models of spread of epidemics are widely used in different fields of studies and directly related to everyone's life. They are well used in medicine for the purpose of studying an epidemiological process, analysis and forecasting outbreaks of such infections as influenza, cholera, HIV/AIDS, syphilis and others [7, 15, 20, 21]; in computer science, for studying the spread of computer viruses [6, 18]; in psychology, for studying how individuals could change their behaviour in response to others (for example, an applause, contagious yawning, the use of social networks such as MySpace, Facebook and others) [3, 16]; in marketing, for modelling a viral distribution of advertising [10, 23, 24]; in economics and finance, in order to study the financial crisis transmission, contagion in banking [5, 26, 29]; and even in science fiction, to describe the propagation of zombies attacks [14, 17, 22].

The economy field has gained a special spotlight in recent years. This is due to the global economic decline during the early 21st century. In terms of overall impact, it was the worst global downturn since Second World War, by the conclusions of the International Monetary Fund. The crisis touched every country and it has led to the financial sector's collapse in the world economy, the effects of which can be seen and felt till today [2, 4, 8].

Several scientists, from all over the world, have proposed their theories about the development of the financial crisis and methods to prevent it. However, this topic has not been studied yet completely. Indeed, there is a lack of consensus among scientists, and financial crises continue to occur from time to time. These arguments support the importance, relevance and necessity to continue to study such phenomena.

This paper is focused on studies of contagion in banking market using an epidemiological approach through mathematical modelling. Initially, it is important to understand that the contemporary economy is an open system, which is based on direct and inverse, vertical and horizontal linkages, and can be developed successfully only with effective management of these relations at both macro and micro levels [13].

Banks serve the needs of market participants, and constitute the horizontal linkages in the market environment. They do not just form their own resources, but also provide domestic accumulation of funds for the national economy development. Therefore, banks organize the monetary-lending process (moving money from lenders to borrowers) and issue banknotes. Banks are thus, constantly, faced with many risks in their activity. As an example, the credit risk (risk of not fulfilling liabilities to the credit institution by a third party) may occur during the loan or other equivalent actions, which are reflected on the balance sheet and could have the off-balance character. The occurrence of this risk can cause infection and even bank failure. That could trigger the contagion of other banks and the emergence of a banking crisis.

The risk of contagion in the financial sector presents a serious threat for a country's economy. Failure of one particular institution in the financial sector threatens the stability of many other institutions. This situation is called a systemic risk [11].

Moreover, the spread of one type of crisis is able to initiate the development of other types of crises as well. For example, the external debt crisis could undermine the stability of banks; problems in the banking sector can launch a debt crisis; and the banking crisis, in its turn, often precedes currency crisis. Thus, the risky lending and loan defaults, generally, precede crisis in banking [25]. Banking crises often are accompanied by panic among population. It usually happens when many banks suffer runs simultaneously, as people have suspicions and mistrust to banks, and they suddenly try to withdraw their money from bank accounts.

Banks are directly at the centre of the financial system and it is easy to understand that a crisis in banking is one of the most serious type of financial crisis. Therefore, it is very important to study how the crisis spreads in banking market, to find its basic laws and methods to control it. A good understanding of their propagation mechanism makes possible to find and propose suitable policy interventions, which can most effectively reduce their contagious spread. This is the main goal of our research. Namely, we show that the application of epidemiological models is able to describe well the nature and character of the contagion spread and its behaviour over time. Our analysis identifies which measures to adopt and when they must be taken in order to prevent effects and serious negative consequences for a particular bank, and for economy as a whole.

The paper is structured as follows. In Sect. 12.2, we introduce the main bank concepts into the epidemic model. Various scenarios of contagion and the results of simulations are presented in Sect. 12.3. An optimal control problem is then formulated in Sect. 12.4, where the control simulates the European Central Bank. In Sect. 12.5, the main conclusions are carried out.

12.2 The SIR Mathematical Model

The availability of interactions between banks in a financial system confirms the possibility of a contagion occurrence. Contagion, by definition, refers to the idea that any type of financial crisis may spread from one subject (financial institution) to another. As an example, if a large number of bank customers are withdrawing their funds, then it provokes a bank run that may spread from a few banks to many others within the country, or even spreading from one country to another. Currency crisis is one type of financial crisis, which also may spread from one country to another. Even sovereign defaults and stock market crashes spread across countries. Such processes of contagion in economy are very similar to the disease propagation in a population (from one individual to another). This similarity allows us to consider contagion in such type of financial sectors, as banking, using the same mathematical models of infection spreading as used in epidemiology.

There are many kinds of epidemiological models of spread of infection, which differ in the different assumptions about the disease's nature and character, about the structure of population that is under consideration, and other constraints on which the model is based. In our work, we assume that contagion can be transmitted from an infected bank to another one, which has not yet been infected, and after recovery the bank produces immunity for a long time. Our assumptions about the process of contagion spreading in the banking sector are very similar to the characteristics of many childhood diseases, including chickenpox, measles, rubella and others, after which a strong immunity is produced. In order to simulate an epidemiological process of such diseases, there is a well-known suitable model, called SIR, which was first introduced in the works of Kermack and McKendrick [12]. The model is named SIR for the reason that population is divided into three classes: class S of susceptible, those who are susceptible to the disease; class I of infected, those who are ill/infected; class R of recovered, those who are recovered and have become immune to the disease.

Mathematically, the SIR model is a multi-parameter system of ordinary differential equations, where each equation defines the current state of health of an investigated object. This model is relevant and frequently used in current research. Our work is also based on its application to the transmission of a contagious disease between banks. The model must be, however, adapted to the banking sector, in order to have a deeper understanding of how the process of infection transmission is carried out, and especially how banks, as financial institutions, become contagious.

The earlier formulated rules of bank transitions from one condition to another lead to the following system of differential equations:

$$\begin{cases} \dfrac{dS(t)}{dt} = -\beta S(t)I(t), \\ \dfrac{dI(t)}{dt} = \beta S(t)I(t) - \gamma I(t), \\ \dfrac{dR(t)}{dt} = \gamma I(t), \end{cases} \tag{12.1}$$

$t \in [0, T]$, subject to the initial conditions

$$S(0) = S_0, \quad I(0) = I_0, \quad R(0) = R_0. \tag{12.2}$$

The first equation defines the number of banks that leave the group S by becoming a target of market speculation at time t; the second equation defines the number of contagious banks at time t; and the third equation defines the number of banks that have recovered from the crisis at time t, where $t \in [0, T]$, $T > 0$.

Model (12.1)–(12.2) depends on two parameters. Parameter β denotes the contagion spreading rate. It represents the strength of contagion, where a susceptible bank, that is in contact with a single infected bank, changes its state to infected with probability β. The second parameter, γ, denotes the speed of recovery. It represents the resistance to contagion. Therefore, an infected bank changes its state to recover with probability γ.

A susceptible bank from group S can obtain contagion if it has a relationship with a contagious bank from group I, and if it has not enough money in reserve to cover possible risk losses. In other words, the bank may be contaminated, if it has not enough strong "health" in order to resist an epidemic. According to the standard SIR methodology, we are assuming that all banks are similar, and both parameters β and γ are constant for the entire sample. However, in reality, these data are unique for each bank and depend on the force of contagion of the affected bank and the financial stability of the susceptible. Dynamics of the strength of banks is not taken into consideration in the present study.

It is assumed that N, the number of institutions during the period of time under study, is fixed throughout the contamination time (in our simulations, $N = 169$ — see Sect. 12.3). It means that $S(t) + I(t) + R(t) = N$ for all $t \in [0, T]$. Moreover, we assume that the initial conditions (12.2) at the beginning of the period under study satisfy

$$S(0) \gg I(0) > 0 \quad \text{and} \quad R(0) = 0. \tag{12.3}$$

12.3 SIR Model Simulations with Real Data

The contagion and recovery propagation rates used in our paper, β and γ, follow the statistics and empirical findings of Philippas, Koutelidakis and Leontitsis [19]. The statistical data is taken, with respect to the year 2012, for the 169 largest European banks with total assets over than 10 bn Euro. The values of parameters β and γ were calculated in [19] by assuming that all banks of a country are infected; then, using Monte Carlo simulations, the parameters β and γ were tracked for each bank in a country; and aggregated in such a way they represent the average over simulations, with respect to the country in which the bank originates. Therefore, using the information that the total number of banking institutions (N) is equal to 169, and assuming that only one bank is contagious at the initial time, the values of initial conditions for the SIR model were obtained: $S(0) = 168$, $I(0) = 1$, and $R(0) = 0$. The parameters β and γ are presented for each bank with respect to the country in which the bank originates.

In our work, in order to present contrast scenarios of contagion spreading among European banks, we have chosen three countries that have completely different values of parameters β and γ:

1. in the first scenario, the initially contagious bank is located in Portugal;
2. in the second scenario, the group I of infected consists only of a bank from Spain;
3. while in the third scenario the contamination begins from a United Kingdom's bank.

Such countries were chosen from [19]. However, it should be mentioned that our scenarios are different from those of [19]. Indeed, in [19] they begin with one or more random banks among three different levels of assets. In contrast, here we do not look to the assets level of the banks and, instead, we chose countries by their contagion and recovery rates, respectively β and γ. Moreover, in our cases all the three scenarios begin with exactly one random infected bank in the chosen country, while in [19] more than one is allowed.

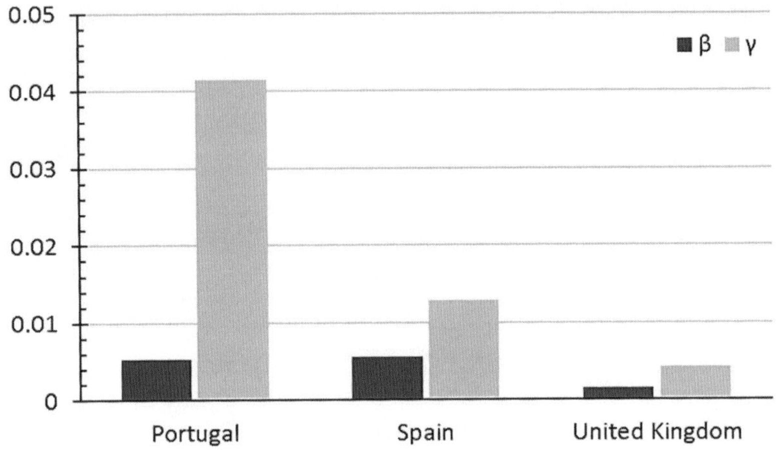

Fig. 12.1 Summary statistics for β and γ based on the data of [19]

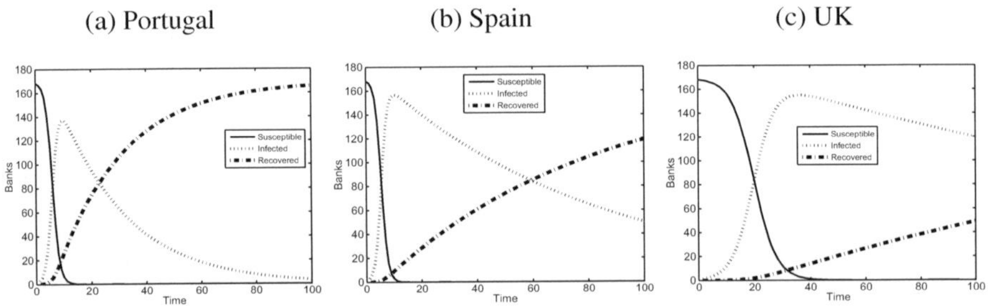

Fig. 12.2 The SIR contagion risk model (12.1)–(12.2) with parameters β and γ as in Fig. 12.1, $S(0) = 168$, $I(0) = 1$, $R(0) = 0$ and $T = 100$

Data of the contagion spreading rate and the speed of recovery for these three scenarios are shown in Fig. 12.1 and the behaviour of the contagion risk model (12.1)–(12.2), during the same time scale of $T = 100$ days, is shown in Fig. 12.2.

In the first scenario, contagion spreads rapidly. In a short span of time, contagion reaches its peak and then swiftly decreases. The recovery process takes fast. See Fig. 12.2a. Regarding the second scenario, Fig. 12.2b demonstrates that contamination occurs rapidly as in the first scenario. The contagion affects a large number of banks but the recovery process takes longer. In Fig. 12.2c, we see what happens in the third scenario: the spread of contagion occurs in a longer period of time and, consequently, the recovery process goes slowly too. The graphs of Fig. 12.2 show that the processes of contagion take place in different ways, depending on the country where it begins.

Figure 12.2 also shows that, in case of the first scenario, contagion has almost reached the contagion-free equilibrium ($I(T) = 0$) at the end of 100 days. However, $T = 100$ is not enough to reach the contagion-free equilibrium for the second and

(a) Portugal, $T = 365$ (b) Spain, $T = 450$ (c) UK, $T = 1200$

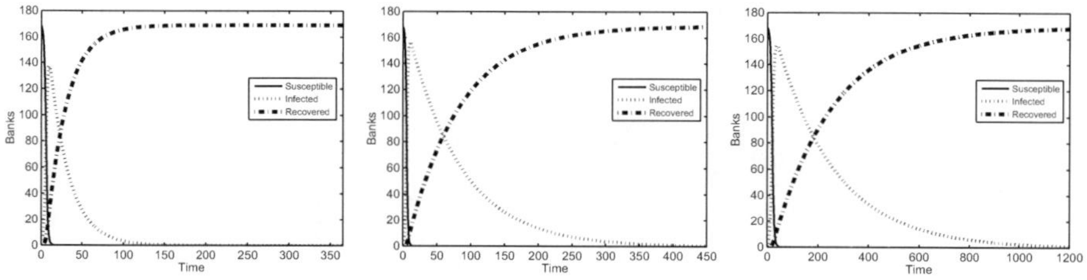

Fig. 12.3 The SIR contagion risk model (12.1)–(12.2) with parameters β and γ as in Fig. 12.1, $S(0) = 168$, $I(0) = 1$ and $R(0) = 0$: illustration of the time needed to achieve the contagion-free equilibrium

third scenarios. In those cases, it makes sense to increase the time. The results of such simulations are reflected in Fig. 12.3.

As seen from Fig. 12.3a, in case of bank contagion starting in Portugal, less than in half a year the contagion spreading stops. Figure 12.3b demonstrates that a spreading of contagion starting in Spain will only stop after a full year. The third scenario is the most severe: the contagion will disappear and the banks will be recovered only after three years (see Fig. 12.3c).

The reason behind the differences found, for the three scenarios considered, rely in the different economic relevance, in the global banking market, of the countries where contagion begins. So, if one of the large and important banks of United Kingdom will be the starting point in the spread of contagion, then the global banking market will experience serious problems, the negative effects of which will continue during a long period of time. In other words, United Kingdom is one of the main financial markets to have into account at European level. On the other hand, Portugal has a less powerful and influential economy on an European scale.

12.4 Optimal Control

Given the horizontal linkages in market environment, all banks are equally needed to be under financial supervision. Such supervision is necessary in order to prevent global contamination and avoid serious consequences due to spread of contagion between banks. This partially explains why the Central European Bank exists. The Central Bank is a supervisory competent authority that implements vertical connections among interacting commercial banks. Since one of the main goals is to avoid the wide dissemination of a contagion, we pose the following optimal control problem:

$$\min \mathscr{J}[I(\cdot), u(\cdot)] = I(T) + \int_0^T bu^2(t)dt \longrightarrow \min \tag{12.4}$$

subject to the control system

$$\begin{cases} \dfrac{dS(t)}{dt} = -\beta S(t)I(t), \\ \dfrac{dI(t)}{dt} = \beta S(t)I(t) - \gamma I(t) - u(t)I(t), \\ \dfrac{dR(t)}{dt} = \gamma I(t) + u(t)I(t), \end{cases} \tag{12.5}$$

where $S(0) = S_0$, $I(0) = I_0$, $R(0) = R_0$, satisfying (12.3), and the weight $b > 0$, are fixed. In our simulations, the initial values of the state variables are given as discussed in Sect. 12.3, while the weight b, associated with the cost of control measures, is taken to be 1.5, motivated by the value of possible recapitalization with state funds considered in [19]. The first term of the objective functional (12.4) reflects the fact that the control organization should take care about the number of contagious institutions $I(T)$ at the final time T. The integral represents the general cost of financial assistance necessary to prevent the spread of contagion and economic decline in the period $[0, T]$. The control $u(\cdot)$ is a Lebesgue function with values in the compact set $[0, 1]$: $0 \leq u(t) \leq 1$, with $u(t)$ denoting the rate at which assistance will be provided to contagious banks, that is, it is the ratio between the financial support from the Central Bank at time t and the financing needed by the banks at that time. In this way, $u(t) = 1$ means full support from the Central Bank at time t (all money needed by the banks is being covered by Central Bank), while $u(t) = 0$ means no financial lending or recapitalization from the Central Bank at time t.

In order to use the BOCOP optimal control solver [1, 9], the optimal control problem in Bolza form (12.4)–(12.5) is rewritten in the following equivalent Mayer form:

$$\min \mathscr{J}[S(\cdot), R(\cdot), Y(\cdot)] = N - S(T) - R(T) + Y(T) \longrightarrow \min \tag{12.6}$$

subject to

$$\begin{cases} \dfrac{dS(t)}{dt} = \beta S^2(t) + \beta S(t)(R(t) - N), \\ \dfrac{dR(t)}{dt} = \gamma(N - S(t) - R(t)) + u(t)(N - S(t) - R(t)), \\ \dfrac{dY(t)}{dt} = bu^2(t), \\ S(0) = S_0, \quad R(0) = 0, \quad Y(0) = 0, \\ u(t) \in [0, 1]. \end{cases} \tag{12.7}$$

The optimal control problem (12.6)–(12.7) is approximated by BOCOP into a nonlinear programming problem (NLP) using a standard time discretization. The NLP problem is then solved by the open-source Ipopt optimization solver [27], using

(a) without control (b) with optimal control (c) the extremal control $u(t)$

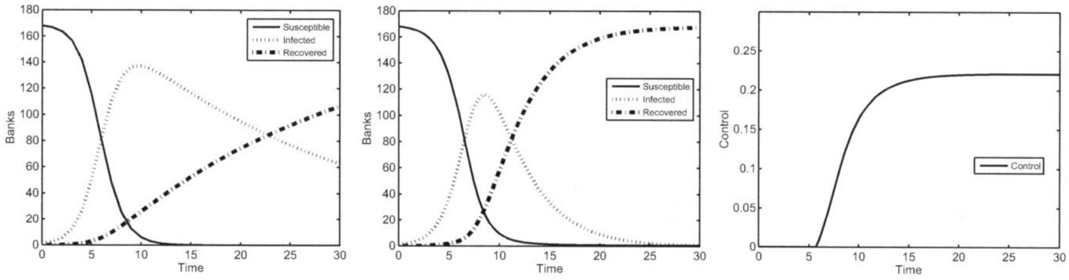

Fig. 12.4 Contagion risk from Portugal (Scenario 1) with and without optimal control

(a) without control (b) with optimal control (c) the extremal control $u(t)$

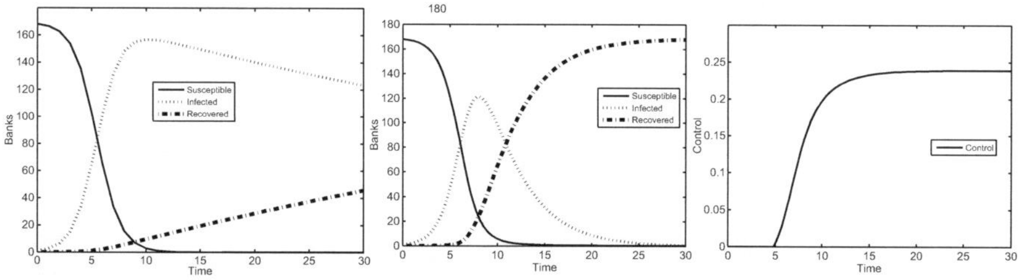

Fig. 12.5 Contagion risk from Spain (Scenario 2) with and without control

sparse exact derivatives computed by the Adol-C (Automatic Differentiation by OverLoading in C++) package [28]. Figures 12.4, 12.5 and 12.6 show the results of simulations of our SIR bank contagion risk model, with and without optimal control, for a period of 30 days ($T = 30$) for the first, second and third scenarios of Sect. 12.3, respectively.

Figure 12.4a shows that if we have no interest to stop contagiousness or just do not want to spend any money on it, that is, $u(t) \equiv 0$, then, in case of the first scenario, the number of contagious banks at final time $T = 30$ will be equal to 64. If the control $u(t)$ is chosen as the solution of our optimal control problem, then in one month the number of contagious banks will become smaller than $I = 2$, see Fig. 12.4b.

In case of the second scenario, Fig. 12.5a shows that the number of contagious without control is equal to 124. In contrast, the number of contagious banks at time $T = 30$ using optimal control, taking into account the costs associated with the control interventions during all the period [0, 30], is equal to 2 (see Fig. 12.5b). On the other hand, after one month, the number of banks that already recovered from the risks are less than fifty without control; while with optimal control measures this number more than tripled.

Finally, when the contagion begins from United Kingdom (third scenario), the number of contagious banks at final time $T = 30$ is 149 without control (Fig. 12.6a) with that number decreasing to less than 4 when the Central Bank applies optimal control (see Fig. 12.6b).

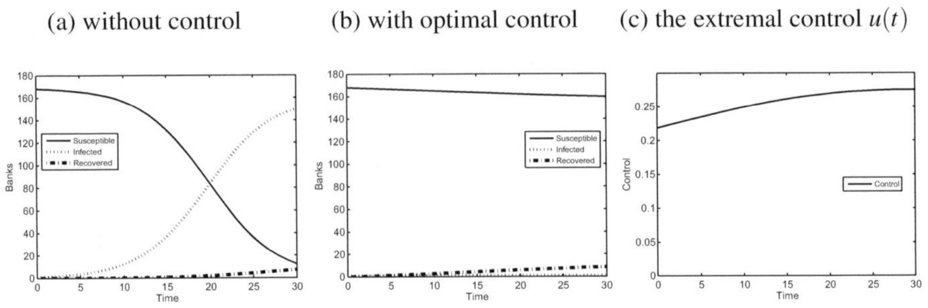

Fig. 12.6 Contagion risk from UK (Scenario 3) with and without control

Table 12.1 Number of contagious banks $I(T)$, $T = 30$ days

Scenario #	No control measures	With optimal control
Scenario 1 (bank contagion starting in Portugal)	64	2
Scenario 2 (bank contagion starting in Spain)	124	2
Scenario 3 (bank contagion starting in UK)	149	4

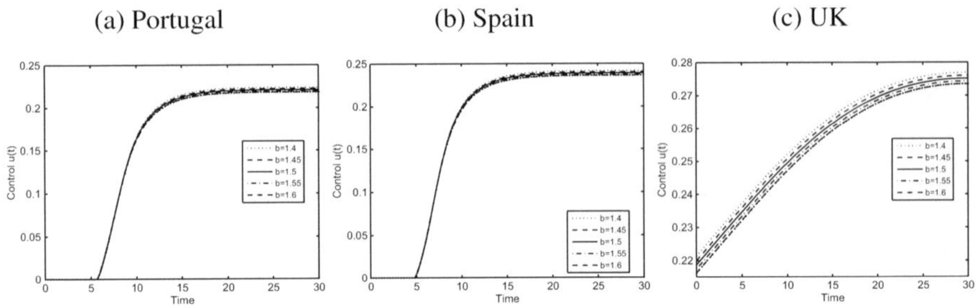

Fig. 12.7 The optimal control $u(t)$ for different values of the weight b

The results of the three scenarios just reported, with respect to the number of contagious banks $I(T)$, are summarized in Table 12.1 for a better comparison.

It is worth to mention that the qualitative results of the paper do not change with the particular choice of the weight b: the extremal curves change continuously with the change of parameter b. For illustrative purposes, we show in Fig. 12.7 the optimal control $u(t)$, $t \in [0, 30]$ days, for different values of the weight b. We see that an increase of the value of the parameter b corresponds to a decrease of the control u, meaning that the financial support from the Central Bank diminishes when we penalize the use of the control by increasing the value of b. The role of parameter b is more visible in Scenario 3 than in Scenarios 1 and 2.

12.5 Conclusions

We investigated the dynamic behaviour of contagiousness in the banking sector, at the macroeconomic level, using a SIR epidemic model and optimal control. The features of contamination were shown to depend on the parameter values of the transmission and recovery rates, as well as on the country for which the process of infection begins. The scale of negative consequences for three different scenarios of bank risk contagion were identified. Banks at risk from countries with greater financial influence in Europe tend to propagate more severely the banking crisis; at the same time, the recovery period is longer. An optimal control problem was proposed, in order to reduce the number of contagious banks, to prevent large-scale epidemic contagiousness, and to avoid serious financial and economic consequences.

Acknowledgements This research was supported in part by the Portuguese Foundation for Science and Technology (FCT – Fundação para a Ciência e a Tecnologia), through CIDMA – Center for Research and Development in Mathematics and Applications, within project UID/MAT/04106/2013. Kostylenko is also supported by the Ph.D. fellowship PD/BD/114188/2016. We are very grateful to the authors of [19] for providing us the parameter values that they have obtained in their work; and to two anonymous referees for valuable remarks and comments, which significantly contributed to the quality of the paper.

References

1. F. Bonnans, P. Martinon, D. Giorgi, V. Grelard, S. Maindrault, O. Tissot, *BOCOP 2.0.4 User Guide*. (Inria-Saclay, 2015)
2. R.J. Caballero, On the macroeconomics of asset shortages. NBER Working Paper No. 12753 (2006)
3. J. Cannarella, J.A. Spechler. Epidemiological modeling of online social network dynamics. Preprint (2014)
4. B. Davis. What's a global recession? The Wall Street Journal (2013)
5. A. Garas, P. Argyrakis, C. Rozenblat, M. Tomassini, S. Havlin, Worldwide spreading of economic crisis. N. J. Phys. **12**(11), 113043 (2010)
6. X. Han, Q. Tan, Dynamical behavior of computer virus on internet. Appl. Math. Comput. **217**(6), 2520–2526 (2010)
7. M.B. Hooten, J. Anderson, L.A. Waller, Assessing north american influenza dynamics with a statistical SIRS model. Spat. Spatio-temporal Epidemiol. **1**(2–3), 177–185 (2010)
8. International Monetary Fund: World Economic Outlook: Crisis and Recovery. World Economic and Financial Surveys (2009)
9. Inria Saclay Team Commands. BOCOP: an open source toolbox for optimal control. http://bocop.org (2017)
10. K. Kandhway, J. Kuri, How to run a campaign: optimal control of SIS and SIR information epidemics. Appl. Math. Comput. **231**, 79–92 (2014)
11. G.G. Kaufman, K.E. Scott, What is systemic risk, and do bank regulators retard or contribute to it? The Indep. Rev. **7**(3), 371–391 (2003)
12. W.O. Kermack, A.G. McKendrick, A contribution to the mathematical theory of epidemics. R. Soc. Lond. A **115**(772), 700–721 (1927)
13. O.O. Kostylenko, O.M. Nazarenko, Modelling and identification of macroeconomic system dynamic interindustry balance. Mech. Econ. Regul. **1**(63), 76–86 (2014)

14. H.P. Langtangen, K.A. Mardal, P. Røtnes, in *Escaping the Zombie Threat by Mathematics*, Chapter18 (University of Chicago Press, 2013)
15. D. Li, W. Ma, Asymptotic properties of a HIV-1 infection model with time delay. J. Math. Anal. Appl. **335**(1), 683–691 (2007)
16. R.P. Mann, J. Faria, D.J.T. Sumpter, J. Krause. The dynamics of audience applause. J. R. Soc. Interface **10**(85), 20130466 (2013)
17. P. Munz, I. Hudea, J. Imad, R.J. Smith, in *When zombies attack!: mathematical modelling of an outbreak of zombie infection*. (Nova Science Publishers, 2009), p. 133–150
18. Y. Öztürk, M. Gülsu, Numerical solution of a modified epidemiological model for computer viruses. Appl. Math. Model. **39**(23–24), 7600–7610 (2015)
19. D. Philippas, Y. Koutelidakis, A. Leontitsis, Insights into european interbank network contagion. Manag. Financ. **41**(8), 754–772 (2015)
20. A. Rachah, D.F.M. Torres, Modeling, dynamics and optimal control of Ebola virus spread. Pure Appl. Funct. Anal. **1**(2), 277–289 (2016)
21. A. Rachah, D.F.M. Torres, Optimal control strategies for the spread of Ebola in West Africa. J. Math. Anal. **7**(1), 102–114 (2016)
22. H.S. Rodrigues, Application of SIR epidemiological model: new trends. Intern. J. Appl. Math. Inf. **10**, 92–97 (2016)
23. H.S. Rodrigues, M.J. Fonseca, Viral marketing as epidemiological model, in *Proceedings of the 15th International Conference on Computational and Mathematical Methods in Science and Engineering*, p. 946–955 (2015)
24. H.S. Rodrigues, M.J. Fonseca, Can information be spread as a virus? Viral marketing as epidemiological model. Math. Methods Appl. Sci. **39**(16), 4780–4786 (2016)
25. M. Simkovic, Competition and crisis in mortgage securitization. Indiana Law J. **88**(1), 213–217 (2013)
26. M. Toivanen, Contagion in the interbank network: An epidemiological approach. Bank of Finland Research Discussion Paper No. 19 (2013)
27. A. Wächter, L.T. Biegler, On the implementation of an interior-point filter line-search algorithm for large-scale nonlinear programming. Math. Program. **106**(1), 25–57 (2006)
28. A. Walther, A. Griewank, Getting started with ADOL-C, in *Combinatorial Scientific Computing, Chapman-Hall CRC Computational Science* ed. by U. Naumann, O. Schenk, pp. 181–202 (2012)
29. Y. Wu, Reputation risk contagion and control of rural banks in China based on epidemic model. Comput. Model. New Technol. **18**(7), 229–235 (2014)

Chapter 13
A Generator of Nonregular Semidefinite Programming Problems

Eloísa Macedo and Tatiana Tchemisova

Abstract Regularity is an important property of optimization problems. Various notions of regularity are known from the literature, being defined for different classes of problems. Usually, optimization methods are based on the optimality conditions, that in turn, often suppose that the problem is regular. Absence of regularity leads to theoretical and numerical difficulties, and solvers may fail to provide a trustworthy result. Therefore, it is very important to verify if a given problem is regular in terms of certain regularity conditions and in the case of nonregularity, to apply specific methods. On the other hand, in order to test new stopping criteria and the computational behaviour of new methods, it is important to have an access to sets of reasonably-sized nonregular test problems. The paper presents a generator that constructs nonregular Semidefinite Programming (SDP) instances with prescribed irregularity degrees and a database of nonregular test problems created using this generator. Numerical experiments using popular SDP solvers on the problems of the database are carried out and permit to conclude that the most popular SDP solvers are not efficient when applied to nonregular problems.

Keywords Semidefinite programming · Regularity · Constraint qualification
Good behaviour · Generator of nonregular sdp problems

13.1 Introduction

Semidefinite programming is an active area of research due to its many applications in combinatorial, convex, and robust optimization, computational biology, systems and control theory, sensor network location, and data analysis, among others [1].

E. Macedo
TEMA, Department of Mechanical Engineering, University of Aveiro, Aveiro, Portugal
e-mail: macedo@ua.pt

T. Tchemisova (✉)
CIDMA, Department of Mathematics, University of Aveiro, Aveiro, Portugal
e-mail: tatiana@ua.pt

A. I. F. Vaz et al. (eds.), *Operational Research*, Springer Proceedings
in Mathematics & Statistics 223, https://doi.org/10.1007/978-3-319-71583-4_13

SDP refers to convex optimization problems where a linear function is minimized subject to constraints in the form of linear matrix inequalities (LMIs).

The most efficient methods for solving SDP problems are based on the first-order necessary optimality conditions, also called Karush–Kuhn–Tucker-type (KKT) conditions [23], which in turn are derived under some special assumptions on the feasible set of the problem, the regularity conditions [5, 10, 23]. Regularity plays an important role in characterizing optimality of feasible solutions, guaranteeing the efficiency of numerical methods and stability of solutions. There exist different notions of regularity, such as Constraint Qualification (CQ) [1, 18, 23], well-posedness [7, 11], or good behaviour in the sense of Pataki [16], which were recently proved to be closely related to each other [15, 16].

The Slater condition, which consists in existence of strictly feasible solutions, is a widely used CQ in SDP, and many authors assume in their works that this condition holds. However, in practice, there are many SDP problem instances that fail to satisfy the Slater condition, i.e., are nonregular (e.g., [3, 7, 8, 11, 21]). In the absence of regularity, theoretical and numerical difficulties may occur. Although in these cases some special regularization techniques (e.g., preprocessing [3], presolving [9], self-dual embedding [5]) can be applied, in practice, the SDP solvers may still run into numerical difficulties. In fact, the popular SDP solvers do not check the regularity of problems, consequently, trustworthiness of results is not guaranteed while solving nonregular problems. Therefore, it is very important to verify if a given SDP problem is regular in some sense before passing it to a solver. In [14, 15], we presented a detailed description of two numerical procedures to check regularity of SDP problems in terms of the fulfilment of the Slater condition and conducted several numerical experiments, which show that many problems from the known SDPLIB database [2] are nonregular. Nevertheless, this database has been widely used for testing the performance and robustness of SDP software, which works under assumption of problem's regularity.

As it was pointed out in [7, 17], it would be important to have a library of problems with a particular structure, irregular or even infeasible instances, to develop and test new stopping criteria for SDP methods, and create more efficient solvers. In [13], an algorithm for generating infeasible SDP instances was presented. A generator of *hard* SDP instances, for which the strict complementary fails, was proposed in [22]. In this light, it is also important to create libraries of nonregular SDP problems, or develop procedures that permit to generate nonregular SDP problem instances to evaluate the computational behaviour of new methods, in particular those specially conceived for solving the nonregular problems.

The main purpose of the paper is to describe an algorithm for generating SDP problems failing the Slater condition, and present a generator of nonregular SDP problem instances that was implemented in MATLAB. We also present a collection of nonregular SDP instances with a particular structure, encoded in standard format.

The paper is organized as follows. Section 13.1 hosts the Introduction. In Sect. 13.2, some notation and definitions are introduced. The regularity notions, such as the Slater condition and good behaviour, as well as their relationships are discussed in Sect. 13.3. In Sect. 13.4, we present procedures to verify regularity of SDP problems in terms of the fulfilment of the Slater condition and determine the level of their (ir)regularity. Section 13.5 is devoted to a particular class of nonregular SDP problems. A generator of nonregular SDP problems with prescribed irregularity degree is presented. The collection of nonregular SDP test problems called NONREGSDP and numerical experiments are described in Sect. 13.6. The final Sect. 13.7 contains conclusions.

13.2 Linear Semidefinite Programming Problem

Given $s \in \mathbb{N}$, $\mathscr{S}(s)$ denotes the space of the $s \times s$ real symmetric matrices equipped with the trace inner product given by $tr(\mathbf{AB}) = \sum_{i=1}^{s} \sum_{j=1}^{s} a_{ij} b_{ji}$, for $\mathbf{A}, \mathbf{B} \in \mathscr{S}(s)$, and $\mathscr{P}(s) \subset \mathscr{S}(s)$ denotes the cone of $s \times s$ positive semidefinite symmetric matrices. Consider the following SDP problems:

$$\min_{x \in \mathbb{R}^n} c^T x \text{ s.t. } \mathscr{A}(x) \preceq 0, \tag{13.1}$$

$$\max_{\mathbf{Z} \in \mathscr{S}(s)} \text{tr}\,(\mathbf{A}_0 \mathbf{Z}) \text{ s.t. } -\text{tr}\,(\mathbf{A}_i \mathbf{Z}) = c_i, \forall i = 1, \ldots, n, \ \mathbf{Z} \succeq 0, \tag{13.2}$$

where x is the primal vector variable, $\mathbf{Z}$ is the dual matrix variable, $c \in \mathbb{R}^n$ and $\mathscr{A}(x)$ is a matrix-valued function defined as $\mathscr{A}(x) := \sum_{i=1}^{n} \mathbf{A}_i x_i + \mathbf{A}_0$, where $\mathbf{A}_i \in \mathscr{S}(s)$, $i = 0, 1, \ldots, n$.

Without loss of generality, we can assume that the matrices $\mathbf{A}_i$, $i = 1, ..., n$, are linearly independent. Problem (13.1) is called primal, and (13.2) is its dual. We denote the primal and dual feasible sets of (13.1) and (13.2), by $\mathscr{X} = \{x \in \mathbb{R}^n : \mathscr{A}(x) \preceq 0\}$ and $\mathscr{Z} = \{\mathbf{Z} \in \mathscr{P}(s) : -\text{tr}\,(\mathbf{A}_i \mathbf{Z}) = c_i, i = 1, \ldots, n\}$, respectively.

13.3 Regularity in Semidefinite Programming

The most common regularity notions are given in terms of some special conditions on the feasible sets or on the constraint functions. Constraint qualifications are special conditions that guarantee that the first-order necessary optimality conditions – the KKT optimality conditions – are satisfied. An optimization problem is often called regular if certain CQ is satisfied [10], and nonregular, otherwise.

Duality results are fairly subtle in SDP, requiring regularity of the problem in some sense. It is well known that as well as in Linear Programming, in SDP the weak duality holds for any pair of primal and dual feasible solutions $x \in \mathscr{X}$ and $\mathbf{Z} \in \mathscr{Z}$ of problems (13.1) and (13.2), i.e., $p = c^T x \geq \text{tr}(\mathbf{A}_0 \mathbf{Z}) = d$. Let p^* and d^*

denote the optimal values of the SDP problems (13.1) and (13.2), respectively. The difference $p^* - d^*$ is called duality gap. In SDP, to guarantee the vanishing of the duality gap some additional assumptions have to be made. An often used sufficient condition to ensure zero duality gap is the existence of a strictly feasible solution. This condition is called strict feasibility or the Slater regularity condition [5].

Definition 13.1 The constraints of the problem (13.1) satisfy the Slater (regularity) condition if the interior of its feasible set $\mathscr{X}$ is nonempty, i.e., $\exists\, \bar{x} \in \mathbb{R}^n : \mathscr{A}(\bar{x}) \prec 0$.

If assume that the primal optimal value is finite and the Slater condition holds, then strong duality holds, i.e., the duality gap is zero, and the dual optimal value can be attained [5]. The strong duality plays an important role in the numerical solving of SDP problems. However, it can fail in the absence of the Slater condition and either a dual optimal solution may not exist or the duality cap may be not zero. Therefore, solvers may run into numerical difficulties and not be able to provide trustworthy solutions.

In the literatures, there are other notions of regularity for SDP problems, such as well-posedness and good behaviour. The well-posedness of a problem is related to its behaviour in terms of (in)feasibility under small perturbations [5, 7, 11]. The good behaviour of a SDP problem is related to the fulfilment of the strong duality property [16]. More specifically, assuming that a SDP problem is feasible, the following definition was introduced in [16].

Definition 13.2 The SDP problem in the form (13.1) is said to be well-behaved, if strong duality holds for all objective functions. Otherwise, the problem is said to be badly-behaved.

A SDP problem is well-behaved in the sense of Pataki [16] if strong duality holds, which can be ensured if a regularity condition, such as the Slater condition, holds. Therefore, the good behaviour of a SDP problem is closely related to the Slater condition. The following result was proved in [16] (Corollary 1).

Proposition 13.1 *If the constraints of the SDP problem* (13.1) *satisfy the Slater condition, then the problem is well-behaved.*

On the basis of this proposition, we can conclude that if the SDP problem (13.1) is badly-behaved, then it does not satisfy the Slater condition.

13.4 Testing and Measuring Regularity in SDP

Different approaches to verify regularity of SDP problems have been proposed in the literature. In terms of well-posedness, two characterizations are known, one based on the Renegar condition number [7], and another based on a rigorous upper bound on the primal optimal value [11]. In [16], a characterization of good behaviour in the sense of Pataki is described and we will briefly discuss it later. In what follows, we suggest our original approach to verify regularity in terms of the fulfilment of the Slater condition.

13.4.1 Subspace of Immobile Indices and Irregularity Degree of SDP Problems

The following definition was given in [12].

Definition 13.3 Given the linear SDP problem (13.1), the subspace of $\mathbb{R}^s$ defined by

$$\mathcal{M} := \left\{ l \in \mathbb{R}^s : l^T \mathcal{A}(x) l = 0, \forall x \in \mathcal{X} \right\} \tag{13.3}$$

is called the subspace of immobile indices.

On the basis of this definition, and the results in [12], we can prove the following theorem.

Theorem 13.1 *The SDP problem* (13.1) *satisfies the Slater condition if and only if the subspace of immobile indices* $\mathcal{M}$ *is null, i.e.,* $\mathcal{M} = \{0\}$.

Proof First, let us reformulate the linear SDP problem (13.1) in the equivalent form:

$$\min c^T x \text{ s.t. } l^T \mathcal{A}(x) l \leq 0, \forall l \in L := \{ l \in \mathbb{R}^s : \|l\|_2 = 1 \}, \tag{13.4}$$

where the set L is an (infinite) index set. This problem has an infinite number of constraints, and thus, is a convex Semi-Infinite Programming (SIP) problem. Notice that the feasible sets of (13.4) and (13.1) coincide: $\left\{ x \in \mathbb{R}^n : l^T \mathcal{A}(x) l \leq 0, \forall l \in L \right\} = \{ x \in \mathbb{R}^n : \mathcal{A}(x) \preceq 0 \} = \mathcal{X}$.

It was proved in [12] that the SIP problem (13.4) satisfies the Slater condition, i.e., $\exists \bar{x} \in \mathcal{X} : l^T \mathcal{A}(\bar{x}) l < 0, \forall l \in L$, if and only if the *set of immobile indices* given by $L^* = \left\{ l \in L : l^T \mathcal{A}(x) l = 0, \forall x \in \mathcal{X} \right\}$ is empty. Evidently, $L^* = L \cap \mathcal{M}$ and then, the subspace $\mathcal{M}$ of immobile indices is null if and only if L^* is empty.

Since the problems (13.1) and (13.4) are equivalent, then they satisfy or not the Slater condition, simultaneously. Therefore, (13.1) satisfies the Slater condition if and only if $\mathcal{M}$ is null. ■

The connection established between the subspace of immobile indices and the Slater condition permits us to introduce a measure of nonregularity (or irregularity) for SDP problems, which we will call here the *irregularity degree* of a SDP problem.

Definition 13.4 The dimension of a basis of the subspace $\mathcal{M}$ of immobile indices for the SDP problem (13.1), denoted by s^*, is called irregularity degree of this problem.

This definition permits to classify SDP problems in the form (13.1) taking into account the dimension s^* of the subspace $\mathcal{M}$ as follows:

- if $s^* = 0$, then the problem is regular, i.e., the Slater condition holds;
- if $s^* = 1$, then the problem is nonregular, with minimal irregularity degree;
- if $s^* = s$, then the problem is nonregular, with maximal irregularity degree.

In fact, for a given SDP problem, the nonvanishing dimension of a basis of the subspace of immobile indices can be considered as a *certificate* of nonregularity, i.e., it proves the failure of the Slater condition.

13.4.2 Testing Regularity and Determining the Irregularity Degree

In [12], an algorithm DIIS (Determination of the Immobile Index Subspace) was proposed to find a basis of the subspace $\mathscr{M}$. Here, we will show that the DIIS algorithm can be used to check weather the Slater condition is satisfied for a given SDP problem.

Given a feasible SDP problem in the form (13.1), the DIIS algorithm constructs a basis of the subspace $\mathscr{M}$ of immobile indices which is formed by s^* vectors $m_i \in \mathbb{R}^s$, $i = 1, \ldots, s^*$ obtained by (13.3). The vectors of this basis form a matrix $\mathbf{M}$. It can be shown that the rank of $\mathbf{M}$ is equal to the irregularity degree of the problem, that in turn permits to conclude about the regularity of the problem. For the sake of completeness, we present the algorithm here. At the kth iteration, let I^k denote a set of indices and M^k a set of vectors which, at the end of the algorithm, will form the basis of $\mathscr{M}$.

The procedure of the Algorithm 1 is constructed so that:

- if the Slater condition holds, then the algorithm stops at the first iteration with $k = 1, \mathscr{M} = \{0\}$ and $s^* = 0$;
- if the Slater condition fails to hold, then the algorithm returns a basis $\mathbf{M}$ with $\text{rank}(\mathbf{M}) = s^* > 0$.

The main task on each iteration of this algorithm consists in solving the system of quadratic equations (13.5). At the k-iteration, this system has $p_k + |I^k|$ vector variables (and $s(p_k + |I^k|)$ scalar variables) and $n + 2 + p_k \times |I^k|$ equations. Notice that one iteration is enough to verify if a given SDP problem is regular in terms of the Slater condition and in this case, one has to solve a system with s vector variables and $n + 2$ equations.

In [15], we have developed two MATLAB numerical tools:

- `SDPreg`, verifies regularity by performing a single iteration of the Procedure 1;
- `DIISalg`, determines the irregularity degree of SDP problems, performing all iterations of the Procedure 1.

These tools are available from the authors upon request. These presolving tools should be run before solving any SDP problem, in order to foresee either a standard SDP solver may be applied for the numerical solving of the given problem. In the case the test indicates that the given SDP problem is irregular, to ensure trustworthiness of solution, some special procedure should be applied.

Algorithm 1 Testing regularity and determining the irregularity degree of SDP problems

input : n, number of variables in the SDP problem;
s, dimension of the constraint matrices;
$\mathbf{A}_j$, $j = 0, 1, \ldots, n$, $s \times s$ symmetric real constraint matrices of a SDP problem of form (13.1).
output: `status`, classification of the problem as regular or nonregular;
s^*, irregularity degree value.

set $k := 1$, $I^1 := \emptyset$ *and* $M^1 := \emptyset$.
repeat
given $k \geq 1$, I^k, $M^k = \{m_1, m_2, ..., m_{|I^k|}\}$ *with* $m_i \in \mathbb{R}^s$, $i \in I^k$:
set $p_k := s - |I^k|$ *and solve the system*

$$\begin{cases} \sum_{i=1}^{p_k} l_i^T \mathbf{A}_j l_i + \sum_{i \in I^k} \gamma_i^T \mathbf{A}_j m_i = 0, \ j = 0, 1, \ldots, n, \\ \sum_{i=1}^{p_k} \|l_i\|^2 = 1, \\ l_i^T m_j = 0 \ , \ j \in I^k, \ i = 1, \ldots, \ p_k, \end{cases} \quad (13.5)$$

w.r.t. the variables $l_i \in \mathbb{R}^s$, $i = 1, \ldots, p_k$, $\gamma_i \in \mathbb{R}^s$, $i \in I^k$.
until *system (13.5) is inconsistent*;
if *system (13.5) is inconsistent,* **then** stop
else *given the solution* $\{l_i \in \mathbb{R}^s, i = 1, \ldots, p_k, \gamma_i \in \mathbb{R}^s, i \in I^k\}$ *of (13.5):*
construct the maximal subset of linearly independent vectors of the set $\{l_1, \ldots, l_{p_k}\}$;
rename its vectors as $\{\xi_1, \ldots, \xi_{s_k}\}$, *where* s_k *is the number of linearly independent vectors in* $\{l_1, \ldots, l_{p_k}\}$ *given* $\{\xi_1, \ldots, \xi_{s_k}\}$, *update:*
$\Delta I^k := \{|I^k| + 1, \ldots, |I^k| + s_k\}$,
$m_{|I^k|+i} := \xi_i$, $i = 1, \ldots, s_k$,
$M^{k+1} := M^k \cup \{m_j, \ j \in \Delta I^k\}$,
$I^{k+1} := I^k \cup \Delta I^k$
end
set $k := k + 1$ *given* M^k: *construct* $\mathbf{M}$, *whose columns are the vectors from* M^k;
compute $s^* := \text{rank}(\mathbf{M})$.
if $k = 1$ **then** *set* `status`*: Regular*
| **else** `status`*: Nonregular*
end
return `status`, Irregularity degree = s^*.

13.4.3 Testing Regularity in Terms of Good Behaviour

In [16], the following characterizations of the badly-behaved SDP problems were proposed.

Theorem 13.2 ([16], Theorem 2) *The SDP problem* (13.1) *is badly-behaved if and only if there exists a matrix* $\mathbf{V}$, *which is a linear combination of the matrices* $\mathbf{A}_i$, *for* $i = 0, \ldots, n$, *of the form*

$$\mathbf{V} = \begin{bmatrix} \mathbf{V}_{11} & \mathbf{V}_{12} \\ \mathbf{V}_{12}^T & \mathbf{V}_{22} \end{bmatrix}, \quad (13.6)$$

where $\mathbf{V}_{11}$ *is a* $(r \times r)$ *symmetric matrix,* $\mathbf{V}_{22}$ *is a* $((s-r) \times (s-r))$ *positive semidefinite matrix and* $\mathbf{V}_{12}$ *is a* $((s-r) \times r)$ *matrix such that the range space of* $\mathbf{V}_{12}^T$*, denoted here by* $\mathscr{C}(\mathbf{V}_{12}^T)$*, is not contained in* $\mathscr{C}(\mathbf{V}_{22})$.

Theorem 13.3 ([16], Theorem 4) *The SDP problem* (13.1) *is badly-behaved if and only if it has a reformulation in the form*

$$\begin{aligned} &\min\ c^T x \\ &\text{s.t.}\ \sum_{i=1}^{k} x_i \begin{bmatrix} \mathbf{F}_i & \mathbf{0} \\ \mathbf{0} & \mathbf{0} \end{bmatrix} + \sum_{i=k+1}^{n} x_i \begin{bmatrix} \mathbf{F}_i & \mathbf{G}_i \\ \mathbf{G}_i^T & \mathbf{H}_i \end{bmatrix} \preceq \begin{bmatrix} \mathbf{I}_r & \mathbf{0} \\ \mathbf{0} & \mathbf{0} \end{bmatrix} = \mathbf{S}, \end{aligned} \tag{13.7}$$

where

1. $\mathbf{S}$ *is the maximum rank slack matrix,* $\mathbf{S} = \begin{bmatrix} \mathbf{I}_r & \mathbf{0} \\ \mathbf{0} & \mathbf{0} \end{bmatrix}$*, where* r *is an integer taking values between* 1 *and* $s-1$*,* $\mathbf{I}_r$ *is the identity matrix of order* r *and* $\mathbf{0}$ *is the null matrix of suitable dimensions;*
2. *the matrices* $\begin{bmatrix} \mathbf{G}_i \\ \mathbf{H}_i \end{bmatrix}$*, for* $i = k+1, \ldots, n$*, are linearly independent;*
3. $\mathbf{H}_n \succeq 0$.

According to [16], the matrices $\mathbf{S}$ and $\mathbf{V}$ provide a *certificate* of the bad behaviour of a SDP problem in the form (13.1).

Since it can be shown that a badly-behaved problem does not satisfy the Slater condition, then we can use the developed numerical tool `SDPreg` to verify the good behaviour of a given SDP problem.

13.5 A Generator of Nonregular SDP Instances

As it was remarked in [7, 15, 17], to develop and test new numerical SDP methods, it is important to have libraries of nonregular SDP problems, as well as problems with a particular structure or infeasible SDP instances. In this section, we propose a generator of nonregular SDP problem instances with certain predefined properties.

Based on the Theorems 13.2 and 13.3 formulated in the previous section, we can describe a class of nonregular SDP problems with a desired irregularity degree s^*, $1 \le s^* \le s-1$, and the optimal value $p^* = 0$ as a class of problems in the form (13.1) that satisfy the following conditions:

1. the integers s and n are as follows: $s \ge 2$, $1 \le n \le \frac{s(s+1)}{2}$;
2. c is a n-dimensional vector: $c = \begin{bmatrix} 1 & 0 & \cdots & 0 \end{bmatrix}^T$;

3. $\mathbf{A}_0 := -\left[\begin{array}{c|c} \mathbf{D}_r & \mathbf{0} \\ \hline \mathbf{0} & \mathbf{0} \end{array}\right]_{s\times s}$, where $r = 1, \ldots, s-1$, and $\mathbf{D}_r = \mathrm{diag}(\beta_1, \cdots, \beta_r)$ with $\beta_i \in \mathbb{R}^+, i = 1, \ldots, r$,
4. the matrices $\mathbf{A}_i,\ i = 1, \ldots, n$, have the form

$$\mathbf{A}_i = \left[\begin{array}{c|c} \mathbf{F}_i & \mathbf{G}_i \\ \hline \mathbf{G}_i^T & \mathbf{H}_i \end{array}\right]_{s\times s},\quad i = 1, \ldots, n. \tag{13.8}$$

   Here for $i = 1, \ldots, n$, matrices $\mathbf{F}_i$ are symmetric: $\mathbf{F}_i \in \mathscr{S}(r)$; matrices $\mathbf{H}_i \in \mathscr{S}(s-r)$, have null diagonal, $\mathbf{H}_1$ being a null matrix: $\mathbf{H}_1 := \mathbf{0} \in \mathscr{S}(s-r)$. A non vanishing matrix $\mathbf{G}_1$ has the form $\mathbf{G}_1 = \sum_{j=1}^{r(s-r)} \alpha_j \mathbf{T}_j$, where $\alpha_j \in \mathbb{R}$, and $\mathbf{T}_j \in T$, $j = 1, \ldots, r(s-r)$, T being the canonical basis of $\mathbb{R}^{r\times(s-r)}$. Matrices $\mathbf{G}_i \in \mathbb{R}^{r\times(s-r)}$, $i = 2, \ldots, s-1$, are linearly independent and are chosen in the form of multiples of the matrices from T. For $i \geq s$, we set $\mathbf{G}_i := \mathbf{0} \in \mathbb{R}^{r\times(s-r)}$.

We can then outline an algorithm for generating nonregular SDP instances as follows.

The following theorem states the main properties of the algorithm.

Theorem 13.4 *Given positive integers* s, $n \leq \frac{s(s+1)}{2}$ *and* s^* *with* $1 \leq s^* \leq s-1$ *as input in the Algorithm 2, the following properties hold for any problem of the form* (13.1) *generated by the Algorithm 2:*

1. *the generated problem is feasible;*
2. *any feasible solution is optimal with* $x_1 = 0$ *and the corresponding optimal value is* $p^* = 0$*;*
3. *the Slater condition is not satisfied.*

Proof A problem generated by the Algorithm 2 is a SDP problem of the form (13.1). It is feasible, since it admits the trivial solution. By construction, the constraint matrices $\mathbf{A}_i, i = 1, \ldots, n$, have the form (13.8) and have at least s^* zeros on the same entries of the main diagonal, while $\mathbf{A}_0$ has exactly s^* zeros. Additionally, for $i = 1, \ldots, n-s$, the matrices $\mathbf{A}_i$ are linearly independent. Thus, the constraint matrix of the problem will have s^* zeros on the diagonal. Since the matrices $\mathbf{G}_i$, $i = 2, \ldots, n-s$, and $\mathbf{G}_1$ form a linearly independent set, using the property that if any diagonal entry is zero, then the corresponding row and column are also full of zeros, it follows that any feasible solution has $x_1 = 0$. Hence, it is easy to see that all feasible solutions are optimal and the optimal value is $p^* = 0$.

Since any problem generated by the Algorithm 2 is badly-behaved [16], it follows from Proposition 13.1 that it does not satisfy the Slater condition. ■

Algorithm 2 Generating SDP instances with pre-specified irregularity degree s^*

input : n, number of variables in the SDP problem;
s, dimension of the constraint matrices;
s^*, desired irregularity degree.
output: $\mathbf{A}_i$, $i = 0, \dots, n$, constraint matrices;
c, vector of coefficients of an objective function.

compute $r = s - s^*$
choose an arbitrary $(r \times r)$ *diagonal matrix* $\mathbf{D}_r$ *with* r *positive entries*
set the $(s \times s)$ *matrix* $\mathbf{A}_0$ *to* $\mathbf{A}_0 = -\left[\begin{array}{c|c} \mathbf{D}_r & \mathbf{0} \\ \hline \mathbf{0} & \mathbf{0} \end{array}\right]$
generate random symmetric $(r \times r)$ *matrices* $\mathbf{F}_i$, $i = 1, \dots, n$
obtain the canonical basis of $\mathbb{R}^{r \times s^*}$, $T = \{\mathbf{T}_j, j = 1, \dots, rs^*\}$
choose the matrix $\mathbf{G}_1 \neq \mathbf{0} \in \mathbb{R}^{r \times s^*}$ *such that* $\mathbf{G}_1 = \sum_{j=1}^{rs^*} \alpha_j \mathbf{T}_j$, *for* $\mathbf{T}_j \in T$ *and arbitrary coefficients* $\alpha_j \in \mathbb{R}$, $j = 1, \dots, rs^*$
for $i = 2, \dots, s$ **do**
| *choose matrices* $\mathbf{G}_i \in \mathbb{R}^{r \times s^*}$ *such that* $\mathbf{G}_i = \alpha \mathbf{T}$, *for some* $\mathbf{T} \in T$, $\alpha \in \mathbb{R}$, *and matrices* $\mathbf{G}_i$, $i = 1, \dots, s$, *are linearly independent*
end
for $i > s$ **do**
| $\mathbf{G}_i := \mathbf{0}$
end
set $\mathbf{H}_1 := \mathbf{0}$
choose arbitrary $\mathbf{H}_i \in \mathcal{S}(s^*)$, $i = 2, \dots, n$, *having a null diagonal*
for $i = 1, \dots, n$ **do**
| $\mathbf{A}_i = \left[\begin{array}{c|c} \mathbf{F}_i & \mathbf{G}_i \\ \hline \mathbf{G}_i^T & \mathbf{H}_i \end{array}\right]$
end
set $c_1 := 1$ *and* $c_i := 0$, *for* $i = 2, \dots, n$
return $\mathbf{A}_i$, $i = 0, 1, \dots, n$, and c.

13.5.1 Implementation Details of the Nonregular SDP Instance Generator `nonregSDPgen`

We have implemented the Algorithm 2 in MATLAB programming language, since many SDP solvers are either coded in MATLAB, or have interface with MATLAB. The resulting function is called `nonregSDPgen` and generates nonregular SDP instances with a pre-specified irregularity degree, s^*, from 1 up to $s - 1$. In the steps of the Algorithm 2, one has to generate random symmetric $(r \times r)$ matrices $\mathbf{F}_i$, $i = 1, \dots, n$. We have implemented a procedure to obtain such matrices as linear combinations of elements of the canonical basis of $\mathcal{S}(r)$. The generated instances have a specific structure and have integer entries in their constraint matrices. The `nonregSDPgen` function returns a nonregular SDP instance written in dat-s format in a new file, whose name should be pre-specified by users. In the MATLAB environment, the user starts by choosing the parameters `n`, `s` and `d`, which correspond to the number of variables of the SDP problem, dimension of the constraint matrices and desired irregularity degree, respectively. The name for the new file that will be created to store the generated data in sparse SDPA format [24], e.g., `examplename.dat-s`, should be specified as well. The basic calling statement structure of the `nonregSDPgen` function is

```
> nonregSDPgen(n,s,d,'examplename.dat-s')
```

The `nonregSDPgen` will create a new dat-s file with a nonregular SDP instance of a pre-specified irregularity degree, which can be used by any SDP solver that requires this input format.

13.6 NONREGSDP: A Nonregular SDP Database

For numerical testing, it is important to have access to collections of test problems "for comparing the performance and robustness of software for solving these optimization problems. Such comparisons have led to significant improvements in the speed and robustness of optimization software" [2]. The SDPLIB [2] is a library of linear SDP test problems with a wide range of sizes, which is usually used to test the performance of solvers. In [7], it is mentioned that it would be interesting to have "a reasonably-sized set of SDP problem instances that might be better suited to empirically examine issues related to the computational behaviour of algorithms for SDP". Since the performance of SDP solvers may be compromised when the Slater condition fails to hold, thus, it makes sense to have a collection of moderate-sized nonregular SDP instances, that is, failing the Slater condition. In this light, we have created a new SDP database and conducted computational experiments.

13.6.1 NONREGSDP

We have generated 100 nonregular SDP instances using the routine `nonregSDPgen` and we have called this collection of test problems NONREGSDP. The current version of this new database is available from the author upon request. The NONREGSDP database is a moderate-sized set of SDP problem instances that can be used for testing the behaviour of SDP algorithms and new stopping criteria. The SDP problems from NONREGSDP were obtained for different values of n and s, with n varying from 1 to 12, s from 2 to 30, and with irregularity degree d varying from 1 up to 29. We have tested the instances from NONREGSDP with our MATLAB function `DIISalg` in order to confirm the irregularity degree of the SDP instances. Table 13.1 provides detailed information on the new SDP library. The column "*Problem*" contains the instance's name, and the parameters n, s, and d refer to the number of variables, the dimension of the constraint matrices and the irregularity degree value, respectively.

Table 13.1 SDP instances from the NONREGSDP database

Problem	*n*	*s*	*d*	*Problem*	*n*	*s*	*d*	*Problem*	*n*	*s*	*d*	*Problem*	*n*	*s*	*d*
nonreg1	2	2	1	*nonreg26*	4	2	1	*nonreg51*	3	5	4	*nonreg76*	12	25	12
nonreg2	3	3	1	*nonreg27*	1	3	1	*nonreg52*	7	4	1	*nonreg77*	12	25	24
nonreg3	3	3	2	*nonreg28*	1	3	2	*nonreg53*	7	4	2	*nonreg78*	2	18	4
nonreg4	4	4	1	*nonreg29*	2	3	1	*nonreg54*	7	4	3	*nonreg79*	9	23	1
nonreg5	4	4	2	*nonreg30*	2	4	1	*nonreg55*	2	20	2	*nonreg80*	9	23	11
nonreg6	4	4	3	*nonreg31*	1	4	1	*nonreg56*	4	21	1	*nonreg81*	9	23	22
nonreg7	5	4	3	*nonreg32*	1	4	3	*nonreg57*	2	30	29	*nonreg82*	4	17	2
nonreg8	3	4	2	*nonreg33*	2	4	1	*nonreg58*	3	11	9	*nonreg83*	4	17	12
nonreg9	6	2	2	*nonreg34*	2	4	2	*nonreg59*	10	15	10	*nonreg84*	10	30	1
nonreg10	1	4	2	*nonreg35*	2	4	3	*nonreg60*	1	27	25	*nonreg85*	10	30	5
nonreg11	5	10	1	*nonreg36*	3	4	1	*nonreg61*	10	30	29	*nonreg86*	10	30	10
nonreg12	5	10	2	*nonreg37*	3	4	3	*nonreg62*	6	24	11	*nonreg87*	10	30	15
nonreg13	5	10	3	*nonreg38*	5	4	1	*nonreg63*	5	13	10	*nonreg88*	10	30	20
nonreg14	5	10	4	*nonreg39*	5	4	2	*nonreg64*	5	13	1	*nonreg89*	10	30	25
nonreg15	5	10	5	*nonreg40*	1	5	1	*nonreg65*	12	30	29	*nonreg90*	1	30	1
nonreg16	5	10	6	*nonreg41*	1	5	2	*nonreg66*	12	30	1	*nonreg91*	1	30	10
nonreg17	5	10	7	*nonreg42*	1	5	3	*nonreg67*	2	25	5	*nonreg92*	1	30	20
nonreg18	5	10	8	*nonreg43*	1	5	4	*nonreg68*	7	28	2	*nonreg93*	1	30	29
nonreg19	5	10	9	*nonreg44*	2	5	1	*nonreg69*	7	28	7	*nonreg94*	8	21	1
nonreg20	2	10	1	*nonreg45*	2	5	2	*nonreg70*	7	28	12	*nonreg95*	8	21	8
nonreg21	12	10	1	*nonreg46*	2	5	3	*nonreg71*	7	28	19	*nonreg96*	8	21	15
nonreg22	6	4	1	*nonreg47*	2	5	4	*nonreg72*	7	28	27	*nonreg97*	8	21	20
nonreg23	6	4	3	*nonreg48*	3	5	1	*nonreg73*	12	11	10	*nonreg98*	12	30	22
nonreg24	1	2	1	*nonreg49*	3	5	2	*nonreg74*	12	11	2	*nonreg99*	12	30	8
nonreg25	3	2	1	*nonreg50*	3	5	3	*nonreg75*	12	25	1	*nonreg100*	12	30	17

13.6.2 Numerical Results and Discussion

In this section, we used 54 instances from the NOREGSDP database to test the computational behaviour of the popular SDP solvers SDPT3 [20] and SeDuMi [19]. All computations were performed on a computer with an Intel Core i7-2630QM processor CPU@2.0GHz, with Windows 7 (64 bits) and 12 GB RAM, using MATLAB (v.7.12 R2013a). We tried to solve some generated instances using two different solvers available on the package CVX [4], SDPT3 and SeDuMi, and the default precision or tolerance values.

The numerical results of the tests are displayed in the Tables 13.2 and 13.3. In these tables, the first column contains the NONREGSDP instance's name. The next three columns contain the number of variables, n, the dimension of the constraint matrices, s, and the desired irregularity degree, d, respectively. The fifth column presents the computed irregularity degree, s^*, obtained using the `DIISalg` function. The last columns of the Tables 13.2 and 13.3 contain the outputs of the SDP solvers SDPT3 and SeDuMi, respectively, where *iter* is the number of iterations, *time* is the computational time, *val* is the returned optimal value, p^* and d^* are the primal and dual optimal values, respectively, *gap* is the actual duality gap, and *Solver's Report* stands for observations which are (warning) output messages returned by solvers. The symbol $*$ in the last column of the tables means that the solver solved the dual problem to get the solution of the given (primal) SDP problem. The lack of results in the tables correspond to the cases when the solvers were not able to provide such results.

While solving the generated nonregular SDP problems, one of the first observations we can make from the experiments is that the number of warning messages delivered by the SDPT3 solver is quite higher than that by SeDuMi. Another observation is that for these nonregular instances the solvers chose to solve the dual problem instead of the given primal one for almost all tested SDP instances.

Observing the Table 13.2, we can see that for 7 generated instances the returned value p^* was quite far from the true one, which is zero. In terms of the returned optimal value *val*, we can see that SDPT3 provided wrong values for 13 instances (i.e., NaN - not a number; $-$Inf - unbounded; or values far from the true optimal ones). We can also see that the most accurate optimal value p^* was computed for the problem *nonreg29* with $p^* = 7.0321e - 14$. However, since the solver has chosen to solve the dual problem, the returned optimal value *val* was $-3.7952e - 7$.

As can be seen from this table, in 19 out of 54 instances the solver SDPT3 returned warning messages related to numerical issues. For all the 18 nonregular SDP instances with $n \geq s$, the solver ran into numerical difficulties and returned wrong solutions or values far from the true optimal values. The exceptions are the problems *nonreg4*, *nonreg6*, *nonreg21* and *nonreg38*, whose computed values can be considered roughly close to (the optimal) zero.

No general assertion about correlation between the level of nonregularity and the number of iterations used by SDPT3 can be made. It may be due to the use of the dual to solve the given problem. However, there are some examples supporting that large

Table 13.2 Numerical results using `DIISalg` and SDPT3 on SDP instances from NONREGSDP (computation time is in seconds)

Problem	n	s	d	`DIISalg`	SDPT3						
				s^*	it	Time	val	p^*	d^*	Gap	Solver's report
nonreg1	2	2	1	1	16	0.19	$-1.1775e-3$	$-1.1775e-3$	0.0000	$-1.18e-3$	Solved
nonreg2	3	3	1	1	24	0.24	NaN	$-2.4329e-2$	$8.7691e-9$	$-2.38e-2$	progress is bad; Failed
nonreg3	3	3	2	2	58	0.61	−Inf				∗ primal problem is suspected of being infeasible; Unbounded
nonreg4	4	4	1	1	24	0.26	$-1.7315e-7$	$1.0218e-7$	$1.7315e-7$	$-7.10e-8$	∗ Solved
nonreg5	4	4	2	2	31	0.35	$-8.7322e-7$	$4.9006e-7$	$8.7322e-7$	$-3.83e-7$	∗ Inaccurate; Solved
nonreg6	4	4	3	3	34	0.36	$-1.2851e-7$	$7.0894e-8$	$1.2851e-7$	$-5.76e-8$	∗ Solved
nonreg7	5	4	3	3	26	0.28	$-3.8859e-5$	$1.9393e-5$	$3.8859e-5$	$-1.95e-5$	∗ Inaccurate; Solved
nonreg8	3	4	2	2	24	0.26	$-1.1372e-6$	$6.4885e-7$	$1.1372e-6$	$-4.88e-7$	∗ Inaccurate; Solved
nonreg9	6	2	2	2	19	0.31	NaN	$7.3161e-6$	$8.0461e-4$	$-7.97e-4$	∗ Failed
nonreg10	1	4	2	2	22	0.19	$-1.1464e-7$	$6.6920e-8$	$1.1464e-7$	$-4.77e-8$	∗ Solved
nonreg11	5	10	1	1	28	0.62	$-2.5503e-8$	$2.3136e-8$	$2.5503e-8$	$-2.37e-9$	∗ Solved
nonreg12	5	10	2	2	22	0.26	$-7.1756e-7$	$6.4886e-7$	$7.1756e-7$	$-6.87e-8$	∗ Inaccurate; Solved
nonreg13	5	10	3	3	21	0.26	$-8.7671e-7$	$8.1750e-7$	$8.7671e-7$	$-5.92e-8$	∗ lack of progress in infeas; Inaccurate; Solved
nonreg14	5	10	4	4	28	0.33	$-2.2118e-8$	$1.5133e-8$	$2.2118e-8$	$-6.98e-9$	∗ Solved
nonreg15	5	10	5	5	27	0.30	$-2.0518e-8$	$1.5921e-8$	$2.0518e-8$	$-4.60e-9$	∗ Solved

(continued)

Table 13.2 (continued)

Problem	n	s	d	`DIISalg`	SDPT3						
				s^*	it	Time	val	p^*	d^*	Gap	Solver's report
nonreg16	5	10	6	6	28	0.31	$-2.0014e-8$	$1.4456e-8$	$2.0014e-8$	$-5.56e-9$	* Solved
nonreg17	5	10	7	7	27	0.30	$-1.9767e-8$	$1.3181e-8$	$1.9767e-8$	$-6.59e-9$	* Solved
nonreg18	5	10	8	8	30	0.30	$-3.6391e-8$	$2.1580e-8$	$3.6391e-8$	$-1.48e-8$	* Solved
nonreg19	5	10	9	9	32	0.34	$-2.7487e-8$	$1.4789e-8$	$2.7487e-8$	$-1.27e-8$	* Solved
nonreg20	2	10	1	1	24	0.23	$-2.5293e-8$	$2.3063e-8$	$2.5293e-8$	$-2.23e-9$	* Solved
nonreg21	12	10	1	1	23	0.26	$-3.4148e-8$	$2.9786e-8$	$3.4148e-8$	$-4.36e-9$	* Solved
nonreg22	6	4	1	1	29	0.31	NaN	$-1.2664e-9$	$1.6762e-2$	$-1.65e-2$	* progress is bad; Failed
nonreg23	6	4	3	3	21	0.20	$-6.9621e-5$	$1.8884e-6$	$6.9621e-5$	$-6.77e-5$	* Inaccurate; Solved
nonreg24	1	2	1	1	17	0.25	$-4.6124e-6$	$2.4798e-6$	$4.6124e-6$	$-2.13e-6$	* progress in duality gap has deteriorated; Inaccurate; Solved
nonreg25	3	2	1	1	21	0.25	$-1.4449e-3$	$-1.4449e-3$	0.0000	$-1.44e-3$	progress is bad; Inaccurate; Solved
nonreg26	4	2	1	1	22	0.26	−Inf				progress is bad; dual problem is suspected of being infeasible; Unbounded
nonreg27	1	3	1	1	32	0.29	$-2.2120e-7$	$1.2689e-7$	$2.2120e-7$	$-9.43e-8$	* Solved
nonreg28	1	3	2	2	31	0.29	$-5.2713e-7$	$2.8473e-7$	$5.2713e-7$	$-2.42e-7$	* Solved
nonreg29	2	3	1	1	51	0.41	$-3.7952e-7$	$7.0321e-14$	$3.7952e-7$	$-3.80e-7$	* Solved
nonreg30	2	4	1	2	30	0.29	$-3.3806e-7$	$1.8231e-7$	$3.3806e-7$	$-1.56e-7$	* Solved
nonreg31	1	4	1	1	19	0.22	$-5.9690e-8$	$4.1990e-8$	$5.9690e-8$	$-1.77e-8$	* Solved

(continued)

Table 13.2 (continued)

Problem	*n*	*s*	*d*	DIISalg	SDPT3						
				s^*	it	Time	val	p^*	d^*	Gap	Solver's report
nonreg32	1	4	3	3	26	0.19	$-4.3577e-7$	$2.4237e-7$	$4.3577e-7$	$-1.93e-7$	∗ Solved
nonreg33	2	4	1	1	36	0.31	$-2.2030e-7$	$1.2831e-7$	$2.2030e-7$	$-9.20e-8$	∗ Solved
nonreg34	2	4	2	2	29	0.25	$-1.6806e-7$	$9.5636e-8$	$1.6806e-7$	$-7.24e-8$	∗ Solved
nonreg35	2	4	3	3	35	0.34	$-1.4537e-7$	$8.9635e-8$	$1.4537e-7$	$-5.57e-8$	∗ Solved
nonreg36	3	4	1	1	19	0.22	$-3.2653e-8$	$2.6820e-8$	$3.2653e-8$	$-5.83e-9$	∗ Solved
nonreg37	3	4	3	3	32	0.30	$-3.4000e-7$	$1.8926e-7$	$3.4000e-7$	$-1.51e-7$	∗ progress is bad; Solved
nonreg38	5	4	1	1	30	0.34	$-2.6551e-7$	$1.5231e-7$	$2.6550e-7$	$-1.13e-7$	∗ Solved
nonreg39	5	4	2	2	20	0.39	$-1.6548e-5$	$1.1077e-5$	$1.6548e-5$	$-5.47e-6$	∗ lack of progress in infeas; Solved
nonreg40	1	5	1	1	21	0.23	$-2.0888e-7$	$6.0000e-1$	$6.0000e-1$	$-3.31e-8$	∗ Solved
nonreg41	1	5	2	2	21	0.18	$-5.4662e-8$	$3.5468e-8$	$5.4662e-8$	$-1.92e-8$	∗ Solved
nonreg42	1	5	3	3	25	0.30	$-1.1985e-7$	$6.8445e-8$	$1.1985e-7$	$-5.14e-8$	∗ Solved
nonreg43	1	5	4	4	29	0.23	$-2.8877e-7$	$1.5750e-7$	$2.8876e-7$	$-1.31e-7$	∗ Solved
nonreg44	2	5	1	1	21	0.21	$-5.5529e-8$	$3.7617e-8$	$5.5529e-8$	$-1.79e-8$	∗ Solved
nonreg45	2	5	2	2	28	0.29	$-4.9874e-8$	$3.3628e-8$	$4.9874e-8$	$-1.62e-8$	∗ Solved
nonreg46	2	5	3	3	32	0.34	$-1.3129e-7$	$7.4843e-8$	$1.3129e-7$	$-5.65e-8$	∗ Solved
nonreg47	2	5	4	4	30	0.20	$-1.7913e-7$	$9.8066e-8$	$1.7913e-7$	$-8.11e-8$	∗ Solved
nonreg48	3	5	1	1	24	0.26	$-1.4151e-7$	$8.6130e-8$	$1.4151e-7$	$-5.54e-8$	∗ Solved
nonreg49	3	5	2	2	27	0.27	$-7.2936e-8$	$4.5615e-8$	$7.2936e-8$	$-2.73e-8$	∗ Solved

(continued)

Table 13.2 (continued)

Problem	n	s	d	`DIISalg`	SDPT3						
				s^*	it	Time	val	p^*	d^*	Gap	Solver's report
nonreg50	3	5	3	3	30	0.32	$-1.1485e-7$	$5.9662e-8$	$1.1485e-7$	$-5.52e-8$	$*$ Solved
nonreg51	3	5	4	4	30	0.32	$-2.4319e-7$	$1.3273e-7$	$2.4319e-7$	$-1.10e-7$	$*$ Solved
nonreg52	7	4	1	1	56	0.56	NaN	$8.0698e-4$	$6.8436e-2$	$-6.33e-2$	$*$ lack of progress in dual infeas; Failed
nonreg53	7	4	2	2	21	0.27	$-1.0291e-3$	$6.4224e-6$	$1.0291e-3$	$-1.02e-3$	$*$ Inaccurate; Solved
nonreg54	7	4	3	3	28	0.42	NaN	$5.6348e-7$	$1.9355e-3$	$-1.93e-3$	$*$ Failed

values of the irregularity degree correlate well with large number of iterations of the solver (e.g., *nonreg40–nonreg43*, *nonreg44–nonreg47*, *nonreg48–nonreg51*).

From Table 13.3, it can be observed that SeDuMi reported 5 warning messages about numerical problems on solving the given SDP instances.

While the results provided by SDPT3 permit to consider many of them to be rather close to the true optimal value, notice that the results from SeDuMi can not be considered so good. Moreover, SeDuMi had never reported that it failed to solve some instances. A closer analysis on the results presented in the Table 13.3 permits to conclude that there are significant discrepancies between the computed optimal values and the true ones, even when the solver has reported "Solved". See, for example, the problems *nonreg2*, *nonreg3*, *nonreg7*, *nonreg22*, *nonreg25*, *nonreg26*, *nonreg29*, *nonreg33*, *nonreg52*. Notice that the closest value to zero in *val* is $-8.1019e-6$ for the problem *nonreg15*.

Regarding the computed value for p^*, only for the problem *nonreg3* SeDuMi had returned zero, and for almost all other instances, the computed optimal values are fairly far from the true ones. The closest value to zero corresponds to the problem *nonreg22*. Based on the results presented in the Table 13.3, there is no empirical evidence that there exists some correlation between the level of nonregularity and the number of iterations, or computational time spent by SeDuMi.

It is worth mentioning that in both tables, the problem *nonreg40* is particularly nasty, since both solvers behaved poorly, returning similar values for p^* and d^* (close to 0.6), and a different optimal value *val* of the given problem, which should be zero.

The following table summarizes the results obtained in this section (Table 13.4).

Based on the numerical results presented in this section, we can conclude that they support the conclusion that standard SDP solvers applied to nonregular problems may be unable to provide accurate solutions.

13.7 Conclusions

In this paper, we presented an algorithm for generating nonregular SDP instances with a pre-specified irregularity degree. We have implemented this algorithm in MATLAB by the function `nonregSDPgen`. The routine `nonregSDPgen` is very simple to use and returns a dat-s file containing the generated nonregular SDP instance, that in turn can be used as input in popular SDP solvers. By construction, all the generated instances are feasible and have optimal value equal to zero. We have generated nonregular SDP instances and formed a new SDP database with nonregular SDP problems called NONREGSDP. The NONREGSDP library is described in the Table 13.1. This collection of nonregular SDP test problems was used to evaluate the performance and robustness of two popular SDP solvers.

The numerical experiments showed that the tested SDP solvers do not have a reliable behaviour on nonregular SDP instances. Although SeDuMi uses a self-dual embedding technique to regularize the nonregular problem, many examples showed that it may still return inaccurate solutions.

Table 13.3 Numerical results using `DIISalg` and SeDuMi on SDP instances from NONREGSDP (computation time is in seconds)

Problem	n	s	d	`DIISalg`	SeDuMi						
				s^*	iter	Time	val	p^*	d^*	Gap	Solver's report
nonreg1	2	2	1	1	5	0.30	$-1.6672e-1$	$-1.6672e-1$	0.0000	$-1.67e-1$	Run into numerical problems; Solved
nonreg2	3	3	1	1	25	0.30	$-7.1391e-2$	$-7.1391e-2$	$-1.1324e-9$	$-7.14e-2$	Solved
nonreg3	3	3	2	2	25	0.50	$-4.5511e-2$	0.0000	$4.5511e-2$	$-4.55e-2$	$*$ Solved
nonreg4	4	4	1	1	23	0.30	$-9.1231e-5$	$7.4726e-5$	$9.1231e-5$	$-1.65e-5$	$*$ Solved
nonreg5	4	4	2	2	18	0.20	$-9.4062e-5$	$6.3427e-5$	$9.4062e-5$	$-3.06e-5$	$*$ Solved
nonreg6	4	4	3	3	16	0.20	$-5.1708e-5$	$3.1247e-5$	$5.1708e-5$	$-2.05e-5$	$*$ Solved
nonreg7	5	4	3	3	20	0.30	$-2.6713e-4$	$1.6394e-4$	$2.6713e-4$	$-1.03e-4$	$*$ Solved
nonreg8	3	4	2	2	19	0.20	$-2.2489e-5$	$1.7561e-5$	$2.2489e-5$	$-4.93e-6$	$*$ Solved
nonreg9	6	2	2	2	16	0.40	$-1.1971e-1$	$1.8696e-6$	$1.1971e-1$	$-1.20e-1$	$*$ Run into numerical problems; Inaccurate; Solved
nonreg10	1	4	2	2	19	0.30	$-4.0186e-5$	$3.4643e-5$	$4.0186e-5$	$-5.54e-6$	$*$ Solved
nonreg11	5	10	1	1	23	0.50	$-2.8726e-5$	$1.8988e-5$	$2.8726e-5$	$-9.74e-6$	$*$ Solved
nonreg12	5	10	2	2	22	0.50	$-1.4633e-5$	$9.3842e-6$	$1.4633e-5$	$-5.25e-6$	$*$ Solved
nonreg13	5	10	3	3	23	0.40	$-1.0469e-5$	$6.8688e-6$	$1.0469e-5$	$-3.60e-6$	$*$ Solved
nonreg14	5	10	4	4	22	0.30	$-4.3341e-5$	$2.8588e-5$	$4.3341e-5$	$-1.47e-5$	$*$ Solved
nonreg15	5	10	5	5	22	0.30	$-8.1019e-6$	$5.0566e-6$	$8.1019e-6$	$-3.04e-6$	$*$ Solved
nonreg16	5	10	6	6	21	0.30	$-8.6966e-6$	$7.1936e-6$	$8.6966e-6$	$-1.50e-6$	$*$ Solved
nonreg17	5	10	7	7	19	0.20	$-9.4683e-6$	$8.2214e-6$	$9.4683e-6$	$-1.25e-6$	$*$ Solved

(continued)

Table 13.3 (continued)

Problem	*n*	*s*	*d*	`DIISalg`	SeDuMi						
				s^*	iter	Time	val	p^*	d^*	Gap	Solver's report
nonreg18	5	10	8	8	20	0.20	$-1.0959e-5$	$8.8137e-6$	$1.0959e-5$	$-2.14e-6$	∗ Solved
nonreg19	5	10	9	9	18	0.20	$-8.6502e-6$	$5.5011e-6$	$8.6502e-6$	$-3.15e-6$	∗ Solved
nonreg20	2	10	1	1	25	0.40	$-1.7112e-5$	$1.5546e-5$	$1.7112e-5$	$-1.57e-6$	∗ Solved
nonreg21	12	10	1	1	24	0.30	$-5.0062e-5$	$3.6993e-5$	$5.0062e-5$	$-1.31e-5$	∗ Solved
nonreg22	6	4	1	1	27	0.40	$-7.9842e-2$	$-1.5632e-9$	$7.9842e-2$	$-7.98e-2$	∗ Solved
nonreg23	6	4	3	3	19	0.40	$-6.6666e-2$	$2.0036e-7$	$6.6666e-2$	$-6.67e-2$	∗ Run into numerical problems; Inaccurate; Solved
nonreg24	1	2	1	1	18	0.30	$-1.7451e-5$	$1.1614e-5$	$1.7451e-5$	$-5.84e-6$	∗ Solved
nonreg25	3	2	1	1	26	0.40	$-4.1919e-2$	$-4.1919e-2$	0.0000	$-4.19e-2$	Solved
nonreg26	4	2	1	1	26	0.30	$-9.0094e-3$	$-9.0094e-3$	0.0000	$-9.01e-3$	Solved
nonreg27	1	3	1	1	18	0.20	$-5.0650e-5$	$4.1677e-5$	$5.0650e-5$	$-8.97e-6$	∗ Solved
nonreg28	1	3	2	2	17	0.20	$-6.5905e-5$	$4.7263e-5$	$6.5905e-5$	$-1.86e-5$	∗ Solved
nonreg29	2	3	1	1	24	0.30	$-1.4064e-2$	$2.3344e-9$	$1.4064e-2$	$-1.41e-2$	∗ Solved
nonreg30	2	4	1	2	19	0.30	$-1.7670e-5$	$1.1010e-5$	$1.7670e-5$	$-6.66e-6$	∗ Solved
nonreg31	1	4	1	1	20	0.20	$-3.2979e-5$	$2.9286e-5$	$3.2979e-5$	$-3.69e-6$	∗ Solved
nonreg32	1	4	3	3	15	0.10	$-2.8308e-5$	$1.6821e-5$	$2.8308e-5$	$-1.15e-5$	∗ Solved
nonreg33	2	4	1	1	22	0.30	$-1.7365e-4$	$1.0958e-4$	$1.7365e-4$	$-6.41e-5$	∗ Solved
nonreg34	2	4	2	2	19	0.30	$-3.2226e-5$	$2.7211e-5$	$3.2226e-5$	$-5.01e-6$	∗ Solved
nonreg35	2	4	3	3	18	20	$-8.0299e-5$	$5.2231e-5$	$8.0299e-5$	$-2.81e-5$	∗ Solved
nonreg36	3	4	1	1	20	0.20	$-2.5493e-5$	$2.2435e-5$	$2.5493e-5$	$-3.06e-6$	∗ Solved

(continued)

Table 13.3 (continued)

Problem	n	s	d	`DIISalg`	SeDuMi						
				s^*	iter	Time	val	p^*	d^*	Gap	Solver's report
nonreg37	3	4	3	3	16	0.20	$-4.2555e-5$	$2.6337e-5$	$4.2555e-5$	$-1.62e-5$	∗ Solved
nonreg38	5	4	1	1	22	0.30	$-4.1273e-5$	$3.4208e-5$	$4.1273e-5$	$-7.06e-6$	∗ Solved
nonreg39	5	4	2	2	18	0.20	$-3.7746e-5$	$2.4570e-5$	$3.7746e-5$	$-1.32e-5$	∗ Solved
nonreg40	1	5	1	1	22	0.20	$-2.8475e-5$	$6.0001e-1$	$6.0002e-1$	$-1.00e-5$	∗ Solved
nonreg41	1	5	2	2	19	0.20	$-1.5858e-5$	$1.3866e-5$	$1.5858e-5$	$-1.99e-6$	∗ Solved
nonreg42	1	5	3	3	20	0.20	$-4.7861e-5$	$3.9980e-5$	$4.7861e-5$	$-7.88e-6$	∗ Solved
nonreg43	1	5	4	4	18	0.20	$-2.1816e-5$	$1.3399e-5$	$2.1816e-5$	$-8.42e-6$	∗ Solved
nonreg44	2	5	1	1	21	0.20	$-3.4246e-5$	$3.1213e-5$	$3.4246e-5$	$-3.03e-6$	∗ Solved
nonreg45	2	5	2	2	20	0.20	$-1.4613e-5$	$1.2855e-5$	$1.4613e-5$	$-1.76e-6$	∗ Solved
nonreg46	2	5	3	3	19	0.20	$-7.0759e-5$	$5.4775e-5$	$7.0759e-5$	$-1.60e-5$	∗ Solved
nonreg47	2	5	4	4	16	0.20	$-1.7131e-5$	$1.1590e-5$	$1.7131e-5$	$-5.54e-6$	∗ Solved
nonreg48	3	5	1	1	23	0.20	$-3.6285e-5$	$3.2545e-5$	$3.6285e-5$	$-3.74e-6$	∗ Solved
nonreg49	3	5	2	2	19	0.20	$-3.3603e-5$	$2.9011e-5$	$3.3603e-5$	$-4.59e-6$	∗ Solved
nonreg50	3	5	3	3	20	0.20	$-6.7126e-5$	$5.1655e-5$	$6.7126e-5$	$-1.55e-5$	∗ Solved
nonreg51	3	5	4	4	18	0.20	$-1.8706e-5$	$1.1907e-5$	$1.8706e-5$	$-6.80e-6$	∗ Solved
nonreg52	7	4	1	1	29	0.40	$-2.2805e-1$	$7.0992e-9$	$2.2805e-1$	$-2.28e-1$	∗ Solved
nonreg53	7	4	2	2	24	0.50	$-3.0310e-2$	$5.2935e-8$	$3.0310e-2$	$-3.03e-2$	∗ Run into numerical problems; Inaccurate; Solved
nonreg54	7	4	3	3	17	0.40	$-2.8461e-1$	$1.3761e-6$	$2.8461e-1$	$-2.85e-1$	∗ Run into numerical problems; Inaccurate; Solved

Table 13.4 Summary of computational behaviour of SDPT3 and SeDuMi evaluated on 54 instances from NONREGSDP

Solver	Primal problem solved	Dual problem solved	Accurate solutions			Report of	
			10^{-4}	10^{-6}	10^{-8}	Failures	Warnings
SDPT3	4	50	47	41	2	5	19
SeDuMi	4	50	47	6	4	0	5

It should be noticed that it was not the aim of the paper to compare or test the efficiency of SDP solvers. We used two popular SDP solvers, SDPT3 and SeDuMi to analyse the solutions of nonregular SDP problems, and the testes showed that these solvers have not succeeded to find accurate solutions in many cases.

This work reinforces the needed of developing new SDP methods/solvers particularly suitable for nonregular problems, that is failing the Slater condition. Our future work will be dedicated to such a study based on the new CQ free optimality conditions for SDP problems formulated in [12]. Comparison of SDP optimization software on the basis of advanced metrics such as number of function evaluation, ratio of one solver's runtime to the best runtime and performance profiles (see [6] and the references therein) can be another interesting topic of study. To fulfill such comparison, one use the collections of benchmark SDP problems, including the NONREGSDP library.

Acknowledgements The authors would like to thank the anonymous referees for their suggestions and valuable comments that have helped to improve the paper. This work was supported by Portuguese funds through the CIDMA - Center for Research and Development in Mathematics and Applications, and the Portuguese Foundation for Science and Technology (FCT - Fundação para a Ciência e a Tecnologia), within project UID/MAT/04106/2013.

References

1. M.F. Anjos, J.B. Lasserre (eds.), *Handbook of Semidefinite, Conic and Polynomial Optimization: Theory, Algorithms, Software and Applications*, vol. 166, International Series in Operational Research and Management Science (Springer, US, 2012)
2. B. Borchers, Sdplib 1.2, a library of semidefinite programming test problems. Optim. Methods Softw. **11**(1–4), 683–690 (1999)
3. Y. Cheung, S. Schurr, H. Wolkowicz, Preprocessing and reduction for degenerate semidefinite programs. Comput. Anal. Math. Springer Proc. Math. Stat. **50**, 251–303 (2013)
4. Inc. CVX Research. Cvx: Matlab software for disciplined convex programming, version 2.0 (2012)
5. E. de Klerk, *Aspects of Semidefinite Programming - Interior Point Algorithms and Selected Applications*, vol. 65, Applied Optimization (Kluwer, 2004)
6. D.E. Dolan, J.J. Moré, Benchmarking optimization software with performance profiles. Math. Program. Ser. A **91**, 201–213 (2002)

7. R.M. Freund, F. Ordó nez, K.C. Toh, Behavioral measures and their correlation with ipm iteration counts on semi-definite programming problems. Math. Program. **109**(2), 445–475 (2007)
8. G. Gruber, F. Rendl, Computational experience with ill-posed problems in semidefinite programming. Comput. Optim. Appl. **21**, 201–212 (2002)
9. G. Gruber, S. Kruk, F. Rendl, H. Wolkowicz, Presolving for semidefinite program without constraint qualifications. Technical report, CORR 98-32, University of Waterloo, Waterloo, Ontario (1998)
10. B. Hernández-Jiménez, M.A. Rojas-Medar, R. Osuna-Gómez, A. Beato-Moreno, Generalized convexity in non-regular programming problems with inequality-type constraints. J. Math. Anal. Appl. **352**, 604–613 (2009)
11. C. Jansson, D. Chaykin, C. Keil, Rigorous error bounds for the optimal value in sdp. SIAM J. Numer. Anal. **46**(1), 180–200 (2007)
12. O.I. Kostyukova, T.V. Tchemisova, Optimality criterion without constraint qualification for linear semidefinite problems. J. Math. Sci. Springer, US **182**(2), 126–143 (2012)
13. M. Liu, G. Pataki, Exact duals and short certificates of infeasibility and weak infeasibility in conic linear programming. Math. Program. Ser. A, 1–46 (2017)
14. E. Macedo, in *Testing Regularity on Linear Semidefinite Optimization Problems* (Springer International Publishing, Cham, 2015), pp. 213–236
15. E. Macedo, Numerical study of regularity in Semidefinite Programming and applications. Ph.D. thesis, University of Aveiro, Portugal, 2016
16. G. Pataki, Bad semidefinite programs: they all look the same. SIAM J. Optim. **27**(1), 146–172 (2017)
17. I. Polik, T. Terlaky, New stopping criteria for detecting infeasibility in conic optimization. Optim. Lett. Springer **3**(2), 187–198 (2009)
18. M.V. Solodov, Constraint qualifications, *Encyclopedia of Operations Research and Management Science* (Wiley, New York, 2010)
19. J.F. Sturm, Using SeDuMi 1.02, a MATLAB toolbox for optimization over symmetric cones. Optim. Methods Softw. **11**, 625–653 (1999)
20. R.H. Tutuncu, K.C. Toh, M.J. Todd, Solving semidefinite-quadratic-linear programs using sdpt3. Math. Program. Ser. B **95**, 189–217 (2003)
21. H. Waki, M. Nakata, M. Muramatsu, Strange behaviors of interior-point methods for solving semidefinite programming problems in polynomial optimization. Comput. Optim. Appl. Springer **53**(3), 823–844 (2012)
22. H. Wei, H. Wolkowicz, Generating and measuring instances of hard semidefinite programs. Math. Program. **125**(1), 31–45 (2010)
23. H. Wolkowicz, R. Saigal, L. Vandenberghe (eds.), *Handbook of Semidefinite Programming: Theory, Algorithms, and Applications* (Kluwer Academic Publishers, Boston, 2000)
24. M. Yamashita, K. Fujisawa, M. Kojima, Implementation and evaluation of sdpa 6.0 (semidefinite programming algorithm 6.0). Optim. Methods Softw. **18**, 491–505 (2003)

Chapter 14
Bus Fleet Management Optimization Using the Augmented Weighted Tchebycheff Method

William Emiliano, Lino Costa, Maria do Sameiro Carvalho and José Telhada

Abstract This paper presents a multi-objective optimization model for the buses fleet management problem. This model is solved using the Augmented Weighted Tchebycheff method. The aim is to minimize three objective functions, Z_1 (CO_2 emissions), Z_2 (other types of emissions) and Z_3 (total costs), for a bus fleet that uses four types of buses: diesel, electric bus, electric bus of fast charging, and Compressed Natural Gas (CNG). A public transport (PT) company of Joinville, Brazil, where it operates three different PT lines, owns the fleet. The respective data was modelled and optimized using the MS Excel solver. Results provided interesting insights concerning the most adequate strategy for bus selection according with public transport line characteristics and taking into account trade-off between costs and emissions. The results indicate that optimal solutions include the diesel in the Itinga line and the CNG in the South line. The electric bus is more adequate in the South-North line due to the large number of stops and low average speed. However, when the costs are disregarded, in some scenarios, the best option is the electric bus for all lines.

Keywords Augmented weighted tchebycheff · Buses fleet management problem · Ghg emissions

14.1 Introduction

The public transport sector has been testing new technologies, for example, battery electric buses (BEs). In Choi et al. [2], it is reported a test with a few BEs in the city of Seoul in South Korea. The study of Aber [1] investigates the replacement of all conventional bus fleet by electric vehicles in New York city (USA). In Los Angeles (USA), twelve new BEs (fast charging) were tested and compared with Compressed Natural Gas (CNG) buses [7]. In all of these viability studies for the use

W. Emiliano (✉) · L. Costa · M. do Sameiro Carvalho · J. Telhada
Algoritmi Research Center, Universidade do Minho, Braga, Portugal
e-mail: williammachado16@gmail.com

A. I. F. Vaz et al. (eds.), *Operational Research*, Springer Proceedings in Mathematics & Statistics 223, https://doi.org/10.1007/978-3-319-71583-4_14

of new technologies, different methods were applied, such as economic methods [5], simulation [16], optimization and others [6].

The optimization of a fleet of buses involves multiple conflicting objectives that are related to economical, environmental and sustainability issues. Therefore, it is important that the decision-making process takes into account different trade-offs.

In this paper, it is proposed a multi-objective model to optimize a fleet of buses using the Augmented Weighted Tchebycheff method [18]. This approach was applied to optimize a fleet of buses of the city of Joinville [11] considering four types of buses and three different types of lines with different kilometrages, average speed and number of passengers, among other features. In this problem, the goal is to minimize, simultaneously, the costs, the CO_2 emissions and Other Types of Emissions (OTE) (CO (Carbon Monoxide), SO_2 (Sulfur Dioxide), NO_x (Nitrogen Oxide), PM_{10} (Particulate Matter) and $PM_{2.5}$). This approach allows obtaining multiple compromise solutions that are different trade-offs between the three objectives. Thus, decision makers can perceive the trade-offs and select the solution that fits better their preferences taking into accounts the company's policies and/or current law. Ercan et al. [6] have also applied the MINMAX method to optimize a multi-objective model of a fleet of buses. However, to the best of our knowledge, there is no study addressing the use of the Augmented Weighted Tchebycheff [18] applied to the optimization of bus fleets. This study comes to fill this gap in the literature. The advantages and drawbacks of these approaches are also discussed. More generally, this study aims to assist managers in the decision-making process for public transport fleet management.

This paper is organized as follows. Section 14.2 reports the literature review in this research area of application; Sect. 14.3 presents the proposed models; Sect. 14.4 presents the application methods and case study; Sect. 14.5 discusses the results of its application to a case study; Sect. 14.6 ends the paper by presenting a summary of key findings.

14.2 Literature Review

Alternative fuel vehicles, such as hybrid vehicles, electric and hydrogen-powered have emerged as a new solution to comply with the environmental legislation, as suggested in some countries. Therefore, the integration of this new technology with conventional vehicles is necessary to facilitate its implementation. For example, Feng and Figliozzi [8] provide a detailed sensitivity analysis for the definition of a fleet of diesel or/and electric commercial vehicles. Comparisons between electric and conventional vehicles have also been discussed in the public transport sector. Some studies strongly relate the increase of GHG (Greenhouse Gas) emissions with public transport, especially in major world cities. Therefore, the use of alternative fuels has gained prominence in the literature. In 2013, about 63.5% of the bus fleet in USA consisted of diesel buses, accounting for 1% of total GHG emissions of the country.

Based on these data, some reports were developed as in Aber [1], that analyze the replacement of all conventional buses (diesel, hybrid and CNG) by a BE fleet in New York city (USA). In Eudy et al. [7], a comparison between CNG and BE (fast charging) in a public transport operator in USA is conducted. Similar analysis is done by Choi et al. [2], comparing CNG versus BEs in South Korea. All these studies reported a reduction of emission levels in the PT by replacing diesel buses with alternative fuel buses.

For this type of problem, some models have been proposed. Emiliano et al. [5] apply a Life-Cycle Cost model to analyze the best type of bus (diesel versus electric) for three different types of lines. Their results suggest the use of a fleet composed of 90% of electric buses when considering the impact of revenues. Feng and Figliozzi [9] use a Mixed Integer Linear Programming (MILP) model for determining the optimal composition of a homogeneous fleet (diesel and hybrid vehicles) over a horizon of 100 years. The results show that emissions and operating costs can be reduced by using hybrid vehicles only, but a government subsidy of at least 66% is necessary to ensure the economic viability of the project, due the high cost of purchasing hybrid buses.

Research applying multi-objective (MO) optimization for buses fleet is limited in the literature [13]. However, there are works applying MO optimization for other means of transport. Desai et al. [4] apply a multi-objective genetic algorithm to a MO model to minimize fuel consumption and HC (emissions of hydrocarbons), CO, NO_x, for managing a fleet of electric vehicles. Siddiqi et al. [17] propose a memory-efficient MO optimization algorithm to search for optimal path selection functions (OPS) in electric vehicles (EVs), using the MO model shortest path (MOSP). In Qiang et al. [15], a MO model is developed to train formation, the train counts, as well as operating periods, considering factors such as the carrying capacity, organizational requirements of traffic, the benefits of the corporation, passenger behavior and others. In Mishra et al. [13], a MO optimization model is proposed to maximize of TSWARL (Total System Weighted Average Remaining Life) and minimize NPC (Net Present Cost) of a buses fleet for maintaining the quality measure minimum standards.

A MO optimization model for fleet replacement is proposed and applied in Ercan et al. [6]. Six types of buses are considered: Diesel, Hybrid, BE, Biodiesel, CNG and LNG (Liquefied Natural Gas). These buses were tested in three distinct lines and the goal was to find the best composition of the fleet (fuel type) in each row (scenario). In this case, the results depend solely on the characteristics of each scenario and can be operated by a fleet of up to three types of buses. A MINIMAX scalarization approach based on the minimization of the relative distance between a candidate solution and the ideal solution is used. They conclude that electric buses will always be the best option for lines with a large number of stops and low average speed.

14.3 Augmented Weighted Tchebycheff Methods

The main goal of the MO optimization is to obtain the set of Pareto-optimal solutions that correspond to different tradeoffs between objectives. In this study, scalarization methods can be used to obtain an approximation to the Pareto-optimal set [12]. It should be noted that these approaches can be extended into an interactive form as presented in Hwang and Masud [10] and also in Ercan et al. [6]. During the search, the decision-maker indicates which objective value(s) must be improved and revises the aspiration levels of the corresponding goals and the problem is automatically modified with additional constraints.

In this work, the Augmented weighted Tchebycheff method [18] is applied to solve the fleet management multi-objective optimization model. This method has the advantage that it can converge to non-extreme final solutions and may also be applicable to integer linear and nonlinear multiple objective problems [18]. The Augmented Weighted Tchebycheff method can be formulated as:

$$\min \max[w_i|f_i(x) - z_i^\star|] + \rho \sum_{i=1}^{M} |f_i(x) - z_i| \tag{14.1}$$

where w_i are the weighting coefficients for objective i, $z_i^\star$ are the components of a reference point, and ρ is a small positive value [3]. The solutions of the optimization problem defined by Eq. 14.1 are Pareto-optimal solutions. This non-linear problem can be reformulated as a linear problem, as follows:

$$\begin{aligned} \min \quad & \lambda + \rho \sum_{i=1}^{M} (f_i(x) - z_i^\star) \\ \text{s.t.} \quad & w_i(f_i(x) - z_i^\star) - \lambda \leq 0. \end{aligned} \tag{14.2}$$

In this study, the optimization problem formulated in Eq. 14.2 was used to solve the fleet management multi-objective optimization problem. The objectives of this fleet management problem are to minimize, simultaneously, CO_2 emissions, other types of emissions (OTE) (CO, SO_2, NO_x, PM_{10}, $PM_{2.5}$), and the total cost of the life cycle.

The model considers a set V of different types of buses (diesel, electric and CNG buses). The buses operate in a set L of different lines that correspond to different driving conditions. The goal is to determine the number of buses P_{ij} of type $i \in V$ for a driving condition $j \in L$ that simultaneously minimize the CO_2 emissions, the other types of emissions, and the total cost.

The CO_2 emissions objective (Z_1) includes several components:

$$Z_1(P_{ij}) = \sum_{i=1}^{V} \sum_{j=1}^{L} (tc_{ij} + uc_{ij} + fc_{ij}) P_{ij} \tag{14.3}$$

where tc_{ij} are the CO_2 tailpipe emissions, uc_{ij} are the CO_2 emissions for manufacturing, infrastructure, maintenance, and battery replacement process, and fc_{ij} are the CO_2 emissions due to fuel generation for bus type $i \in V$ and driving condition $j \in L$.

The objective with respect to other types of emissions (Z_2) have several components:

$$Z_2(P_{ij}) = \sum_{i=1}^{V}\sum_{j=1}^{L}(tp_{ij} + up_{ij} + fp_{ij})P_{ij} \tag{14.4}$$

where tp_{ij} are the OTE tailpipe emissions, up_{ij} are the OTE emissions for manufacturing, infrastructure, maintenance, and battery replacement process, and fp_{ij} are the OTE emissions due to fuel generation for bus type $i \in V$ and driving condition $j \in L$.

The total cost objective (Z_3) considers different costs:

$$Z_3(P_{ij}) = \sum_{i=1}^{V}\sum_{j=1}^{L}(ic_i + if_i + mr_{ij} + op_{ij} - re_{ij})P_{ij} \tag{14.5}$$

where $ic_i$is the initial cost (purchase) and infrastructure cost for bus type ($i \in V$), and mr_{ij}, op_{ij} and re_{ij} are, respectively, the lifetime maintenance and repair costs, the lifetime operation costs, and the lifetime revenue, for bus type $i \in V$ in driving condition$j \in L$.

The goal is to minimize, simultaneously, the three objective functions $Z_1(P_{ij})$, $Z_2(P_{ij})$, and $Z_3(P_{ij})$ expressed, respectively, by Eqs. 14.3–14.5. Applying the Augmented Weighted Tchebycheff defined in Eq. 14.2, the mathematical formulation of the obtained scalarized problem can be expressed by:

$$\min \quad Q + \rho \sum_{i=1}^{3} |Z_i(P_{ij}) - z_i^{\star}| \tag{14.6}$$

$$\text{s.t.} \quad \sum_{i=1}^{V} P_{ij} = N_j, \forall j \in L \tag{14.7}$$

$$\beta_{co_2}(Z_1(P_{ij}) - z_1^{\star}) - Q \leq 0 \tag{14.8}$$

$$\beta_{OTE}(Z_2(P_{ij}) - z_2^{\star}) - Q \leq 0 \tag{14.9}$$

$$\beta_{COST}(Z_3(P_{ij}) - z_3^{\star}) - Q \leq 0 \tag{14.10}$$

$$P_{ij} \geq 0 \text{ and } P_{ij} \text{ integer } \forall i \in V, \forall j \in L \tag{14.11}$$

$$Q \geq 0 \tag{14.12}$$

In this formulation, the variable Q should be minimized, where ρ is a small positive value (Eq. 14.6). The decision variables P_{ij} are the number of buses of type $i \in V$ for a driving condition $j \in L$. The first constraint (Eq. 14.7) ensures that the total number of buses for each driving condition $j \in L$ is exactly N_j, i.e., the fleet size is exactly N_j buses per line. The weighted deviation constraints with respect to each objective are expressed by Eqs. 14.8–14.10. Equation 14.11 imposes that P_{ij} are non-negative and integer. Equation 14.12 constrains the slack variable Q to non-negative values.

The parameters of this problem are: N_j that is the number of buses for a driving condition $j \in L$, and βco_2, β_{OTE} and β_{COST} that are, respectively, the weights for CO_2 emissions, other types of emissions, and total cost. The weights are uniformly varied in the interval ranging from 0 to 1 such as $\beta co_2 + \beta_{OTE} + \beta_{COST} = 1$.

14.4 Case Study Application

The fleet management model and multi-objective optimization method was applied to the urban transportation system of the city of Joinville, located in the northern state of Santa Catarina in southern Brazil. Currently, the city has two licensee's companies to operate public transport services, where one operates in the north and the other in the southern region of the city. Both companies have a fleet of 364 diesel buses with different sizes and carrying about 3 million passengers per month. In this study, three different lines in the southern region of the city were chosen. South-North line (SN) is a DSL type line and it has Passengers by Kilometer Index (PKI) of 0.42. It aims to carry passengers from the southern district station to the north district station. This line has 37.5 km, 176 stop points and average operation speed of 9.77 km/h. The Itinga line (IT) is a type of SL and aims to transport passenger of Itinga district (estimated population of 6,847 inhabitants, IPPUJ [11]) to the south district station. This line has 17.1 km, 32 stop points, a PKI of 1.3 and average operation speed of 20.54 km/h. The South-Central line (ST) is type of CSL responsible for transporting passengers between the south district station and the central station. This line has 11.54 km, 40 stop points, a PKI of 1.35 and average operation speed of 18.34 km/h. It will be assumed for purposes of this study that we wish to purchase and operate of a fleet with 20 buses by line. This amount represents the number of buses required to operate these lines. In 2015, one BE of BYD Chinese Company has been tested in the city, to analyze its economic viability. The results to date were not been disclosed by the operating company. So, to analyze this viability of implementing a bus alternative fuel, this study considers four types of buses of different technologies (diesel, electric and CNG), as presented in Table 14.1. This data includes operating costs, maintenance, purchase (purchase bus age 0), generated revenue (ticket sales per line), and GHG emissions (per ton). The data is reported to 12 years, representing the life cycle (maximum age allowed by law) of a bus in Joinville. The inputs used in this study are real data obtained from the company for 2015. Table 14.1 presents the information concerning four types of buses (Diesel bus, BEs I, BEs II and CNG). BEs I and BEs II are battery electric buses with different charging times.

It is possible to observe that the investment costs of BEs are higher when compared to the other types of buses. This is mainly due to the high cost of batteries. In this case, battery replacements are not considered because manufacturers offer a 12-year warranty. This corresponds to the same time the life cycle of a bus in the city of Joinville. BEs II (fast charging) have the highest infrastructure cost, because it is a new technology that enables the loading of a bus in just 10 min [7]. The diesel bus type has greater autonomy without the need for filling or refilling and also consumes less fuel.

Table 14.1 Types of buses analysed in the case study and their characteristics

	Diesel bus	BEs I	BEs II	CNG bus
Bus length (m)	13.2	12	10.7	12.8
Capacity (passengers)	90	80	55	50
Fuel tank capacity	275 L	N/A	N/A	22,204scf at 3,600psi
Battery capacity	N/A	324 kWh	330 kWh	N/A
Autonomy (km) (af_i)	590	250	250	390
Fuel consumption (ef_i)	0.46 L/km	1.28 kW/km	1.34 kW/km	0.52 L/km
Fuel costs ($)	0.70/L	0.12/kWh	0.12/kWh	0.54/L
Initial cost ($) (ic_i)	330,000.00	900,000.00	904,490.00	575,000.00
Infrastructure cost ($) (if_i)	–	22,500.00	349,000.00	80,000.00
Recharging time (min) (rt_i)	–	300	10	–

However, the diesel fuel is more expensive. The bus length and passenger capacity of all types of buses are similar, with exception of BEs II bus type. Operating costs include the salaries of drivers, fuel costs and compulsory insurance. Maintenance costs consider the wages of the maintenance sector, lubricating oil costs, tires, parts and accessories. Costs were computed by multiplying unitary costs per kilometer by monthly kilometrage for each type of bus and line.

The maintenance and operation costs of a bus life cycle (by line type) life are shown in Fig. 14.1. Increases in maintenance and operating costs over the years (such as inflation, bus aging, rising fuel prices and others) were not considered. Revenue from the ticket sales was also taken into consideration. These values are measured by line type in PKI format. The PKI was obtained by multiplying the price of the ticket in 2015 ($ 0.86) by the daily kilometrage traveled by bus type and line type. The total annual kilometrage of the lines considered was of 534,717 km in ST line, 481,228 km in IT line and 423,750 km in SN line. Thus, a bus diesel type (with autonomy 500 km) can travel more kilometers than a BEs with a range of 250 Km and it also needs 5 hours to fully recharge.

All these data were used compute total cost objective Z_3 (Eq. 14.5). In this study, it was not considered any benefit in bus purchasing alternative fuels. The CO_2 emissions objective Z_1 and OTE objective Z_2 were computed considering data from Ercan et al. [6], due to the similarity between the analyzed lines. These data represent a simulation of 12 years of emissions (tons) by type of bus and line, considering total emissions of Fuel Supply (fuel production), Upstream (battery replacement, manufactory, maintenance, repair, and infrastructure) and Tailpipe as shown in Table 14.2. It is possible to identify that CNG bus is the most polluting in terms of CO_2 as well as OTE, as also referred in Aber [1]. The diesel bus type also has a high level of CO_2 emissions (mainly in line SN), much higher than the BEs buses. It should be noted that BEs have the lowest emission rate and for all the types of lines considered. This is because the level of tailpipe pollution is null.

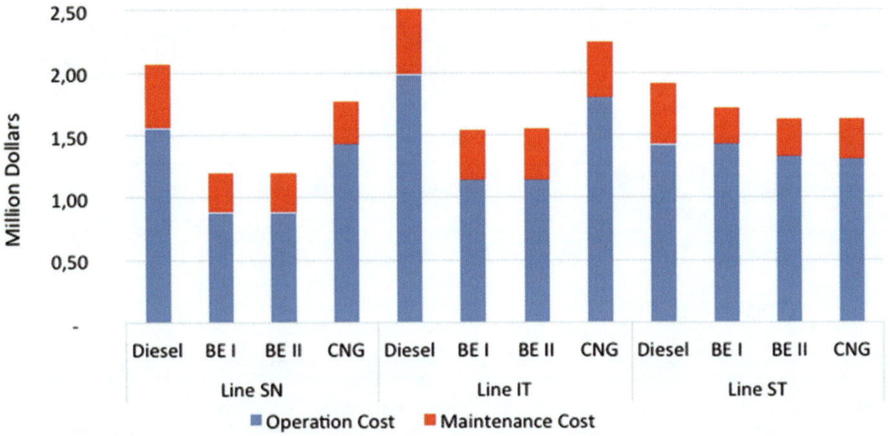

Fig. 14.1 Maintenance and operating costs per line and per type of bus

Table 14.2 Total emissions in the life cycle (12 years) per type of bus and per line (ton)

		Type of emission	
Line	Type of bus	CO_2	OTE
SN	Diesel	2819	10.48
	BEs I	1003	9.05
	BEs II	1003	9.05
	CNG	3035	1522.3
IT	Diesel	2024	8.41
	BEs I	1004	9.05
	BEs II	1004	9.05
	CNG	2053	976.2
ST	Diesel	2007	8.38
	BEs I	1003	9.05
	BEs II	1003	9.05
	CNG	2166	1037.78

14.5 Results and Discussion

The data and model to optimize a fleet of buses of the city of Joinville [11] with four types of buses ($V = 4$) and three different types of lines ($L = 3$) were implemented using MS Excel. This optimization problem has 12 integer non-negative decision variables, one non-negative continuous variable, three equality constraints and three inequality constraints. Optimization was performed using the solver add-in provided by MS Excel using a computer with a AMD A8-6410 APU processor with 8 GB of RAM and a AMD Radeon R5 Graphics (2.00 GHz). In this study, 66 different combinations of uniformly distributed weights (βco_2, β_{OTE} and β_{COST}) were considered. The number of weights combinations considered in this study is higher than the used by Ercan et al. [6]. The average computation time was 8 s. Sensitivity analysis was used to study the different trade-offs between the objectives. Each trade-off corresponds to different fleets constituted by different quantities of each bus type. Table 14.3 presents the 9 non-dominated solutions found using the Augmented Weighted Tchebycheff method. The trade-offs are represented in Fig. 14.2.

The solutions obtained are good compromises that can be considered by the public transport operator. In addition, it is observed that the quantities of buses of each type is directly related to the characteristics of each line (number of stops, average speed, among others). The BEs I is always the best choice for reducing emissions, but its high initial cost and low autonomy (generating less revenue) make it unfeasible in certain scenarios. The BEs low autonomy can be compensated by the use of the new

Table 14.3 Results of the Augmented Weighted Tchebycheff method

	Type of line						Objectives		
	SN line	IT line		ST line					
Solution	BEs I	Diesel	BEs I	Diesel	BEs I	CNG	Z_1 (ton)	Z_2 (ton)	Z_3 ($)
1	20	20				20	103,860	21,105	34,302,310
2	20	20		1		19	103,701	20,075	34,371,411
3	20	20		2		18	103,542	19,046	34,440,513
4	20	19	1	2		18	102,522	19,047	34,678,374
5	20		20		20		60,200	543	47,266,142
6	20	19	1	1		19	102,681	20,076	34,609,273
7	20	19	1			20	102,840	21,105	34,540,172
8	20	20		3		17	103,383	18,017	34,509,614
9	20	20		20			100,680	517	35,684,338

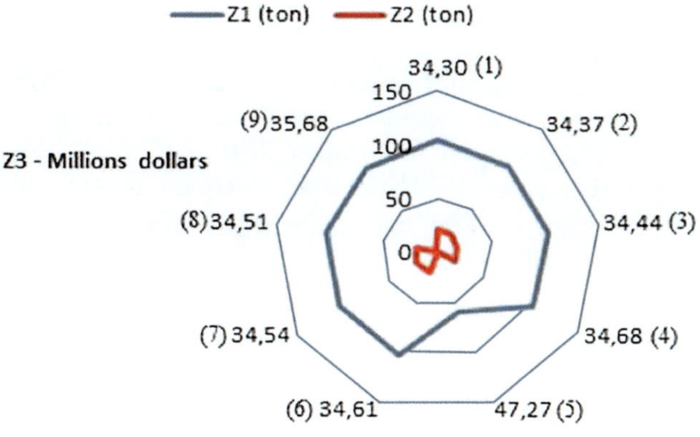

Fig. 14.2 Non-dominated solutions obtained by the Augmented Weighted Tchebycheff method

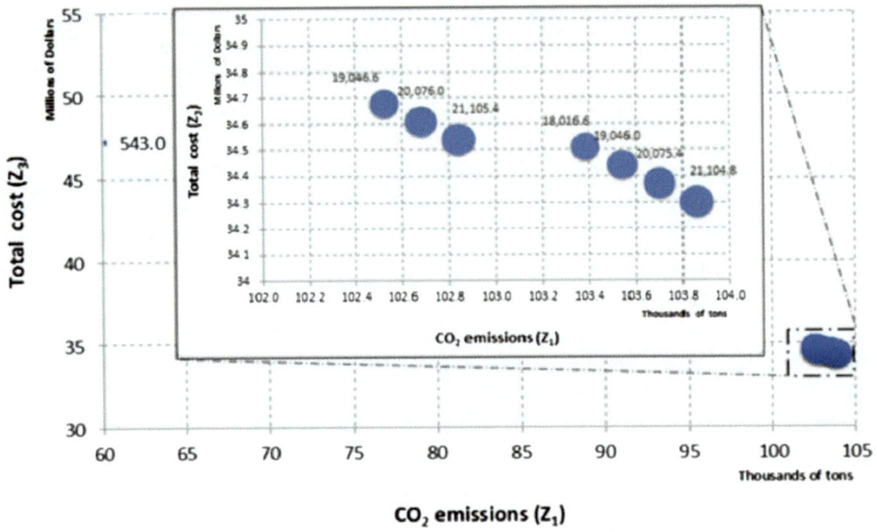

Fig. 14.3 Pareto front obtained by the Augmented Weighted Tchebycheff method

fast charging technology. Costs are still an important issue for operating companies, unless they can benefit of governmental subsidies [9]. However, in general, there is no legislation requiring that the operators replace conventional buses by more ecological vehicles. Thus, the operators may disregard the exchange of technologies due to the high investment required.

Figure 14.3 presents the Pareto front defined by non-dominated solutions for the three objectives. This graph allows perceiving the trade-offs between the nine solutions obtained. There is one solution that has smaller CO_2 emissions; in contrast, they have the largest total cost values. The other solutions have larger CO_2 emissions but smaller total cost values. They represent different trade-offs of other type of emissions.

Figure 14.4 shows the relative magnitude of the three objectives. It can be seen that solution 5 is the solution with the largest total cost (Z_3) and the smallest CO_2 emissions (Z_1). However, solution 5 is worse than solution 9 in terms of OTE values (Z_2), because OTE levels of the diesel buses in these lines (IT and ST) are smaller than the electric buses. Solution 1 has the largest CO_2 emissions (Z_1) and OTE values (Z_2), but the lowest total cost (Z_3).

In terms of the SN line, it is observed the predominance of BEs. The high initial cost for the purchase of BEs do not compromise that optimal solutions include this type of buses. This occurrence is directly related to operating costs since it is a line with a high number of stops and a very low average speed. In this situation, BEs are more suitable than other technologies because factors such as speed and stop number does not affect their energy efficiency, as it was also observed by Ribau et al. [16]. It should be stressed that diesel buses tend to consume more fuel due to these

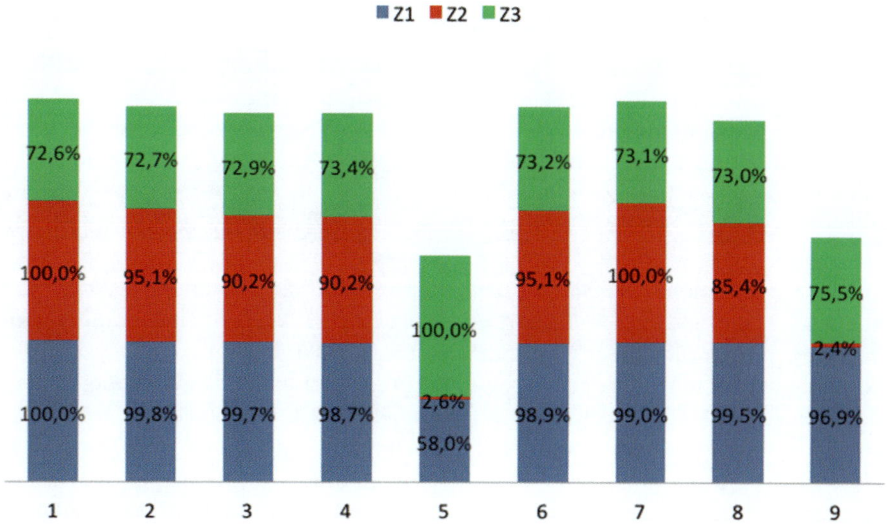

Fig. 14.4 Trade-offs between solutions

factors. In the study conducted by Ercan et al. [6], BEs have been identified as the most adequate in most of the Manhattan line, which has features like SN line.

For IT line, solutions with two types of technologies, diesel buses and BEs were obtained. This is due to IT line features such as a high average speed and the existence of few stops that are specific conditions for the adequate use of the diesel buses. This is remarkable when it is only analyzed the diesel buses costs compared with other technologies. Feng and Figliozzi [9] also observed that considering just costs the best solution is to use diesel buses. However, in cases of changes in fuel prices or subsidies in the purchase price BEs may be an alternative. This is mainly due to the high initial costs of BEs. When costs are disregarded the use of BEs become viable, except for the SN line.

Finally, for the ST line, the obtained solutions involve the highest number of bus types. The CNG bus type was the most selected. This type of buses is selected because the ST line has a higher number of stops and a lower average speed when compared with the IT line. Furthermore, the operation and maintenance costs are lower than diesel, which in this case makes it feasible. If the importance given to pollution levels is increased, diesel buses become a good option, because the CNG technology is the one with the largest OTE emission rate to produce fuel (Fuel Supply) and upstream. For this reason, it is important to consider pollution levels for fuel production and not only the upstream of the tailpipe. The CNG bus type is used due to its low price when compared to diesel bus type. The city of Seoul in South Korea, which replaced all diesel buses to CNG technology in the last decade is an example of this situation [2].

14.6 Conclusions and Further Work

In this work, a multi-objective model to optimize a fleet of buses was proposed. This model was optimized using the Augmented Weighted Tchebycheff method. This approach was used to search for the most efficient fleet of buses to the city of Joinville (Brazil), minimizing costs, CO_2 emissions and other types of emissions. Three different types of lines and four types of buses (Diesel, Electric I, Electric II with fast charging, and CNG) were considered.

The results demonstrate that this method provide efficient solutions that represent different trade-offs. Public transport operators can analyze and select the solution according to their preferences. In addition, it was observed that the choice of a bus is directly related to the characteristics of each line (number of stops, average speed, and among others). BEs I is always the best choice for reducing emissions, but its high initial cost and low autonomy (generating less revenue) make it unfeasible in certain scenarios. The low autonomy can be solved with the use of the new fast charging technology of BEs II, but the high costs are still a problem. However, in Brazil there is no governmental subsidy that encourages the purchase of electric vehicles, or legislation that obliges the operator to replace its conventional buses with alternative fuels. Therefore, the replacement of technologies may be disregarded by the operator due to the need for a high investment.

Further developments of this study will consider the application of different scalarization methods, such those described in Nurjanni et. [14]. It is also intended to develop visualization tools to assist the decision-making process. The inclusion of buses of newer technologies (zero emissions), such as the hydrogen hybrid bus (hybrid and electric motor) will be also addressed.

Acknowledgements This work has been supported by CNPq (National Counsel of Technological and Scientific Development, Brazil) and COMPETE: POCI-01-0145-FEDER-007043 and FCT - Fundação para a Ciência e Tecnologia within the Project Scope: UID/CEC/00319/2013.

References

1. J. Aber, Electric bus analysis for new york city transit. Master's thesis, Columbia University (2016)
2. U.-D. Choi, H.-K. Jeong, S.-K. Jeong, Commercial operation of ultra low floor electric bus for seoul city route, in *2012 IEEE Vehicle Power and Propulsion Conference*, 1128–1133 Oct 2012. https://doi.org/10.1109/VPPC.2012.6422619
3. K. Dächert, J. Gorski, K. Klamroth, An augmented weighted Tchebycheff method with adaptively chosen parameters for discrete bicriteria optimization problems. Comput. Oper. Res. **39**(12), 2929–2943 (2012), ISSN 0305-0548. https://doi.org/10.1016/j.cor.2012.02.021. http://www.sciencedirect.com/science/article/pii/S0305054812000470
4. C. Desai, F. Berthold, S.S. Williamson, Optimal drivetrain component sizing for a plug-in hybrid electric transit bus using multi-objective genetic algorithm, in *2010 IEEE Electrical Power Energy Conference*, 1–5 Aug 2010. https://doi.org/10.1109/EPEC.2010.5697242

5. W.M. Emiliano, P. Afonso, J. Telhada. Income-weighted life-cycle cost model: an application to the case of electric versus diesel buses state of the art, in *APMS International Conference*, (Springer, Berlin, 2016), pp. 1–8
6. T. Ercan, Y. Zhao, O. Tatari, J.A. Pazour, Optimization of transit bus fleet's life cycle assessment impacts with alternative fuel options. Energy 93(1), 323–334 (2015). ISSN 0360-5442, https://doi.org/10.1016/j.energy.2015.09.018. http://www.sciencedirect.com/science/article/pii/S0360544215012104
7. L. Eudy, R. Prohaska, K. Kelly, M. Post, *Foothill Transit Battery Electric Bus Demonstration Results*, NREL (2016). https://doi.org/10.2172/1237304. http://www.osti.gov/scitech/servlets/purl/1237304
8. W. Feng, M. Figliozzi, An economic and technological analysis of the key factors affecting the competitiveness of electric commercial vehicles: a case study from the usa market. Transp. Res. Part C: Emerg. Technol. **26**, 135–145 (2013). ISSN 1613-7159, https://doi.org/10.1016/j.trc.2012.06.007. http://www.sciencedirect.com/science/article/pii/S0968090X12000897
9. W. Feng, M. Figliozzi, Vehicle technologies and bus fleet replacement optimization: problem properties and sensitivity analysis utilizing real-world data. Public Transport **6**(1), 137–157 (2014). ISSN 1613-7159, https://doi.org/10.1007/s12469-014-0086-z
10. C.L. Hwang, *Multiple Objective Decision Making - Methods and Applications*, vol. 164, LNEMS (Springer, Berlin, 1979)
11. Instituto de Pesquisa e Planejamento para o Desenvolvimento Sustentável de Joinville (IPPUJ). https://ippuj.joinville.sc.gov.br/. Accessed 26 Feb 2017
12. K. Miettinen, *Nonlinear multiobjective optimization* (Kluwer Academic Publishers, Boston, 1999)
13. S. Mishra, S. Sharma, T. Mathew, S. Khasnabis, Multiobjective optimization model for transit fleet resource allocation. Transp. Res. Rec. J. Transp. Res. Board **2351**, 1–13 (2013). https://doi.org/10.3141/2351-01
14. K.P. Nurjanni, M.S. Carvalho, L. Costa, Green supply chain design: a mathematical modeling approach based on a multi-objective optimization model. Int. J. Prod. Econ. **183**(Part B), 421–432 (2017), ISSN 0925-5273. https://doi.org/10.1016/j.ijpe.2016.08.028. http://www.sciencedirect.com/science/article/pii/S0925527316302237
15. Z. Qiang, D. Lianbo, G. Wai, Z. Wenliang, Optimization method for train plan of urban rail transit with elastic demands, in *CICTP 2012* (American Society of Civil Engineers, 2012), pp. 1189–1199. https://doi.org/10.1061/9780784412442.180
16. J.P. Ribau, J.M.C. Sousa, C.M. Silva, Reducing the carbon footprint of urban bus fleets using multi-objective optimization. Energy **93**(Part 1), 1089–1104 (2015). ISSN 0360-5442, https://doi.org/10.1016/j.energy.2015.09.112. http://www.sciencedirect.com/science/article/pii/S0360544215013225
17. U.F. Siddiqi, Y. Shiraishi, S.M. Sait, Multi-objective optimal path selection in electric vehicles. Artif. Life Robot. **17**(1), 113–122 (2012). ISSN 1614-7456, https://doi.org/10.1007/s10015-012-0025-5
18. R.E. Steuer, *Multiple Criteria Optimization: Theory* (Computation and Application. John Wiley, New York, 1986)

Chapter 15
Multicriteria Location-Routing Problems with Sectorization

Alberto Martinho, Eduardo Alves, Ana Maria Rodrigues and José Soeiro Ferreira

Abstract Logistic decisions involving the location of facilities in connection with vehicle routing appear in many contexts and applications. Given a set of potential distribution centers (DC) and a group of clients, the choice of which DC to open together with the design of a number of vehicle routes, satisfying clients' demand, may define Location-Routing Problems (LRP). This paper contributes with a new method, the 4-Phase Method (4-PhM), to deal with Capacitated LRP. Relevant advantages of 4-PhM are its generality, the possibilities of handling Multiple-Criteria and of facing large dimension problems. This last aptitude is a consequence of the sectorization phases, which permit a simplification of the solution space. Sectors are constructed by two Simulated Annealing based procedures, and they follow SectorEl, a sectorization approach inspired by electrostatics. In the last phase, the results obtained are evaluated using multicriteria analysis. Here, decision makers play an important role by reflecting preferences in a pairwise comparison matrix of the Analytic Hierarchy Process. Computational results, based on randomly generated instances, confirm the expectations about 4-PhM and its potentiality to deal with LRP.

Keywords Location-routing problems · Sectorization · Multiple-criteria
Electrostatics · Simulated annealing

A. Martinho (✉) · E. Alves · A. M. Rodrigues
INESC TEC - Technology and Science, Campus da FEUP, Rua Dr. Roberto Frias 378, 4200-465 Porto, Portugal
e-mail: albertommmartinho@gmail.com

E. Alves
e-mail: eduardo.alves1901@gmail.com

A. M. Rodrigues
CEOS.PP - Centre for Organisational and Social Studies of P.Porto, Polytechnic of Porto, Rua Jaime Lopes de Amorim, 4465-111 S.Mamede de Infesta, Portugal
e-mail: amr@inesctec.pt,anamaria@iscap.ipp.pt

J. S. Ferreira
Faculty of Engineering and INESC TEC - Technology and Science, University of Porto, Campus da FEUP, Rua Dr. Roberto Frias 378, 4200-465 Porto, Portugal
e-mail: jsf@fe.up.pt

A. I. F. Vaz et al. (eds.), *Operational Research*, Springer Proceedings in Mathematics & Statistics 223, https://doi.org/10.1007/978-3-319-71583-4_15

15.1 Introduction

Combining facility location and vehicle routing is recognized as crucial to improve logistic systems and to reduce global costs. Examples of potential applications are diverse and they are found everywhere: food and drink distribution, waste collection, blood bank locations, technical support or paper distribution. The associated problems are usually entitled Location Routing Problems (LRP), and exact methods and heuristics have been used to deal with them. They may appear with many different characteristics. Reference [9], for instance, organize a classification of LRP according to eleven characteristics: hierarchical level; nature of demand; number of facilities; the size of vehicle fleets; vehicle capacity; facility capacity; facility layer; planning horizon; time restriction; objective function and types of model data.

The aim of this paper is to propose a quite general method to solve Capacitated LRP, the 4-Phase Method (4-PhM), which comprises Pre-sectorization, Sectorization, Routing and Multicriteria evaluation phases. The novelties are associated with the improvement and application of the new approach for sectoring SectorEl, which will be presented in the next section, the inclusion of dedicated heuristics and the articulation with AHP (Analytical Hierarchy Process) to handle multiple criteria. This last phase allows the evaluation of the obtained solutions, according to predefined criteria. Decision makers' preferences play a vital role here, affecting solution fitness. The 4 phases can be executed sequentially but they also can be adopted individually, as modules or parts of other LRP approaches, as it will be explained. It should be emphasized that the main merits of 4-PhM are its generality, the easy way it handles multicriteria and the potential to face real large-scale problems. This last aptitude comes from the incorporation of the sectorization phase, allowing for some hierarchy and simplification of the solution space.

This paper is organized as follows: Sect. 15.2 briefly reviews some concepts related to sectorization, in particular, the use of SectorEl, a new approach inspired by electrostatics and, at the end, multicriteria analysis in connection with AHP. The 4-PhM proposed in this paper is described in Sect. 15.3. Then, Sect. 15.4 provides new instances of location-routing problems and the computational results obtained. Everything is available online. Finally, Sect. 15.5 presents some conclusions.

15.2 Location-Routing Problems: Sectorization and Multiple-Criteria

This section about Location Routing Problems will essentially focus on sectorization and multicriteria analysis. It prepares and anticipates the 4-PhM, described in Sect. 15.3. It is supported by the creation of convenient sectors and by the observation of diverse relevant criteria. Furthermore, as sectorization is grounded on a new approach SectorEl inspired on electrostatics, that requires a brief description.

15.2.1 Location-Routing Problems

Location-Routing Problem (LRP) are a combination of two well-known problems:

- **Facility Location Problems (FLP)** - which consist of choosing the location of facilities or depots (for instance, warehouses), to minimize the cost related to the satisfaction of the demand and,
- **Routing Problems (RP)** - in which the objective is to find a set of circuits, each one for each vehicle, satisfying the demand of each customer, at minimum cost.

There are many types of LRP, depending on various characteristics. For instance, if vehicles and depots have a limited capacity, then the Capacitated LRP (CLRP) is considered. Most of the times, RP involved in LRP are related to node routing problems, named Vehicle RP (VRP). Quite fewer researchers have studied the RP as an arc routing problem, such as, for example, [12].

LRP appear in a large variety of contexts. Reference [10] show a table of LRP applications as vast as food and drink distribution, postbox location or military equipment location. Reference [5] include a survey of LRP models, solution techniques, and some case studies. [12] present another survey where 72 recent papers are analyzed. Both exact methods and heuristics have been used to solve LRP.

The typical CLRP presented (described in [1]) consists of a set of distribution centers (DC), represented by big red circles having a certain capacity, and a set of clients to serve, represented by the small blue squares, (see Fig. 15.1). Given the number of DC to install, the goal is to choose the DC to open and to design routes (that must begin and end at the DC) to serve the clients and respect the capacity of the transportation vehicle. Ultimately, the different criteria used to evaluate the solutions should embody decision maker's opinions and preferences as much as possible.

The 4-PhM has distinct advantages when dealing with CLRP in large-scale due to the stepwise decrease of complexity of the problems attained in each phase.

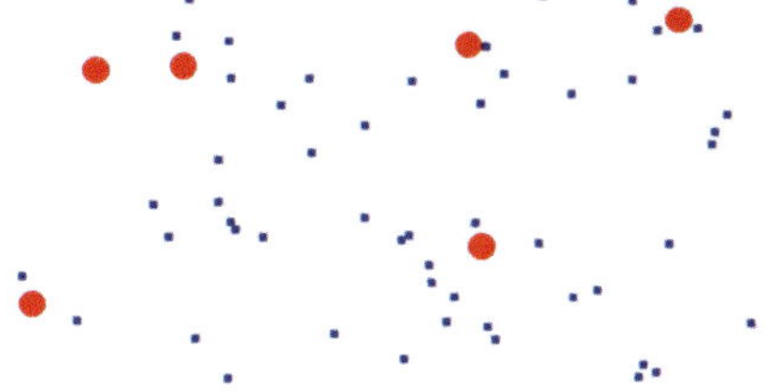

Fig. 15.1 CLRP with 100 clients and 10 DC

15.2.2 Sectorization and Multiple-Criteria

Sectorization means dividing into sectors or parts, a procedure that arises in many contexts and applications, usually to achieve some goal or to facilitate an activity. Most of the time, this division or partition aims at better organizing or simplifying a large problem into smaller sub-problems or promoting groups with similar characteristics. The concept of sectorization is not new. The first publications, [6], discusses the division of a territory applied to political districting.

More recently, the literature refers to many other applications, such as the division of sales territories by different sellers, school districting, salt spreading operations in the winter or waste collection.

Reference [14] present the original SectorEl approach, inspired by Electrostatics, in which Coulomb's Law was used as a metaphor for the "approximation" or "separation" of two points or elements, based on the weight or importance of each point/element and the distance between them.

Coulomb's Law states that:

"The force between a given pair of charges is inversely proportional to the square of the distance between them. [...] is directly proportional to the *quantity* of one charge multiplied by the *quantity* of the other," (in [7]).

Two electrical charged points with charges q_1 and q_2, at a distance of d_{12} will be subject to a force F with intensity given by:

$$F = k \cdot \frac{q_1 \cdot q_2}{d_{12}^2}$$

k represents the Constant of Coulomb and corresponds to $8.99 \times 10^9\,\mathrm{Nm^2C^{-2}}$.

The force is along the straight line joining the charges. If they have the same sign, the electrostatic force between them is repulsive; if they have different signs, the force between them is attractive.

Following this inspiration for the sectorization of a group of n points (and considering $k = 1$), the attraction between each pair of points, i and j, is calculated as

$$a_{ij} = \frac{q_i \cdot q_j}{d_{ij}^2}$$

where q_i and q_j represent the capacity of each point i and j.

Without details, and to conclude, points with greatest attraction should belong to the same sector; see [14]. Adaptations and extensions of the concepts of attraction and repulsion may be contemplated for realistic applications in sectorization.

When multiple-criteria are implied, it is not easy to classify a solution as "good" or "bad". Furthermore, the situation can get more complicate when some criteria appear in a quantitative range and others are manifested on a qualitative scale. Examples of more qualitative nature can be personal desires, experience and the intuition of a decision maker. In order to design sectors or to evaluate their quality (after sectorization) some criteria, quite common in the literature, may be employed:

- Equilibrium - different sectors must contain approximately the same "population" or "amount of work";
- Contiguity - each sector must be composed of "one body", that is to say, it should be possible to "move" between any pair of points in a sector without leaving the sector;
- Compactness - it is difficult to evaluate and it gives the idea of "concentration", that is to say, "U" shapes and dispersed territory should be avoided; instead, round shapes that are more compact should be preferred.

Analytic Hierarchy Process (AHP), developed by T. Saaty, is a well-known and important method to help deal with complex-decision problems. Regarding all criteria, the decision maker can construct a pairwise comparison (PC). Shortly referring to AHP, this comparison is made using a scale of importance in which the values represent how much more one criterion dominates another, (see [16, 17]). The intensity of importance is scaled from 1 to 9 and reciprocal values. For instance, if the intensity of importance when comparing criteria A and B is equal to 7, this means that criterion A has a very strong importance when compared to criterion B or, similarly, the intensity of importance when comparing criterion B with criterion A is equal to $1/7$. So, a pairwise comparison matrix (PCM) is formed. Dividing each element of the matrix by the sum of all the elements of the respective column, the relative weight is normalized. The priority vector is obtained by calculating the average across each row. The elements of the priority vector represent the relative weights associated with different criteria, according to the opinion of the decision maker.

15.3 The 4-Phase Method to Solve CLRP

The 4 phases of 4-PhM are now detailed:

- **Pre-sectorization phase** - A problem such as the one presented in Fig. 15.1: Decide which are the DC to be opened (selected DC must have enough capacity to serve the demand of all clients) and allocate clients to opened DC;
- **Sectorization phase** - For each DC opened, divide its clients into sectors respecting the capacity constraint of each sector;
- **Routing phase** - Define as many transportation routes as the number of sectors. Each route must start and finish in DC and visit all the clients that belong to the same sector;
- **Multicriteria evaluation phase** - Evaluate each solution according to all criteria and the decision maker's "judgments".

Each phase will be described in more detail in the next subsections.

Table 15.1 Importance value (v_{ij}) assigned to the DC i from client j

d_{ij}	$\left[0, \frac{d_{max}}{4}\right]$	$\left]\frac{d_{max}}{4}, \frac{d_{max}}{2}\right]$	$\left]\frac{d_{max}}{2}, \frac{3d_{max}}{4}\right]$	$\left]\frac{3d_{max}}{4}, d_{max}\right]$
v_{ij}	6	3	1	0

15.3.1 Pre-sectorization Phase

Given a set of potential DC locations, the importance of each DC is examined, taking into account the distances between each DC and each client. Thus, a "scoring" system is used where, from each client, each potential DC receives a certain score according to their proximity.

Let us assume n represents the total number of clients and m the number of DC. The importance of the DC i, I_i, $i = 1, \ldots, m$, is calculated by:

$$I_i = \sum_{j=1}^{n} v_{ij}$$

where v_{ij} is the importance value that DC i receives from client j. This value, as represented in Table 15.1, depends on the distance between them, d_{ij}, and on the largest distance between a DC and a client, d_{max}.

After this stage, it is possible to give a certain "probability of opening", P_i, to each DC i. This probability is directly proportional to the relative importance of the DC, I_i:

$$P_i = \frac{I_i}{\sum_{j=1}^{m} I_j}$$

That is to say, a DC with higher importance presents a higher probability of opening. A random process respecting the calculated probabilities selects the DC to open.

After choosing the group of DC to open, the customers are divided between them. The purpose of this pre-sectorization is to allocate clients to a specific, previously selected DC. This distribution of clients among the DC is done by "attraction", that means that a matrix of attraction between each DC i and each client j is used. It is proportional to $a_{ij} = \frac{1}{d_{ij}^2}$, where $q_i = q_j = 1$ and d_{ij} identifies the distance between i and j (see [13]). The clients' allocation to DC also depends on the DC's relative importance. Matrix $B = [b_{ij}]_{\substack{i=1,\ldots,m \\ j=1,\ldots,n}}$ portrays the normalized "attraction" between DC and clients in which, for each (i, j),

$$b_{ij} = \frac{a_{ij} \cdot P_i}{\sum_{i=1}^{m} (a_{ij} \cdot P_i)}$$

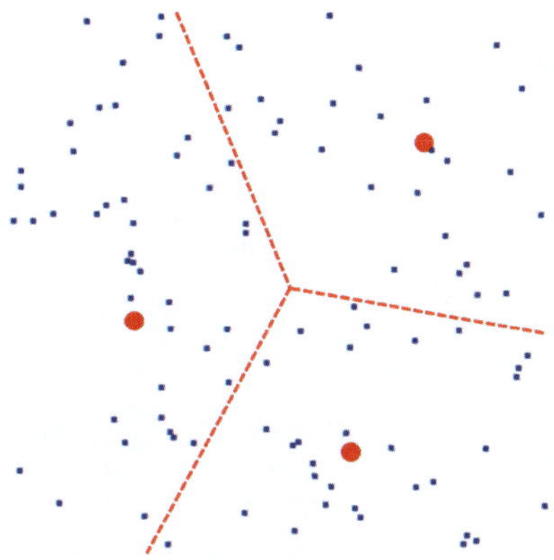

Fig. 15.2 Pre-sectorization phase: allocation of clients to the opened DC

Clients are allocated to the DC from the highest value of B to the lowest.

In Pre-sectorization, the decision agent has the possibility to choose the equilibrium of clients per center $c_{eq}(0 \leq c_{eq} \leq 1)$. So, depending on the number of open DC, f, it is possible to define a maximum number of clients per center n_{max}.

$$n_{max} = \frac{n}{f} \cdot (f - (f - 1)c_{eq})$$

For a set of 50 clients and 2 open DC, if $c_{eq} = 0$, then $n_{max} = 50$, thus a DC could possibly have 50 clients allocated. If $c_{eq} = 1$, then $n_{max} = 25$, thus each DC will have 25 clients allocated. c_{eq} reflects the importance attributed by the decision maker to the balance of clients by DC.

The effect of the pre-sectorization is visible in the red dashed lines in Fig. 15.2.

15.3.2 Sectorization Phase

Until this phase, the clients are allocated to DC. Now, client groups will be partitioned to form sectors. In this context, a sector embodies clients, served by the same DC, and that will be served by the same vehicle. Figure 15.5 illustrates the situation where the sectors are drawn with different colours and symbols.

15.3.2.1 Sectorization with SectorEl

The main idea inspiration behind SectorEl, that is to say, the analogy between the attraction in electrostatic and the "attraction" between points that should belong to the same sector, was the center of two different procedures presented in this paper. The paragraphs that follow are dedicated to different ways to produce sectors

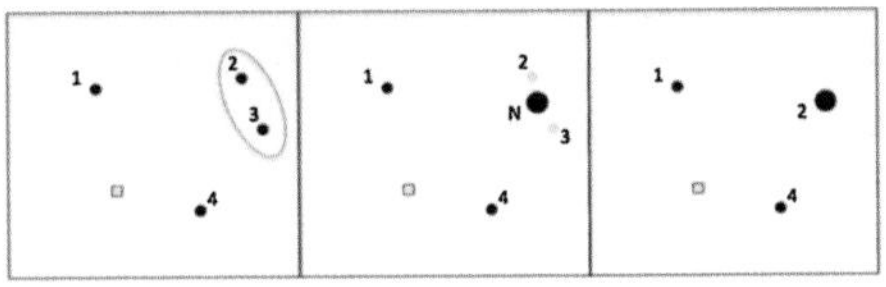

Fig. 15.3 Representation of one iteration in SectorEl-P

where "charges" and "distances" play a significant role. The procedures, both used to sectorize, are SectorEl-Pairs and SectorEl-Seeds.

SectorEl-Pairs (SectorEl-P)

In this procedure, the clients are seen as charged loads and the objective is to produce groups with a high level of "attraction". So, firstly, it is necessary to calculate the "attraction" between each pair of clients i and j with coordinates (x_i, y_i) and (x_j, y_j), respectively, using $a_{ij} = \frac{q_i \times q_j}{d_{ij}^{2-P}}$ where q_i and q_j represent the demand of clients i and j, respectively, d_{ij} the distance between these clients and $P = (n_{ip} - n^*) \times 0.08$ where n_{ip} is the number of initial points (number of clients allocated to the DC), n^* is the current number of sectors (in the first iteration it is equal to the number of clients allocated to the DC and it is decremented at each iteration: $n^* := n^* - 1$) and 0.08 is an adjustment factor according with the problem data type. P increases the denominator at each iteration, in order to counterbalance the continuous increase of the numerator at each iteration, avoiding the "black hole effect". Then an iterative process begins, where the two clients with the highest level of attraction are put together in one imaginary single point, representing the two clients. The coordinates x_N and y_N of this new point N are given by:

$$x_N = \frac{q_i}{q_i + q_j} \cdot x_i + \frac{q_j}{q_i + q_j} \cdot x_j$$

and

$$y_N = \frac{q_i}{q_i + q_j} \cdot y_i + \frac{q_j}{q_i + q_j} \cdot y_j$$

Also, this new point, N, has a "load", q_N, regarding the value of the loads of the two original points, i and j, given by $q_N = q_i + q_j$. Figure 15.3 illustrates this situation.

The clients that originated the new point are marked as "allocated to a sector" in order to not be part of the future calculation of the attraction matrix. This iteration process only stops when it is not possible to put together more pairs of points. This occurs when the junction of any pair of points implies exceeding the limit of the load that is allowed for each sector. This limit is specified by the capacity of the transportation vehicles.

SectorEl-Seeds (SectorEl-S)

SectorEl-S is a procedure based on a seeds algorithm. This kind of algorithm presents a search for solutions simultaneously in multiple locations in space, i.e., in this case,

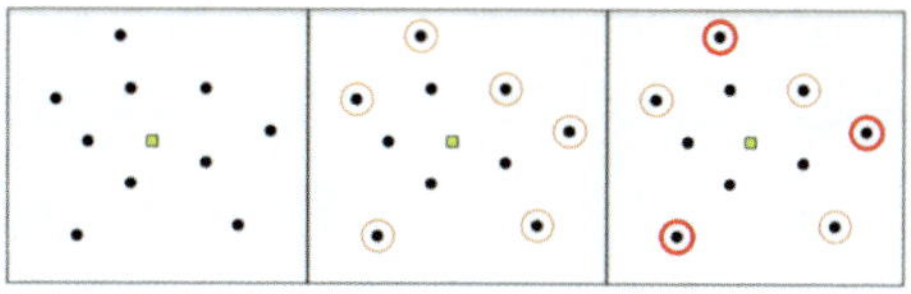

Fig. 15.4 Selecting three clients to be the "seeds" of three sectors

to allocate sectors, SectorEl-S selects one set of clients (seeds), far from each other, to be "initial clients", then these clients are associated to others who have not been associated with any sector. Each time a client is associated to a sector, i, the sector mass center (o_i) is updated and the attraction between the mass center of each sector and each client not allocated yet is calculated. First, to choose the initial clients, it is necessary to determine, for each open DC, how many sectors are required in order to form groups of clients without exceeding the capacity limit of the vehicle. This minimum number of required sectors (n_s) is given by dividing the total load value of N clients associated to the DC by the vehicle capacity (Cv), as shown next.

$$n_s = \frac{\sum_{j=1}^{N} q_j}{0.9 \times C_v}$$

After calculating the minimum number of sectors that are necessary, the initial customers of each sector are chosen. Obviously, the number of clients selected is equal to n_s. The clients to be chosen should, on one hand, be spaced apart from each other, which favors the independence between sectors and, on the other hand, apart from the DC in order to avoid the creation of a very close sector and others sectors too far away from the DC. This choice is made in two stages: first, the distances between clients and the DC, to which they belong, are analyzed and those who present more distant are marked. Then, from the set of clients marked, the number of clients that are actually needed is selected. The number of clients to be marked, in the first step, should be above the minimum number n_s sectors (in this case use $2 \times n_s$) since this set of marked clients is subjected, in the second step, to a new selection. This procedure of choosing clients is represented in Fig. 15.4.

15.3.3 Routing Phase

The objective of the routing phase is to find the shortest path between the clients, who constitute a sector, and the associated DC. The route must pass once in each client's location. Thus, the routing phase requires the solution of k Traveling Salesperson Problem (TSP) (see [4]), where k represents the number of sectors (Fig. 15.5).

If $G = (V, E)$ is a graph, V a set of n vertices and E a set of edges, then it is possible to determine $C = [c_{ij}]$, the distance or cost matrix associated to E, where i

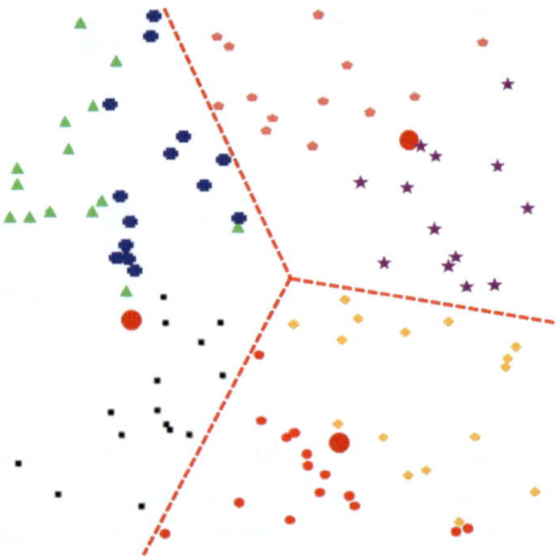

Fig. 15.5 Sectorization phase: creation of client sectors to each opened DC

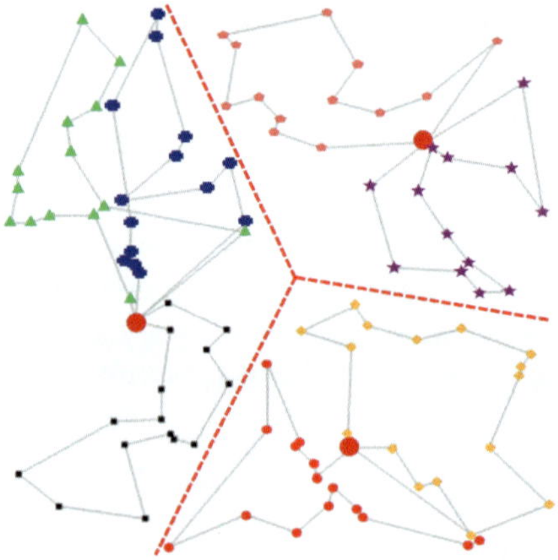

Fig. 15.6 Routing phase: design of routes for each sector

and j correspond to two different locations. "Any sequence of $p + 1$ integers taken from $(1, 2, \ldots, n)$, in which each of the n integers appears at least once and the first and last integers are identical is called a tour" ([2]). An optimal solution is a tour that minimizes the sum of all edge lengths or cost (c_{ij}) included.

It is quite common to use metaheuristics to solve TSP. In this phase of 4-PhM the well-known metaheuristic Simulated Annealing (SA) is applied [8, 11]. SA has its own way of escaping local optima by accepting worst solutions, depending on a certain probability, which decreases as the search for new solutions advances [3, 18].

Figure 15.6 depicts this phase. The routes of the various sectors appear in grey lines.

15.3.4 Multicriteria Evaluation Phase

The analysis of each solution will be held at the end of all iterations and includes the following parameters: load, distance and client equilibrium, compactness and solution cost. It is possible to replace these parameters with others or even add more to this set.

Compactness

Compactness is related to the concentration/density, i.e., it evaluates the concentration of customers around the mass center of the route in the analysis. A value of higher compactness means that the route has its customers close to each other and will hardly be crossed by another route. To better understand the compactness of a solution, this concept was associated with the idea of concentration or density. For each sector i ($i = 1, \ldots, k$) the value of compactness γ_i (see [15]) is defined as:

$$\gamma_i = \frac{q_i}{d(o_i, p_i)}$$

q_i is the capacity/load of the sector i, $i = 1, \ldots, k$, and d(o_i,p_i) represents the Euclidean distance between the mass center of the sector i (o_i) and the farthest point of the mass center, within the same sector (p_i).

To be able to compare the values of compactness of different sector solutions, it is necessary to calculate the compactness coefficient of variation (CV_γ). A higher density in a given sector is characterized by a high γ_i value.

Load Equilibrium

The balance of demand values between routes or load equilibrium can be important if the decision maker demands that all the vehicles travel with similar amounts of merchandize, or if a small "gap" between loads associated with each route is desired. To evaluate the load equilibrium of a sector, i.e., the amount that a vehicle has to carry in each sector, the coefficient of variation (CV_q) is used [15], calculated based on the amounts q_i, $i = 1, \ldots, k$, associated with each of the k sectors.

Distance Equilibrium

Another important criterion is the equilibrium of distances reflecting the "gap" between the distance that each freight vehicle travels. This distance equilibrium reflects, consequently, the travel hours each driver has to do to fulfill his route. To evaluate the balance of distances of a set of sectors, the coefficient of variation (CV_δ) is used. It is calculated for k sectors with the travel distances t_i, $i = 1, \ldots, k$.

Client Equilibrium

The equilibrium of clients per sector is the balance that exists between the number of clients associated with each route. To evaluate the equilibrium of clients of a set of sectors, the coefficient of variation (CV_c) is used. It is calculated for k sectors with a number of clients c_i, $i = 1, \ldots, k$.

For each one of the previous four criteria, it is essential to calculate the coefficient of variation:

$$CV_z = \frac{s'_z}{\overline{z}}$$

where

$$s'_z = \sqrt{\frac{1}{k-1}\sum_{i=1}^{k}(z_i - \overline{z})^2}$$

and z can assume the values of γ, q, δ and c according to the parameter being measured. $\overline{z}$ represents the average value of z.

Solution Cost

The cost related to the CLRP under analysis may have three main sources: the distance traveled, opening a DC and the number of vehicles.

Here, only the cost of the distance traveled for each vehicle and the opening cost of each DC were considered. Therefore, the cost of each solution C is calculated by the sum of the distance traveled in each sector t_i, $i = 1, \ldots, k$, plus the cost of each opened DC O_l, $l = 1, \ldots, f$:

$$C = \sum_{i=1}^{k} t_i + \sum_{l=1}^{f} O_l$$

In order to compare the solutions, the results for each solution must be normalized. Being r_{ab} the result obtained for each criterion b, the solution a is divided by the sum of all solutions, the solutions fit is obtained using the expression:

$$Fit_a = \sum_{b=1}^{5} w_b \times r_{ab}$$

$w = [w_b]_{b=1,\ldots,5}$ is the normalized principal eigenvector of AHP PCM. It is a priority vector that expresses the importance of each b criterion for a certain decision maker.

The lower the cost and the coefficients of variation, the better the solution. Therefore, the solution with the lowest Fit is the best obtained.

Table 15.2 illustrates a hypothetic situation in which 3 criteria and 4 alternatives (4 solutions) are considered. The 4 solutions are obtained using the SectorEl-P procedure and with the same AHP weights. It is, then, possible to score the solutions, the best solution being solution B. The absolute values of solution B are then compared with the absolute values of the solution with the best score of the SectorEl-S procedure, for the same AHP weights. The values are normalized and the Fit is again calculated.

Table 15.2 Multicriteria evaluation phase

Solutions	Criterion 1	Criterion 2	Criterion 3	Fit	Score
Solution A	r_{a1}	r_{a2}	r_{a3}	Fit_a	4
Solution B	r_{b1}	r_{b2}	r_{b3}	Fit_b	1
Solution C	r_{c1}	r_{c2}	r_{c3}	Fit_c	3
Solution D	r_{d1}	r_{d2}	r_{d3}	Fit_d	2

Table 15.3 Pair-wise comparison - DM1

	Compactness	Load equilibrium	Distance equilibrium	Solution cost	Client equilibrium
Compactness	1	6	6	6	3
Load equilibrium	$\frac{1}{6}$	1	1	1	$\frac{1}{2}$
Solution cost	$\frac{1}{6}$	1	1	1	$\frac{1}{2}$
Distance equilibrium	$\frac{1}{6}$	1	1	1	$\frac{1}{2}$
Client equilibrium	$\frac{1}{3}$	2	2	2	1

Table 15.4 Pair-wise comparison - DM2

	Compactness	Load equilibrium	Distance equilibrium	Solution cost	Client equilibrium
Compactness	1	1	$\frac{1}{4}$	$\frac{1}{8}$	1
Load equilibrium	1	1	$\frac{1}{4}$	$\frac{1}{8}$	1
Distance equilibrium	4	4	1	$\frac{1}{2}$	4
Solution cost	8	8	2	1	8
Client equilibrium	1	1	$\frac{1}{4}$	$\frac{1}{8}$	1

Table 15.5 Priority vector

Criteria	DM1	DM2
Compactness	0.5455	0.0667
Load equilibrium	0.0909	0.0667
Distance equilibrium	0.0909	0.2667
Solution cost	0.0909	0.5333
Client equilibrium	0.1818	0.0667

Table 15.6 Computational results - DM1

#DC	Procedure	Fit					
		9 × 3	11 × 3	25 × 5	30 × 5	100 × 10_1	100 × 10_2
2	SectorEl-S	0.613	0.544	0.530	0.510	**0.417**	**0.405**
	SectorEl-P	**0.387**	**0.456**	**0.470**	**0.490**	0.583	0.595
3	SectorEl-S	-	-	-	-	**0.492**	**0.460**
	SectorEl-P	-	-	-	-	0.508	0.540
4	SectorEl-S	-	-	-	-	**0.444**	0.521
	SectorEl-P	-	-	-	-	0.556	**0.479**
5	SectorEl-S	-	-	-	-	0.538	0.591
	SectorEl-P	-	-	-	-	**0.462**	**0.409**

15.4 Computational Results

This section includes the computational tests and results by applying the 4-PhM to CLRP. These tests and results were based on six different sets of locations ("9 × 3", "11 × 3", "25 × 5", "30 × 5", "60 × 6" and "100 × 10"). For the defined points in "60 × 6" and "100 × 10" fifteen instances were created, where the location of the centers is different. This way, the nomenclature of these instances presents the following form: 100 × 10_*i* and 60 × 60_*i*, with $i = 1, \ldots, 15$. Thus, a total of 34 instances (available in www.inescporto.pt/~jsoeiro/CLRP_SectorEl) were generated. The name of the defined instances reflects the client's quantity and the possible existing DC. For example, instance "9 × 3" represents a set of nine clients and three possible DC. All clients have an *id*, cartesian coordinates and load associated. All possible DC have an *id*, cartesian coordinates, load and opening cost associated.

Each instance was tested both for the SectorEl-P and the SectorEl-S. In both procedures, Simulated Annealing was used for the routing problem.

All obtained solutions were evaluated using five criteria. Two decision makers with different priorities were invited: Decision Maker 1 (DM1) and Decision Maker 2 (DM2). DM1 privileges compactness and client equilibrium, the objective being, for instance, to set an approximate number of clients and sales territory for each salesperson (see Table 15.3). DM2 privileges solution cost and distance equilibrium, the objective being to ensure, for example, an approximate traveled distance for each driver of a delivery company (see Table 15.5). The priority vectors are presented in Tables 15.5 and 15.4.

Tables 15.6 and 15.7 contain the solutions obtained after applying the AHP method for DM1 and for DM2, respectively. The solution evaluation of each criteria is presented in Tables 15.8 and 15.9, for DM1 and for DM2, respectively. The values are normalized, as required by the AHP. For instances with a reduced number of customers, no more than 2 distribution centers were considered. In particular, for the instances with 9 and 11 clients, only 3 hypothetical distribution centers were available, of which 2 were chosen.

Table 15.7 Computational results - DM2

#DC	Procedure	Fit					
		9 × 3	11 × 3	25 × 5	30 × 5	100 × 10_1	100 × 10_2
2	SectorEl-S	0.536	0.589	**0.406**	0.520	**0.435**	**0.453**
	SectorEl-P	**0.464**	**0.411**	0.594	**0.480**	0.565	0.547
3	SectorEl-S	-	-	-	-	**0.356**	**0.461**
	SectorEl-P	-	-	-	-	0.644	0.539
4	SectorEl-S	-	-	-	-	0.526	**0.427**
	SectorEl-P	-	-	-	-	**0.474**	0.573
5	SectorEl-S	-	-	-	-	0.536	**0.499**
	SectorEl-P	-	-	-	-	**0.464**	0.501

In the next section we discuss our findings and draw conclusions regarding all these computational results.

15.5 Conclusions

This work proposed the new 4-Phase Method (4-PhM) to deal with Capacitated Location-Routing Problems (CLRP). It is quite general, for it can handle various types of CLRP, with its own additional advantage for large-scale practical problems. In fact, the inclusion of a Sectorization phase may ease a complex situation by conveniently reducing the solution space. Moreover, the Multicriteria Phase approximates the 4-PhM to realistic requirements, which are taken into consideration/evaluation when evaluating the solutions within the various criteria. The paper takes into account and defines five relevant criteria.

Phase 1 divides clients by open Distribution Centres (DC) and a decision maker can restrict a maximum difference between them. Phase 2 designs sectors inside the initial pre-sectors using a new approach inspired in electromagnetism. Two procedures were used: SectorEl-P and SectorEl-S. Each created sector will have one DC and will be assigned one vehicle. Phase 3 generates routes that connect each sector to the respective centre, by resorting to the metaheuristic Simulated Annealing. Finally, Phase 4 compares the best solutions of SectorEl-P and SectorEl-S, according to five criteria. SectorEl-S usually produces solutions with a balanced number of clients in each sector, the major feature distinguishing the two heuristics. SectorEl-P usually produces solutions with lower traveling costs and more compact sectors. AHP allows for the comparison of both procedures in diverse situations. Commonly, SectorEl-P obtains better fit when there are fewer clients for each DC while SectorEl-S obtains better fit when the decision maker privileges the client equilibrium in each sector.

Figures 15.7 and 15.8 comprise two figures each. Each figure presents the convex polygons of each sector.

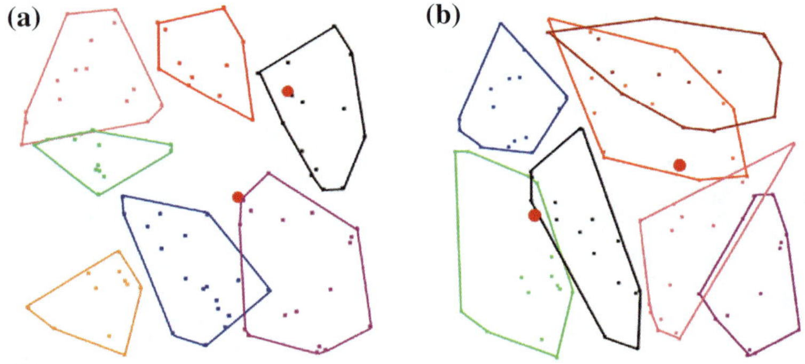

Fig. 15.7 Different solutions obtained for 100 clients and 2 DC applying **a** SectorEl-P for DM2; **b** SectorEl-S for DM2

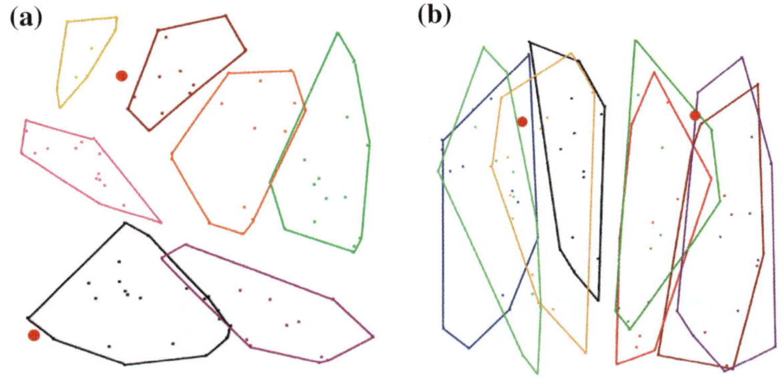

Fig. 15.8 Different solutions obtained for 100 clients and 2 DC applying **a** SectorEl-P for DM1; **b** SectorEl-S for DM1

The two procedures generate different solutions. While SectorEl-P privileges the creation of sectors in regions with high client density (see Fig. 15.7a), SectorEl-S originates sectors that can, sometimes, cover the same area (see Fig. 15.7b). The "geographical" characteristics involved in these procedures allow concluding that, for instance, for a Political District Problem, SectorEl-S would not produce viable solutions, while SectorEl-P could perform very well.

In a scenario where the DC are located in the periphery, it is possible to notice that SectorEl-P produces sectors with a small area that still continues to privilege high client density (see Fig. 15.8a), while SectorEl-S sectors have larger common areas (see Fig. 15.8b). Both procedures allow a large set of options, depending on the decision makers' will. SectorEl-S promotes a greater proximity between vehicles with diverse routes, while SectorEl-P allows deliveries within the same time-space, due to the client's proximity.

Table 15.8 Normalized values - DM1

ID	Vehicles capacity	#DC	Procedure	Compact.	Load Eq.	Distances Eq.	Cost	Clients Eq.	Time(s)	Fit
9 × 3	7000	2	SectorEl-S	0.699	0.565	0.514	**0.464**	**0.500**	21.3	0.613
			SectorEl-P	**0.301**	**0.435**	**0.486**	0.536	**0.500**	18.3	**0.387**
11 × 3	2000	2	SectorEl-S	**0.405**	**0.486**	0.974	0.564	0.763	35.6	0.544
			SectorEl-P	0.595	0.514	**0.026**	**0.436**	**0.237**	20.5	**0.456**
25 × 5	3000	2	SectorEl-S	0.517	0.821	0.582	0.550	**0.387**	462.5	0.530
			SectorEl-P	**0.483**	**0.179**	**0.418**	**0.450**	0.613	559.4	**0.470**
30 × 5	6000	2	SectorEl-S	0.565	**0.482**	0.508	**0.494**	**0.369**	235.8	0.510
			SectorEl-P	**0.435**	0.518	**0.492**	0.506	0.631	659.5	**0.490**
100 × 10_1	9000	2	SectorEl-S	0.511	**0.363**	**0.130**	0.576	**0.225**	17596.1	**0.417**
			SectorEl-P	**0.489**	0.637	0.870	**0.424**	0.775	18155.1	0.583
		3	SectorEl-S	0.608	**0.492**	**0.285**	0.544	**0.220**	10282.7	**0.492**
			SectorEl-P	**0.392**	0.508	0.715	**0.456**	0.780	9850.4	0.508
		4	SectorEl-S	0.511	0.503	**0.279**	0.508	**0.261**	13322.8	**0.444**
			SectorEl-P	**0.489**	**0.497**	0.721	**0.492**	0.739	12459.7	0.556
		5	SectorEl-S	0.586	0.537	**0.495**	0.533	**0.416**	20325.1	0.538
			SectorEl-P	**0.414**	**0.463**	0.505	**0.467**	0.584	20553.7	**0.462**
100 × 10_2	9000	2	SectorEl-S	**0.373**	**0.435**	**0.499**	0.544	**0.371**	17830.7	**0.405**
			SectorEl-P	0.627	0.565	0.501	**0.456**	0.629	16316.3	0.595
		3	SectorEl-S	0.503	0.555	**0.405**	**0.492**	**0.294**	16577.5	**0.460**
			SectorEl-P	**0.497**	**0.445**	0.595	0.508	0.706	16316.4	0.540
		4	SectorEl-S	**0.493**	0.514	0.773	0.529	**0.478**	20595.7	0.521
			SectorEl-P	0.507	**0.486**	**0.227**	**0.471**	0.522	19750.6	**0.479**
		5	SectorEl-S	0.573	0.619	0.680	**0.465**	0.649	22398.9	0.591
			SectorEl-P	**0.427**	**0.381**	**0.320**	0.535	**0.351**	20994.7	**0.409**

Table 15.9 Normalized values - DM2

ID	Vehicles capacity	#DC	Procedure	Compact.	Load Eq.	Distances Eq.	Cost	Clients Eq.	Time(s)	Fit
9 × 3	7000	2	SectorEl-S	0.569	**0.465**	0.656	**0.484**	**0.500**	36.3	0.536
			SectorEl-P	**0.431**	0.535	**0.344**	0.516	**0.500**	33.5	**0.464**
11 × 3	2000	2	SectorEl-S	0.655	**0.464**	0.647	0.567	0.591	68.5	0.589
			SectorEl-P	**0.345**	0.536	**0.353**	**0.433**	**0.409**	37.8	**0.411**
25 × 5	3000	2	SectorEl-S	**0.450**	**0.273**	**0.131**	0.584	**0.168**	443.4	**0.406**
			SectorEl-P	0.550	0.727	0.869	**0.416**	0.832	655.7	0.594
30 × 5	6000	2	SectorEl-S	0.568	0.535	0.570	0.506	**0.367**	229.4	0.520
			SectorEl-P	**0.432**	**0.465**	**0.430**	**0.494**	0.633	447.7	**0.480**
100 × 10_1	9000	2	SectorEl-S	0.725	0.536	**0.220**	0.520	**0.216**	17908.9	**0.435**
			SectorEl-P	**0.275**	**0.464**	0.780	**0.480**	0.784	18855.7	0.565
		3	SectorEl-S	**0.340**	**0.285**	**0.189**	**0.474**	**0.162**	14046.3	**0.356**
			SectorEl-P	0.660	0.715	0.811	0.526	0.838	14688.5	0.644
		4	SectorEl-S	0.532	0.505	**0.496**	0.542	0.530	11414.1	0.526
			SectorEl-P	**0.468**	**0.495**	0.504	**0.458**	**0.470**	10461.8	**0.474**
		5	SectorEl-S	0.623	0.542	0.560	0.523	**0.452**	20635.2	0.536
			SectorEl-P	**0.377**	**0.458**	**0.440**	**0.477**	0.548	21850.2	**0.464**
100 × 10_2	9000	2	SectorEl-S	0.564	0.503	**0.337**	0.517	**0.243**	13256.8	**0.453**
			SectorEl-P	**0.436**	**0.497**	0.663	**0.483**	0.757	13040.9	0.547
		3	SectorEl-S	**0.478**	**0.473**	**0.417**	**0.488**	**0.387**	16513.5	**0.461**
			SectorEl-P	0.522	0.527	0.583	0.512	0.613	16801.1	0.539
		4	SectorEl-S	0.663	**0.382**	**0.308**	**0.485**	**0.253**	18619.4	**0.427**
			SectorEl-P	**0.337**	0.618	0.692	0.515	0.747	17064.0	0.573
		5	SectorEl-S	0.517	0.542	**0.369**	0.540	0.627	20553.4	**0.499**
			SectorEl-P	**0.483**	**0.458**	0.631	**0.460**	**0.373**	21105.2	0.501

Some novelties of the paper include: improving and applying the new approach SectorEl, using pre-sectorization, developing special dedicated heuristics and the articulation with AHP to handle multiple criteria. Moreover, new instances of location-routing problems are provided.

As a conclusion, the general method proposed in this paper, 4-PhM, seems to have great potential to deal with large multicriteria LRP.

Acknowledgements This work is financed by the ERDF – European Regional Development Fund through the Operational Programme for Competitiveness and Internationalisation - COMPETE 2020 Programme within project "POCI-01-0145-FEDER-006961", and by National Funds through the FCT – Fundao para a Cincia e a Tecnologia (Portuguese Foundation for Science and Technology) as part of project UID/EEA/50014/2013.

Appendix A

Tables 15.8 and 15.9 illustrate the normalized values for each considered criterion regarding DM1 and DM2, respectively.

References

1. S. Barreto, C. Ferreira, J. Paixão, B.S. Santos, Using clustering analysis in a capacitated location-routing problem. Eur. J. Oper. Res. **179**(3), 968–977 (2007)
2. M. Bellmore, G.L. Nemhauser, The traveling salesman problem: a survey. ACM Comput. Surv. (CSUR) **16**(3), 538–558 (1968)
3. C. Blum, A. Roli, Metaheuristics in combinatorial optimization: overview and conceptual comparison. ACM Comput. Surv. (CSUR) **35**(3), 268–308 (2003)
4. G. Gutin, A. Yeo, A. Zverovitch, *The Traveling Salesman Problem and its Variations* (Kluwer Academic Publishers, Dordrecht, 2007)
5. A. Hassanzadeh, L. Mohseninezhad, A. Tirdad, F. Dadgostari, H. Zolfagharinia, Location-routing problem. In: *Facility Location. Concepts, Models, Algorithms and Case Studies* (2009), pp. 395–417
6. S.W. Hess, J.B. Weaver, H.J. Siegfeldt, J.N. Whelan, P.A. Zitlau, Nonpartisan political redistricting by computer. Oper. Res. **13**(6), 998–1006 (1965)
7. A.F. Kip, *Fundamentals of Electricity and Magnetism*, 2nd edn. (McGraw-Hill Inc., New York, 1969)
8. S. Kirkpatrick, C.D. Gelatt, M.P. Vecchi, Optimization by simulated annealing. Science **220**(4598), 671–681 (1983)
9. H. Min, V. Jayaraman, R. Srivastava, Combined location-routing problems: a synthesis and future research directions. Eur. J. Oper. Res. **108**(1), 1–15 (1998)
10. G. Nagy, S. Salhi, Location-routing: issues, models and methods. Eur. J. Oper. Res. **177**(2), 649–672 (2007)
11. M. Pirlot, General local search methods. Eur. J. Oper. Res. (1996)
12. C. Prodhon, C. Prins, A survey of recent research on location-routing problems. Eur. J. Oper. Res. **238**(1), 1–17 (2014)
13. A.M. Rodrigues, Sectores e Rotas na Recolha de Resíduos Sólidos Urbanos. Ph.D. thesis, Faculdade de Engenharia da Universidade do Porto, 2014

14. A.M. Rodrigues, J.S. Ferreira, Sectors and routes in solid waste collection, in *Operational Research*, ed. by J.P. Almeida, et al. CIM Series in Mathematical Sciences, vol. 4 (Springer International Publishing, Switzerland, 2015), pp. 353–375
15. A.M. Rodrigues, J.S. Ferreira, Measures in sectorization problems, in *Operations Research and Big Data*, ed. by A.P.F.D. Barbosa Póvoa, J.L. de Miranda. Studies in Big Data, vol. 15 (Springer International Publishing, Switzerland, 2015), pp. 203–211
16. T.L. Saaty, How to make a decision: the analytic hierarchy process. Eur. J. Oper. Res. **48**, 9–26 (1990)
17. T.L. Saaty, Decision making with the analytic hierarchy process. Int. J. Serv. Sci. **1**(1), 83 (2008)
18. G. Zapfel, R. Braune, *Metaheuristic Search Concept: A Tutorial with Applications to Production and Logistics* (Springer Science+Business Media, Heidelberg, 2010)

Chapter 16
A Decomposition Approach for the Long-Term Scheduling of a Single-Source Multiproduct Pipeline Network

William Hitoshi Tsunoda Meira, Leandro Magatão, Susana Relvas, Ana Paula Ferreira Dias Barbosa Póvoa and Flávio Neves Junior

Abstract This paper proposes a decomposition approach combining heuristic algorithms and Mixed Integer Linear Programming (MILP) models to solve the long-term scheduling of a multiproduct pipeline connecting a single-source to multiple distribution centers. The solution considers many operational aspects, such as simultaneous deliveries, pipeline maintenance periods, deliveries of multiple products during the same pumping run, and rigorous inventory control. A long-term scheduling problem from the literature was solved to validate the proposed approach. This problem is composed of a straight pipeline connecting a refinery to 3 distribution centers and transporting 4 different oil derivatives. The approach was able to obtain an operational solution in less than half a minute of CPU time. Moreover, additional tests using the same scenario were executed in order to analyze the performance of the developed decomposition approach.

Keywords Multiproduct pipeline · Scheduling · Decomposition approach
Mixed integer linear programming · Real-world application

16.1 Introduction

Improvements in the transportation activities of the petroleum and its derivatives products to the consumption areas are one of the main challenges in the oil industry, where any increase in performance and better usage of the resources generates considerable companies profits. The transportation is carried mainly through roads,

---

W. H. T. Meira (✉) · L. Magatão · F. Neves Junior
Graduate Program in Electrical and Computer Engineering,
Federal University of Technology - Paraná, Curitiba, Brazil
e-mail: william.meira@tecnico.ulisboa.pt

S. Relvas · A. P. F. D. Barbosa Póvoa
CEG-IST - Centre for Management Studies, Instituto Superior Técnico,
Universidade de Lisboa, Lisboa, Portugal

A. I. F. Vaz et al. (eds.), *Operational Research*, Springer Proceedings
in Mathematics & Statistics 223, https://doi.org/10.1007/978-3-319-71583-4_16

railroads, vessels, and pipelines. The latter is the most used distribution modal, justified by their high volume capacity, reliability, economy, and safety in relation to the others [15]. A pipeline system usually transports more than one product through the same pipeline and connects a single-source (refinery) to one or more distribution centers (DCs). More complex network topologies are also possible, including networks with multiple sources, DCs that can act also as inputs (dual purpose), multiples pipelines that form tree-structures or mesh-structures. This paper focuses on the topology with a single-source and multiple destinations, where a common operation consists of the refinery pumping batches of products into a straight multiproduct pipeline in order to attend the local demand of the connected DCs, with the lowest possible cost.

Authors have proposed many approaches to solve pipeline scheduling, each of them considering different sets of operational constraints, characteristics, and techniques. In the literature, it is common to find knowledge-based search heuristics [15], mathematical programming approaches based on mixed integer linear programming (MILP) [12] or mixed integer non-linear programming (MINLP) formulations [6], and decomposition strategies [4].

On the studied problem, a pipeline with single-source and multiple DCs, one of the first work was presented by Rejowski and Pinto [12] who formulated a discrete MILP model to solve a real-world pipeline network composed of a refinery and multiple DCs. The volume between two consecutive terminals was considered as a pipeline segment and each segment divided into volumes batches. Two models were considered, one with a fixed volume batch size and other with different ones. The objective function was to minimize the costs of storage, pumping, and interface. In order to increase the performance of their previous work, Rejowski and Pinto [13] included special constraints and valid cuts in the model. Still, Rejowski and Pinto [14] proposed a new continuous time MINLP model for the same problem. The new formulation allowed pumping rate variations and also considered some hydraulic characteristics.

Cafaro and Cerdá [1] proposed an MILP continuous formulation for the scheduling of a single-source multiproduct pipeline network with multiple destinations. The objective was to minimize the interface, pumping and inventory costs attending the demand on time. Later on, Cafaro and Cerdá [2] considered a multi-period rolling horizon approach for the long-term scheduling of the same problem and generalized their previous work. Cafaro et al. [3] proposed a two-level hierarchical decomposition. The upper level determines the batches and their pumping sequence during each pumping run, solved by the continuous model presented in [1, 2]. The lower level determines the output operations, the sequence and assigned depot of each lot, and the times when each valve and pump should be turned on/off. The authors presented a continuous-time MILP model in combination with a discrete-event simulation method. Following, the same authors [4] improved their solution for the lower level allowing simultaneous deliveries at two or more distribution centers. As a result, the number of stoppages reduced considerably. In order to obtain a more stable flow rate over the considered horizon, Cafaro et al. [5] developed an MINLP continuous model considering friction loss and tracking power consumption with nonlinear equations.

MirHassani et al. [9] developed a continuous MILP model for the scheduling of a multiproduct pipeline connecting a refinery to several DCs. When the authors addressed a long-term instance, the time horizon was divided into a number of shorter time periods, where each period was determined by a heuristic that calculates the next day a DC would face a shortage. Their approach considered different aspects in the topology treated, such as daily demand due dates, settling periods for quality control of the products, and scheduled shutdown periods for pipeline maintenance.

Mostafaei and Ghaffari-Hadigheh [10] proposed another MILP formulation based on the problem presented by [4], allowing simultaneous deliveries and lots of different products to be delivered to a DC during the same pumping run, a limitation of the [4] formulation. Ghaffari-Hadigheh and Mostafaei [7] developed a monolithic continuous MILP approach for the short-term scheduling and allowing simultaneous deliveries to multiple receiving terminals during a pumping execution. Recently, Zaghian and Mostafaei [16] solved the same problem for the short-term scheduling, however also considering the possibility to deliver multiple products to an active DC while injecting a new batch into the pipeline, similar to [10] consideration.

Every new approach being developed on the scheduling of a multiproduct pipeline with single-source and several destinations has the objective to approximate the models to the operational reality, including more and more aspects of a real-world pipeline network. However, the resulting combinatorial problems have posed significant computational challenges. To overcome this problem some approaches, e.g. Cafaro et al. [4], experimented a structural decomposition and also a temporal decomposition (rolling horizon). In this work, we consider a single-source pipeline distributing to several DCs considering operational aspects as presented in [4] and additional operations and assumptions present in a real-world pipeline network. The following aspects are highlighted: simultaneous deliveries; pipeline maintenance; rigorous inventory control using intermediary inventory levels as detailed in [8, 11]. The demand is assumed as given and the system has to respect the consumption rate that may vary during the time horizon and adapt the pipeline scheduling in order to avoid surplus or shortage problems in the distribution centers. This assumption decreases the liberty to determine how the demand will be attended, thus, consequently, increasing the computational challenge of the proposed approach. To solve the long-term scheduling for this pipeline network, we propose a structural and temporal decomposition strategy, combining heuristics and mixed integer linear programming (MILP).

The remainder of this paper is organized as follows. Section 16.2 describes the problem studied and the assumptions adopted. Section 16.3 describes the decomposition approach developed. Section 16.4 discusses the results for the long-term scheduling of a literature scenario, [9]. The last section presents the conclusions and future work.

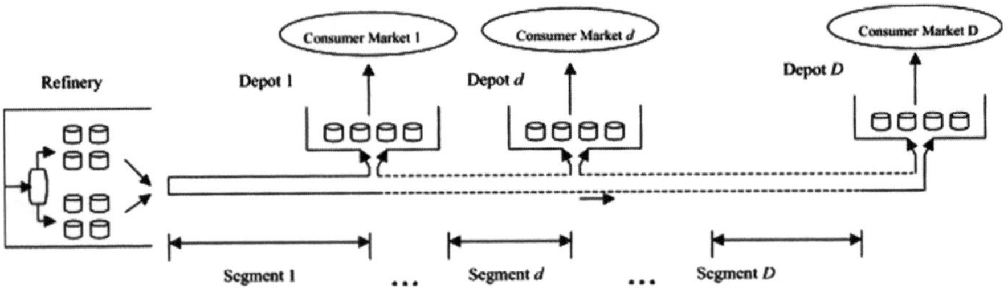

Fig. 16.1 Generic straight pipeline network with a single-source and multiple destinations [13]

16.2 Problem Definition

The problem consists of a straight multiproduct pipeline that connects a single-source (refinery) to multiple destinations (DCs), as generically illustrated in Fig. 16.1. The refinery pumps batches of products, oil derivatives, into the pipeline and the DCs act as removal terminals. The product extracted from a passing batch is stored into a tank. The products are consumed by the local market demand of each DC. The objective of the proposed approach is to schedule the pumping and receiving operations for a long-term period (e.g. 30 days) respecting the operational constraints of the system while minimizing surplus or shortage of inventory for each product in each DC; flow rate fluctuations of each segment and the number of product interfaces.

The developed decomposition approach relies on the following assumptions:

- A unidirectional flow pipeline network connecting a refinery to multiple distribution centers.
- The pipeline is always full of incompressible products and the only way to remove a volume of product to one or more DCs is injecting the same volume into the pipeline.
- The refinery works in an ideal way, that is, the production and storage capacity are sufficient to supply any pumping need into the pipeline.
- Each DC considers all tanks that stores the same product as an aggregate tank.
- The interface between two consecutive batches of different products causes a contamination volume. We consider a fixed penalty cost related to the contamination volume and the necessary changeover operation. However, when the contamination between two products is excessive, the sequence is a forbidden operation.
- The volume of each product batch to be pumped is limited by known lower and upper bounds.
- The volume to be removed from a batch in a delivery (receiving) operation is limited by a known lower bound.
- The admissible minimum and maximum product pumping rate, pipeline segment passage flow rate, and DC receiving flow rate are known.
- Pipeline maintenance may occur. The period of each pipeline maintenance is specified within days in advance and the details are inputs of the system.

- Simultaneous deliveries of products in two or more terminals are allowed.
- Delivery operation of multiple batches into a DC during the same pumping run is allowed.
- Initialization batches (batches inside the pipeline at the beginning of the time horizon), initial inventory volume, inventory levels, demand consumption and horizon size are known parameters of the system. The inventory levels are physical capacity/empty, operational maximum/minimum and target maximum/minimum levels.

The output of the system is the complete detailed scheduling of batch pumping operations and delivery operations during the specified horizon. The solution minimizes surplus/shortage of all inventory levels, respecting the operational constraints of the system.

16.3 Solution Approach

The proposed solution approach for the defined problem is a decomposition strategy that combines heuristics and MILP models. The complete flowchart is presented in Fig. 16.2.

The denominated scenario aggregates all the input parameters for the solution approach. It contains the network information (terminals, pipelines, tanks, and products); the demands of the system for the entire considered horizon; initialization batches; initial inventory at the beginning of the horizon; inventory levels of each product in each terminal; scheduled pipeline maintenance periods; and configuration parameters of the solution approach.

The Allocation and Sequencing Module (ASM) is responsible for allocating the volume, flow rate, product and sequence of the batches to be pumped from the refinery during the considered horizon. Following, the Scheduling Module (SM) determines the detailed scheduling of all the delivery (removal) operations to be performed at each terminal during the considered horizon. After running the two modules, the output is the complete pipeline scheduling, which includes all pumping operations from the refinery and all delivery operations for the complete time horizon.

16.3.1 Allocation and Sequencing Module

The first module to be executed, as observed in the flowchart of Fig. 16.2, is the Allocation and Sequencing Module (ASM). Apart from the solution decomposition, a temporal decomposition using a rolling horizon is applied.

The module receives as input the scenario, preprocesses the received information, then proceeds to the iterative part, which involves solving part of the time horizon at each iteration. During an iteration, the input data processing, the discrete-time

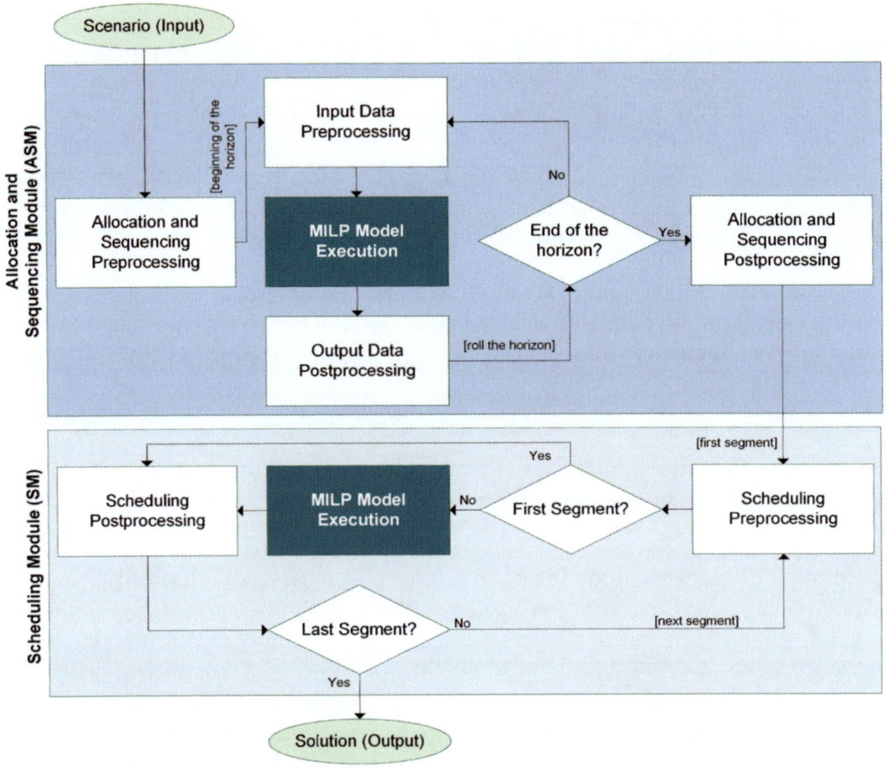

Fig. 16.2 Flowchart of the proposed solution approach

MILP model, and the output data processing are executed. At last, a postprocessing is executed to organize the obtained result, which will be needed as the input of the next module.

The Allocation and Sequencing preprocessing receives the scenario as an input parameter. Then, it applies a network simplification using heuristics in order to approximate some operational aspects that will be better considered at the next module. Two of the simplifications are the estimation of an average flow rate for each segment, and the definition of delivery operations for the initialization batches.

After the applied simplifications, the iterative part of the module is initiated. Here, before each execution of the MILP model, the input data is organized, based on the horizon period being considered, this because some parameters change at every iteration, as initial inventories and initialization batches present in the pipeline. At each iteration, the considered horizon period is constituted by the current period, which we are solving at the moment, and an auxiliary period, whose function is to guide the model, helping to anticipate problems caused, for example, by the sequenced batches, future maintenance periods or demand variations. The model finds a solution for the entire considered period, then the output data processing eliminates the

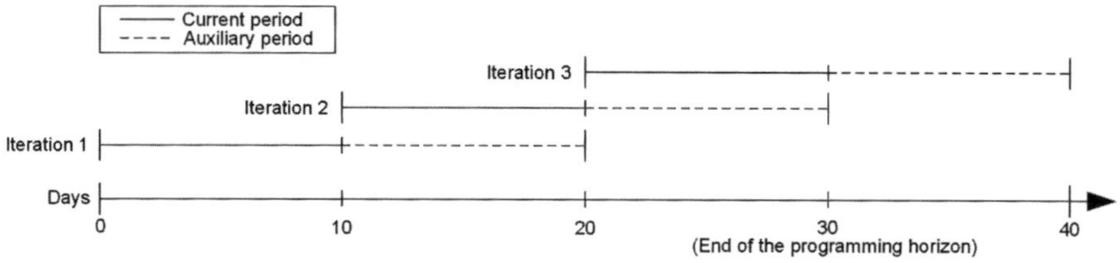

Fig. 16.3 Example of a rolling horizon execution

batches sequenced during the auxiliary period, keeping the solution processed for the current period. The current period is then "rolled" to the next iteration and fixed. Figure 16.3 illustrates the considered period for each iteration, given the following configuration: 30 days of time horizon, 10 days of current period and 10 days of auxiliary period.

The MILP model is the main process of the module. It is a discrete-time model, being the considered horizon divided into uniform intervals of few hours with respect to the whole considered horizon (e.g. 720 h divided into intervals of 12 h). The objective function of the model minimizes three groups of variables: (i) penalty of inventory surplus/shortage of physical capacity/empty, maximum/minimum operational and maximum/minimum target levels; (ii) cost of product interface; and, (iii) an operational penalty associated with pumping flow rate variation, which implies a flow rate with minimal variations and an easier refinery operationalization. The later also aid the execution of the next module.

After all iterations, the postprocessing simply adjusts the allocated and sequenced batches obtained at each iteration, then the execution proceeds to the Scheduling Module.

16.3.2 Scheduling Module

The Scheduling Module (SM) is responsible for programming the delivery operations of each distribution center during the considered horizon. A delivery operation consists of the removal of an amount of volume from a batch passing along a DC. The flowchart of this module is presented in Fig. 16.2. The module applies a structural decomposition, where the scheduling of the delivery operations is determined per pipeline segment.

The execution flow initiates at the first segment, which connects the refinery to the first DC, then moving to the next segment, until the last segment of the network. Each iteration determines the delivery operations of the upstream DC, which justifies why the MILP model is not executed for the first segment, where the upstream DC is the refinery and no delivery operations are considered. In this case, the postprocessing is performed directly, propagating and timing the batches of the entire horizon to the first DC. Since we know the pumping times and flow rate of each batch, we know the behavior of the system in this segment. Then, for the second segment, it is known the moment that each batch is starting to pass along the first DC. However, it is not known yet the flow rate estimated to "enter" in the second segment because the variations of the flow rate are dependent on the delivery operations to be scheduled by the MILP model at the upstream DC. After the MILP model execution, the algorithm is able to determine the flow rate profile; then, it is possible to calculate when each batch will start to pass by the downstream DC. The process continues until the last segment, where the downstream DC receives all the batches.

Before each segment execution, the scheduling preprocessing simplifies the network, in order to be possible to treat just one segment at a time. The simplified network consists of the upstream DC, the actual segment and a representation of the downstream network, combining the downstream DCs and segments. The heuristic used to construct the representation considers aggregate tanks, demands, mean flow rate and other aspects of the remaining network.

The MILP model receives the simplified network and the batch movements entering the current segment. The objective function of the MILP model in the Scheduling Module, in a similar fashion to the ASM, minimizes three groups of variables: (i) penalties of inventory surplus/shortage of physical capacity/empty, maximum/minimum and maximum/minimum target levels; (ii) penalty of a maximum/minimum flow rate violation; and, (iii) an operational penalty associated with pumping flow rate variation, which implies a more constant flow rate and aid the execution of the next segment.

As already said, the postprocessing is responsible for the propagation and timing of the batches passing from the upstream DC to the downstream DC of the current segment. After the last pipeline segment execution, all the delivery operations are determined, meaning that the complete pipeline scheduling for the considered horizon is obtained and the execution is then finished.

16.4 Results and Discussion

The proposed approach was applied to a real case presented by MirHassani et al. [9]. This case studies the scheduling of a straight multiproduct pipeline with 147 km and 14 181 m^3 of volumetric capacity, connecting a refinery to 3 distribution centers (D1, D2, and D3). The scenario considers a horizon of 21 days (3 weeks) and 4 demanded products: jet fuel (P1), gasoline (P2), kerosene (P3) and diesel (P4). The established refinery pumping rate is between 400 and 720 m^3/h. The pipeline begins full of P3, which means a single kerosene initial batch of 14 181 m^3. It is also scheduled a pipeline maintenance period of 12 h starting at the beginning of the third day of the horizon (48–60 h). Complementary data can be found in MirHassani et al. [9].

The programming language used to develop the approach was Java and the two mathematical models were implemented using IBM ILOG CPLEX *Optimization Studio* 12.6. The results were run on a PC platform with an Intel i7-6500 2.5 GHz and 16 GB RAM, using Windows 10 64-bits.

The uniform interval size defined for the execution was initially set to 10 h, and the current and auxiliary periods size in the Allocation and Sequencing Module were 3 and 4 days, respectively. This setting represents a total of 7 iterations for the 21 days horizon during the Allocation and Sequencing Module, as explained in Sect. 16.3.1.

For the set parameters, the complete execution of the solution approach took 23.6 s CPU time (19.1 s during the first module and 4.5 s during the second). MirHassani et al. [9] approach took 442 s of CPU time to obtain a solution, however, it is not valid a comparison since we are using different machines and the objective function is not similar. Moreover, certain aspects considered by [9] are not considered in our approach (e.g. settling periods) and vice versa (e.g. simultaneous deliveries).

The obtained solution planned 22 batches to be pumped by the refinery during the 21 days of the considered horizon. The solution did not present any surplus or shortage of the inventory of each product at each DC and any flow rate violation for each segment. Only penalties related to target and operational level violations, which attempt to maintain the inventory close to the recommended level, occurred. Table 16.1 lists all the batches scheduled and Table 16.2 presents the dimension of each model executed during the Allocation and Sequencing Module and Scheduling Module. According to explanations given in Sects. 16.3.1 and 16.3.2, the ASM and SM models are executed, respectively, 7 and 2 times. In the ASM the 21-day horizon is covered by seven consecutive executions (3 days each execution). In the SM this occurs because of the network topology, where, excluding the first segment, the model executes for the two last segments. The integrality gap of 0 % indicates that the models were executed until optimality. Noteworthy that despite the relatively large dimensions related to the number of variables and constraints, the computational time for the SM models remained within just a few seconds.

Figure 16.4a–l presents the inventory level graphs for each product in each DC during the scheduled horizon. It is interesting to notice that due to the rigorous inventory control, the inventory levels tend to stay almost the entire horizon within the established maximum and minimum target level, as explained in Fig. 16.4.

Additional tests were performed using the same scenario but decreasing the uniform interval size of ASM, in order to analyze the impact on the CPU time and on the quality of the obtained solution. A smaller interval size implies in more events and intervals to be considered by the ASM MILP model and, consequently, increasing its dimension. In relation to the quality of the solution, more intervals mean an increase of the "observation capacity" of the model, which, consequently, tends to reduce problems, such as inventory violations.

The uniform interval size of the ASM was progressively decreased from 10 until 5 h. A running time limit of 3 h (10 800 s) for each MILP model execution was set. Under this running condition, smaller interval sizes were not considered since the optimality (0% of integrality gap) was not reached for some ASM model iterations.

Table 16.3 presents the CPU time and the final integrality gap of each model executed using 10, 9, 8, 7, 6 and 5 h of uniform interval size. Within the considered running time, the ASM model did not reach optimality for the 5-hour-interval in iterations 6 and 7; solutions with, respectively, 13.1 and 5.8% of integrality gap were obtained. The results evidence that the CPU time significantly increases as the interval

Table 16.1 Scheduled batches programmed to be pumped during the horizon

Batch	Start (h)	End (h)	Product	Volume (m^3)	Flow rate (m^3/h)
1	0	20	P1	8640	432
2	20	30	P2	7200	720
3	30	40	P3	4320	432
4	40	48	P4	5760	720
*	48	60	–	–	–
4	60	70	P4	4320	432
5	70	80	P3	4320	432
6	80	110	P2	12000	400
7	110	120	P3	4000	400
8	120	150	P4	12000	400
8	150	160	P4	4163	416
9	160	180	P3	8326	416
10	180	200	P1	8326	416
11	200	220	P2	8326	416
11	220	230	P2	4000	400
12	230	270	P3	16000	400
13	270	320	P4	20000	400
14	320	340	P3	8000	400
15	340	370	P2	12000	400
16	370	380	P3	4000	400
17	380	430	P4	20000	400
18	430	440	P3	4000	400
19	440	470	P4	12000	400
20	470	480	P3	4000	400
21	480	500	P2	8000	400
22	500	504	P1	1600	400

*Pipeline maintenance

Table 16.2 MILP models statistics

Module	Iteration	N° of const.	N° of var.	N° of bin. var.	CPU time (s)	Gap (%)
ASM	1	2716	2092	136	1.4	0
ASM	2	2716	2092	136	5.1	0
ASM	3	2716	2092	136	2.0	0
ASM	4	2716	2092	136	0.7	0
ASM	5	2716	2092	136	3.5	0
ASM	6	2875	2210	144	2.4	0
ASM	7	2716	2092	136	4.0	0
SM	1	19666	8016	2418	1.4	0
SM	2	17347	7254	2132	3.1	0

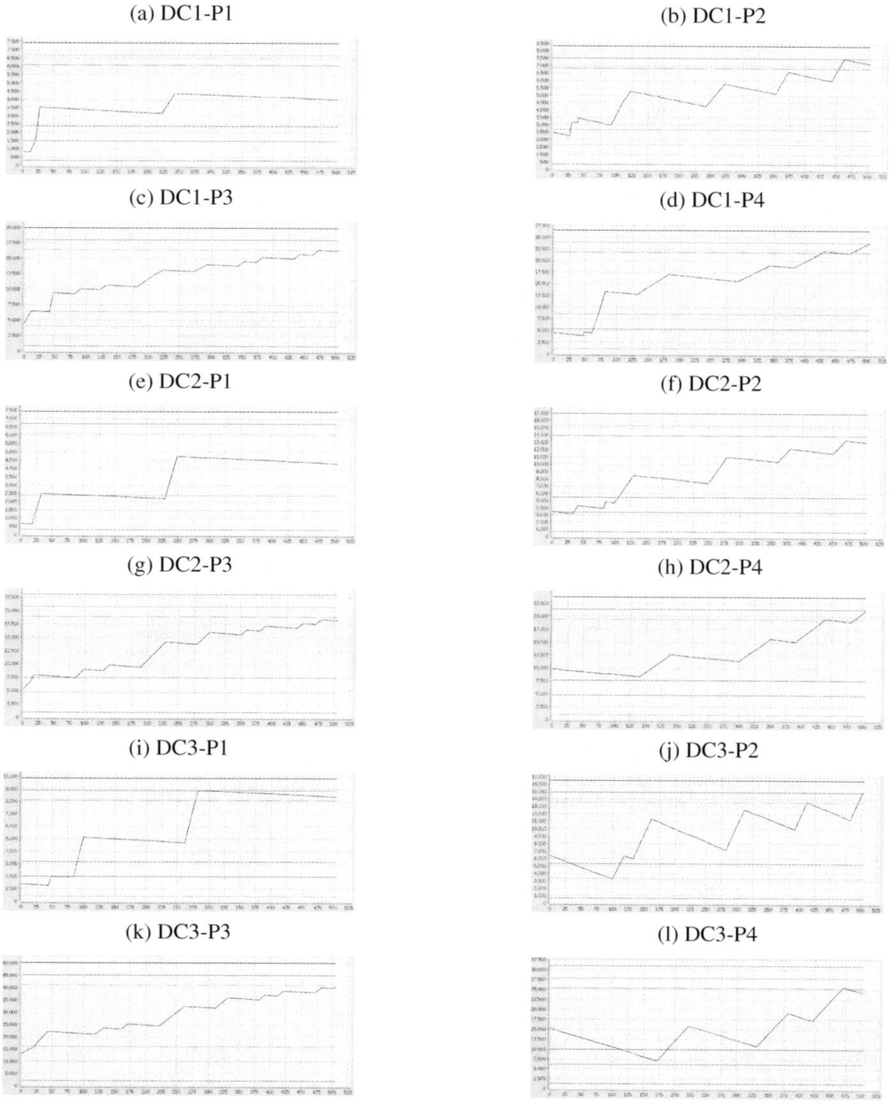

Fig. 16.4 Inventory graphics of every pair DC-product. Where the X-axis is the considered horizon in hours (from 0 to 525 h) and Y-axis is the volume in m^3. The red line represents the inventory at each moment; the dark blue line is the physical capacity level; the outer intermediary lines are the maximum and minimum level; and the inner intermediary lines are the maximum and minimum target level

Table 16.3 CPU time of each model execution for decreasing interval size in ASM

Module	Iteration	Interval duration (h)					
		10	9	8	7	6	5
ASM	1	*1.4 (0)	3.6 (0)	3.7 (0)	4.6 (0)	8.6 (0)	59.8 (0)
ASM	2	5.1 (0)	4.6 (0)	3.7 (0)	7.5 (0)	24.1 (0)	59.4 (0)
ASM	3	2.0 (0)	5.9 (0)	2.5 (0)	28.2 (0)	29.5 (0)	7.1 (0)
ASM	4	0.7 (0)	1.0 (0)	6.7 (0)	65.7 (0)	53.2 (0)	60.8 (0)
ASM	5	3.5 (0)	7.0 (0)	108.4 (0)	59.3 (0)	208.7 (0)	134.5 (0)
ASM	6	2.4 (0)	1.8 (0)	3.6 (0)	99.4 (0)	107.5 (0)	10800.0 (13.1)
ASM	7	4.0 (0)	40.3 (0)	33.8 (0)	527.4 (0)	2470.7 (0)	10800.0 (5.8)
SM	1	1.4 (0)	0.9 (0)	2.5 (0)	3.7 (0)	3.5 (0)	5.8 (0)
SM	2	3.1 (0)	4.6 (0)	5.7 (0)	25.8 (0)	4.4 (0)	9.3 (0)
Total execution time		**23.6**	**69.7**	**170.6**	**821.6**	**2910.2**	**21936.7**

*CPU time in seconds (integrality gap in %)

(a) DC1-P4 with 10 h of interval size (b) DC1-P4 with 5 h of interval size

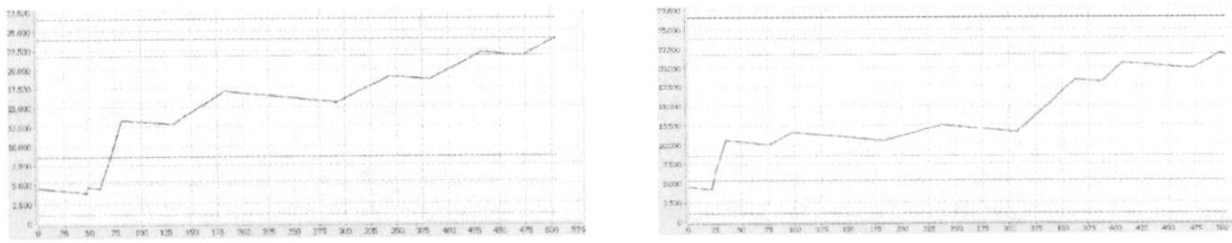

Fig. 16.5 Inventory profile comparison of 10 and 5 h of uniform interval size in ASM

size decreases. For instance, by decreasing the interval size from 10 to 9 h, the total execution time increased more than 3 times, and from 8 to 7 h it increased almost 5 times. However, a better inventory control tends to be evidenced as smaller intervals are considered, as indicated in Fig. 16.5. In this figure, the inventory profile of DC1-P4 for 10-hour-interval size (a) is compared with the same profile obtained for the 5-hour-interval size (b). The differences are observed at the beginning of the curve, where the 5-hour-interval size execution was able to increase the inventory volume faster than the 10-hour-interval size execution, avoiding minimum operational and target violations for a longer period and maintained the inventory between the target levels until the time horizon end, while the first profile (Fig. 16.5a) also presented maximum target level violations at the end of the horizon.

16.5 Conclusion

A decomposition approach using heuristics and MILP models has been presented for the long-term scheduling of a multiproduct pipeline that connects a single-source to multiples distribution centers. The approach determines the sequence of pumping operations and receiving operations to be executed during a certain horizon, attending a continuous demand with a rigorous inventory control. Some important operational aspects are considered, such as pipeline maintenance periods, simultaneous deliveries, and multiple product deliveries to a DC during the same pumping run [10]. The decomposition approach considers two modules. A first module where allocation and sequencing of batches is performed and a second module where the batch deliveries are scheduled. A temporal decomposition was applied during the first module, using the rolling horizon concept to treat the long-term scheduling. In the second module, a structural decomposition is considered, treating the delivery operations one segment at a time.

In order to validate the proposed approach, a real-world case presented by MirHassani et al. [9] was solved. The case consists of a long-term scheduling (21 days) of a multiproduct pipeline connecting a refinery to 3 DCs and transporting 4 different products. The solution was obtained in a low CPU time (23.6 s) and the inventory profiles demonstrated an effective inventory control during the scheduled horizon. The demand consumption was an input parameter determined for each DC and the model had to respect such operational characteristic. Additional tests using the same scenario were also performed in order to analyze the impact on the CPU time and on the quality of the solution while decreasing the uniform interval size in ASM. A fast increase in CPU time was noticed as the interval size decreases, however, a better inventory control was observed, as shown in Fig. 16.5, which indicates an improvement in the quality of the solution. Future work is still necessary in order to improve the current approach, for instance, to consider more complex aspects, such as different system topologies.

Acknowledgements The authors acknowledges the Erasmus Mundus SMART^2 support (Project Reference: 552042-EM-1-2014-1-FR-ERA MUNDUS-EMA2) coordinated by CENTRALESUPÉLEC. The authors would also like to acknowledge financial support from the Brazilian Oil Company PETROBRAS (grant 0050.0066666.11.9) and CAPES - DS.

References

1. D.C. Cafaro, J. Cerdá, Optimal scheduling of multiproduct pipeline systems using a non-discrete MILP formulation. Comput. Chem. Eng. **28**(10), 2053–2068 (2004). https://doi.org/10.1016/j.compchemeng.2004.03.010
2. D.C. Cafaro, J. Cerdá, Dynamic scheduling of multiproduct pipelines with multiple delivery due dates. Comput. Chem. Eng. **32**(4–5), 728–753 (2008). https://doi.org/10.1016/j.compchemeng.2007.03.002

3. V.G. Cafaro, D.C. Cafaro, C.A. Méndez, J. Cerdá, Detailed scheduling of operations in single-source refined products pipelines. Ind. Eng. Chem. Res. **50**(10), 6240–6259 (2011). https://doi.org/10.1021/ie200007a
4. V.G. Cafaro, D.C. Cafaro, C.A. Méndez, J. Cerdá, Detailed scheduling of single-source pipelines with simultaneous deliveries to multiple offtake stations. Ind. Eng. Chem. Res. **51**(17), 6145–6165 (2012). https://doi.org/10.1021/ie202520a
5. V.G. Cafaro, D.C. Cafaro, C.A. Méndez, J. Cerdá, MINLP model for the detailed scheduling of refined products pipelines with flow rate dependent pumping costs. Comput. Chem. Eng. **72**, 210–221 (2015). https://doi.org/10.1016/j.compchemeng.2014.05.012
6. V.G. Cafaro, D.C. Cafaro, C.A. Méndez, J. Cerdá, Optimization model for the detailed scheduling of multi-source pipelines. Comput. Ind. Eng. **88**, 395–409 (2015). https://doi.org/10.1016/j.cie.2015.07.022
7. A. Ghaffari-Hadigheh, H. Mostafaei, On the scheduling of real world multiproduct pipelines with simultaneous delivery. Optim. Eng. **16**, 571–604 (2015). https://doi.org/10.1007/s11081-014-9263-9
8. S.N.B. Magatão, L. Magatão, H.L. Polli, F. Neves-Jr, L.V.R. Arruda, S. Relvas, A.P.F.D. Barbosa-Póvoa, Planning and sequencing product distribution in a real-world pipeline network: an MILP decomposition approach. Ind. Eng. Chem. Res. **51**(12), 4591–4609 (2012). https://doi.org/10.1021/ie2010252
9. S.A. Mirhassani, S. Moradi, N. Taghinezhad, Algorithm for long-term scheduling of multiproduct pipelines. Ind. Eng. Chem. Res. **50**(24), 13899–13910 (2011). https://doi.org/10.1021/ie200101a
10. H. Mostafaei, A. Ghaffari-Hadigheh, A general modeling framework for the long-term scheduling of multiproduct pipelines with delivery constraints. Ind. Eng. Chem. Res. **53**, 7029–7042 (2014). https://doi.org/10.1021/ie4038032
11. H.L. Polli, L. Magatão, S.N.B. Magatão, F. Neves-Jr, L.V.R. Arruda, Collaborative approach based on heuristic algorithm and milp model to assignment and sequencing of oil derivative batches in pipeline networks. Ind. Eng. Chem. Res. **56**(9), 2492–2514 (2017). https://doi.org/10.1021/acs.iecr.6b03516
12. R. Rejowski, J.M. Pinto, Scheduling of a multiproduct pipeline system. Comput. Chem. Eng. **27**, 1229–1246 (2003). https://doi.org/10.1016/S0098-1354(03)00049-8
13. R. Rejowski, J.M. Pinto, Efficient MILP formulations and valid cuts for multiproduct pipeline scheduling. Comput. Chem. Eng. **28**(8), 1511–1528 (2004). https://doi.org/10.1016/j.compchemeng.2003.12.001
14. R. Rejowski, J.M. Pinto, A novel continuous time representation for the scheduling of pipeline systems with pumping yield rate constraints. Comput. Chem. Eng. **32**(4–5), 1042–1066 (2008). https://doi.org/10.1016/j.compchemeng.2007.06.021
15. M. Sasikumar, P. Ravi Prakash, S.M. Patil, S. Ramani, PIPES: a heuristic search model for pipeline schedule generation. Knowl.-Based Syst. **10**(3), 169–175 (1997). https://doi.org/10.1016/S0950-7051(97)00026-9
16. A. Zaghian, H. Mostafaei, An MILP model for scheduling the operation of a refined petroleum products distribution system. Oper. Res. **16**(3), 513–542 (2016). https://doi.org/10.1007/s12351-015-0212-y

Chapter 17
Green Supply Chain Design and Planning: The Importance of Decision Integration in Optimization Models

Bruna Mota, Ana Carvalho, Maria Isabel Gomes and Ana Barbosa Póvoa

Abstract Sustainability is more than ever a central concern when deciding a company's strategy. Stakeholders are continuously pressuring industries to further reduce their environmental impact. In order to achieve that, it is imperative to look for solutions across all of the company's operations adopting a supply chain view. However, supply chains' complexity makes this a challenging task. Analysing the sustainability of such complex systems requires the use of capable tools as are optimization models. Several of these models can be found in the literature, mostly focusing on specific issues or specific decisions in the supply chain. Therefore, a research gap is found in models capable of handling a wider variety of decisions. With this work a mixed integer linear programming model is used to demonstrate the impact of including more or less options/decisions on design and planning decisions, and on the environmental performance of a supply chain. A case-study based on a Portuguese pulp and paper company is analysed. The results obtained for different scenarios are examined.

Keywords Supply chain · Optimization · Sustainability · Decision integration
Life cycle assessment

17.1 Introduction

The need for sustainability has been vastly discussed in the literature, as well as the need to assess it across the supply chains [2]. However, adding sustainability to what is already a complex system often leads to intractable problems. Making use of computational tools such as optimization models becomes necessary. Even then the approach to deal with such problems is often to subdivide the problem into smaller ones. Several works have been proposed that follow this type of approach. However, a research gap is found in integrated modelling approaches, as identified in

B. Mota (✉) · A. Carvalho · A. Barbosa Póvoa
CEG-IST, University of Lisbon, Av. Rovisco Pais, 1049-001 Lisboa, Portugal
e-mail: bruna.mota@tecnico.ulisboa.pt

M. I. Gomes
CMA-FCT, Nova Universtity of Lisbon, Monte de Caparica, 2829-516 Caparica, Portugal

A. I. F. Vaz et al. (eds.), *Operational Research*, Springer Proceedings
in Mathematics & Statistics 223, https://doi.org/10.1007/978-3-319-71583-4_17

several reviews [1, 3, 7]. By integrated modelling approaches we mean models that include decisions at different levels of the supply chain. For instance, the integration of decision variables allowing for capacity definition, supplier selection, technology selection, and transportation network definition over a time horizon, along with location-allocation decisions. By integrating all of these decisions into one model the array of possible solutions widens, leading to the possibility of finding more sustainable solutions.

In this work we aim to demonstrate the importance of such an integrated approach and how this wider integration of options and decisions is reflected on the environmental performance of a supply chain and consequently on supply chain design and planning decisions. To do so a mixed integer linear programming model is used and applied to a case-study based on a Portuguese pulp and paper producer and distributor. Different scenarios simulating the inclusion of more or less options/decisions are analyzed.

This paper is structured as follows. In Sect. 17.2 the problem is defined and the developed model characterized. Section 17.3 presents the case-study. In Sect. 17.4 results are presented and discussed. Final remarks and future work are provided in Sect. 17.5.

17.2 Problem Definition

As stated, the goal of this paper is to examine the extent to which a wider or narrower decision integration level in an optimization model can affect supply chain design and planning decisions as well as the sustainability of the supply chain. ToBLoOM (Triple Bottom Line Optimization Modeling), as described in Mota et al. [6], is used since it is, to the best of our knowledge, the most integrated modelling tool available in the literature. It is a Multi-Objective Mixed Integer Linear Programming (MOMILP) model which includes decisions such as supplier and technology selection, transportation network definition, entity location and capacity definition, and the planning of production, inventory and flow of materials. The environmental pillar is modelled as an objective function using a Life Cycle Assessment method, ReCiPe 2008 [4], on supply chain operations: production, production at suppliers, carbon credits obtained from purchasing raw materials from certified suppliers, intermodal transportation, and entity installation and/or usage. To assess the economic pillar, the Net Present Value (NPV) is modelled as a second objective function. Included are the revenues from sold final products and CO2 allowances (under the European Union's emissions trading system), as well as raw material and production costs, transportation (unimodal or intermodal) and hub handling costs, CO2 allowances costs, and inventory and warehouse renting costs.

In short, **given**: (a) product's demand; (b) bills of materials; (c) a possible superstructure for the supply chain entities; (d) distances between each pair of entities; (e) maximum and minimum supplying flow, storage and production capacities, and installation areas; (f) area occupied per product unit; (g) price per sold product;

(h) weight of each product unit; (i) transportation costs per kilometer and per unit; (j) raw material and production costs; (k) warehouse renting costs; (l) initial stock levels, (m) carbon credits from the usage of certified raw materials; and (n) the environmental impacts of transportation, production processes, production at suppliers (if applicable), and per square meter of entity opened area, **the goal is to determine** (1) the network structure; (2) the installed production technologies; (3) the production and storage levels at each entity; (4) the transportation network; and (5) the flow between entities, **so as to** minimize the global supply chain environmental impact. Different scenarios are studied exploring how the simultaneous decisions' integration in the optimization model affects the supply chain design and planning as well as its environmental performance. Each scenario is solved using the lexicographic approach [5] so as to obtain, for each case, the best economic performance for the minimum environmental impact.

17.3 Case-Study

The case-study of a Portuguese company operating in the pulp and paper industry is analysed in this work, specifically focusing on the supply chain of uncoated wood free paper (UWF paper). The pulp and paper industry is one of the most CO2-intense ones. Strick regulations are imposed and the pressure to continuously reduce the environmental footprint is constant in this sector. It is then important to expand the possible options so as to look for ways to further reduce the environmental impact of the industry's operations. The supply chain structure modelled in ToBLoOM is shown in Fig. 17.1.

Paper production at the factories can be either integrated or non-integrated. In the former, both pulp and paper are produced sequentially, in the same facility. In the latter, pulp is received from suppliers, de-hydrated and used for paper production. The same paper production machine is used in both cases. The source of wood is also a decision. Wood can be obtained from certified or non-certified suppliers. We refer to certified suppliers as those who have their forest management practices

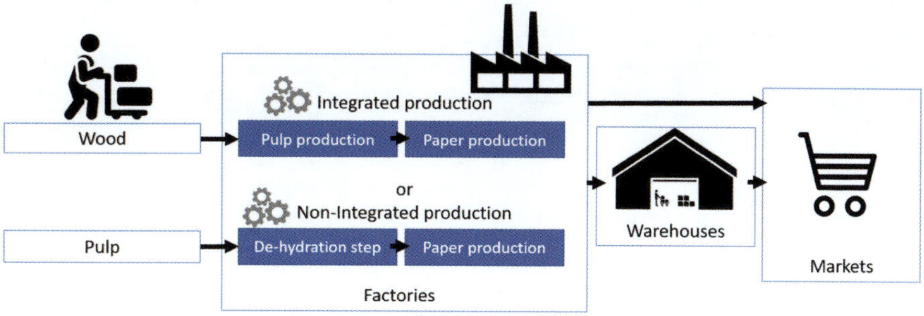

Fig. 17.1 Schematic representation of the UWF paper supply chain

Table 17.1 Potential location for the entities

Wood suppliers		Pulp suppliers	Factories	Warehouses and markets	Rail stations	Seaports
Certified	Non-certified	·	·	·	·	·
Portugal	Portugal	Spain	Figueira da Foz (Portugal)	Portugal	Portugal	Portugal
Brazil	Spain	France	Setubal (Portugal)	Spain	Spain - Algeciras	Spain
Spain	·	Austria	·	France	Spain - Valladolid	United Kingdom
Sweden	·	·	·	United Kingdom	France	Belgium
·	·	·	·	Belgium	Germany	Netherlands
·	·	·	·	Netherlands	·	Germany
·	·	·	·	Germany	·	Italy
·	·	·	·	Italy	·	Greece
·	·	·	·	Sweden	·	Brazil
·	·	·	·	Poland	·	·
·	·	·	·	Greece	·	·

certified, which ensures that forests are sustainably managed in all three pillars of sustainability: economic, environmental and social.

One of the underlying issues around forest certification is reforestation. Hence, carbon credits (reflecting environmental impact reduction) are attributed per ton purchased from certified suppliers, to account for the sequestrated carbon that can be obtained from replacing the cut trees with new ones. For the presented case study certified pulp suppliers are considered. After production, UWF can either be sent directly to the markets or stored at the warehouses.

The locations of factories and potential locations for the remaining entities are provided in Table 17.1. Even though not shown it is important to mention that the supply capacities of the Portuguese and Spanish certified wood suppliers are still too short to meet the needs. Thus it becomes necessary to resort to suppliers further away, to source from non-certified suppliers or to opt for non-integrated production to meet the demand.

Four scenarios are studied to assess the impact of a wider or narrower decision integration level on the design and planning of sustainable supply chains. These differ in terms of the decisions the model will take into account and in terms of the objective function:

- Scenario 1 (base scenario): all supply chain activities and options are considered so as to minimize the total environmental impact;
- Scenario 2: base scenario without the option of non-certified wood;
- Scenario 3: base scenario without the option of sea and rail transportation;
- Scenario 4: all supply chain activities and options are considered so as to maximize NPV.

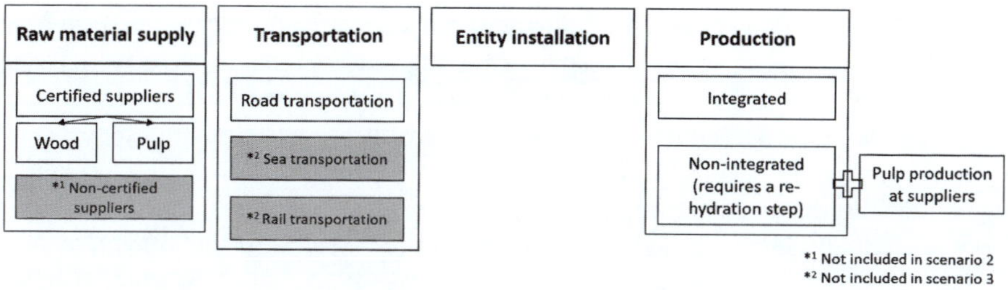

Fig. 17.2 Results obtained across the four scenarios for the economic and environmental indicators

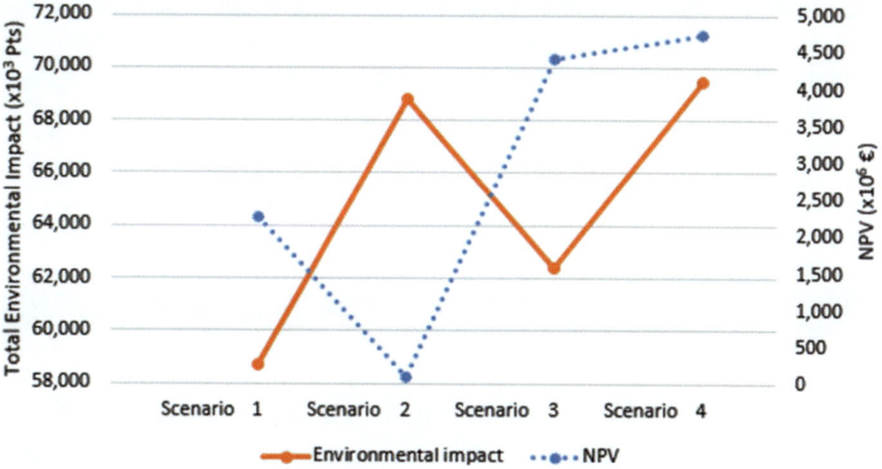

Fig. 17.3 Decisions considered (with environmental impact) in each of the four scenarios

Figure 17.3 depicts the decisions considered in each of the four scenarios. In grey are the decisions that will not be taken into account in scenarios 2 and 3 (Fig. 17.2).

17.4 Results

Results obtained for the total environmental impact (solid line) and for the NPV (dotted line) across the different scenarios are presented in Fig. 17.2.

The minimum environmental impact (58.680×10^3 Pts) is obtained when considering all decisions simultaneously (scenario 1). Interestingly, this supply chain does not correspond to the worst economic performance. However, the NPV is 52.7% (2.246×10^6 Euro versus 4.743×10^6 Euro) lower than the optimal NPV value (scenario 4). Also, very interesting is the fact that optimizing the economical objective function (scenario 4) leads to the worst environmental supply chain structure (18.4% higher than in scenario 1). The reasoning for these results is presented further ahead.

Scenario 2, the one considering only certified suppliers, performs worse in both objectives than scenario 1 (17.3% worse for the environmental performance and 96% worse for the economic performance). Having 100% certified suppliers, a measure that is often presented as environmentally positive in sustainability reports and sustainability policies, may not be so positive after all if considering the collateral damage it may have on the supply chain. In this case, opting for 100% certified suppliers actually increases the environmental impact of transportation and of production. However, one should not forget that it all depends on the availability of certified suppliers, which is still not enough to meet the demand of the pulp and paper industry.

Scenario 3, the one with no intermodal transportation option, performs significantly better economically (96.2%) but worse environmentally (6.4%) compared to scenario 1. This shows that for this case-study, intermodal transportation is better environmentally but not economically as explained further ahead.

Figures 17.4 and 17.5 depict the environmental impact of each activity across the four scenarios, while Table 17.2 presents a summary of selected decisions across the different scenarios. Combined they shed light into the results obtained in Fig. 17.2.

Each scenario is analysed based on the presented results. Note that all comparisons below are made in relation to the base scenario (scenario 1):

- Results show that purchasing pulp, and therefore having a non-integrated production, is avoided when minimizing the environmental impact (scenario 1, Table 17.2). Given the scarcity of certified wood in the proximity of the company's installations (as mentioned above), the purchasing of non-certified wood is preferable to the purchasing of certified pulp and, consequently, an integrated production option is proposed. The minimum amount of certified wood necessary to meet the demand with the non-integrated production technology is totally purchased in Brazil (results not shown).

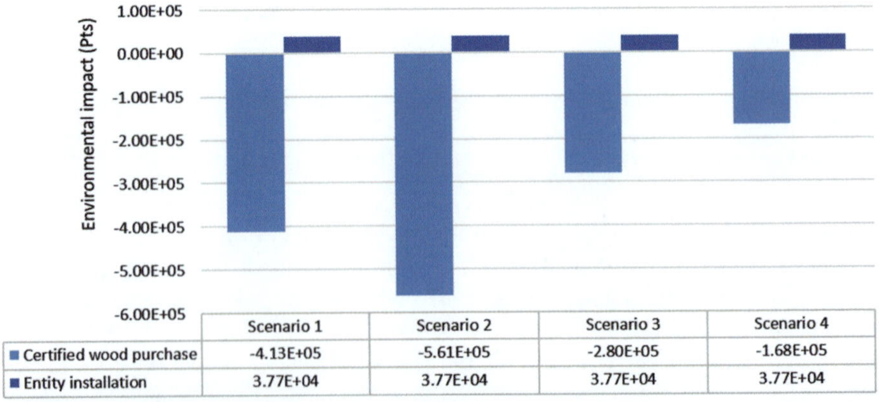

Fig. 17.4 Environmental impact distribution per supply chain activity across the four scenarios: certified wood purchase and entity installation

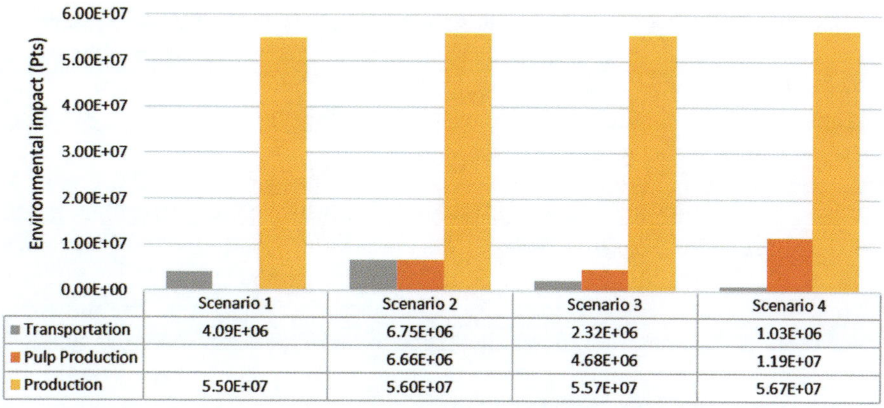

Fig. 17.5 Environmental impact distribution per supply chain activity across the four scenarios: transportation, pulp production at suppliers and paper production

Table 17.2 Summary of selected supply chain decisions obtained across the different scenarios

·	Scenario 1	Scenario 2	Scenario 3	Scenario 4
Certified wood (%)	55.1	100	45.5	40.7
Non-Certified wood (%)	44.9	0	54.5	59.3
Integrated production (%)	100	74.8	82.3	55.2
Non-integrated production (%)	0	25.2	17.7	44.8
Warehouse locations	Portugal	Portugal	Portugal	Portugal
·	Spain	Spain	Spain	Spain
·	France	France	France	United Kingdom
Rail stations	Portugal	Portugal	·	Portugal
·	Spain-Algeciras	Spain-Algeciras	·	Spain-Valladolid
·	Spain-Valladolid	Spain-Valladolid	·	France
·	France	France	·	·
Seaports	Spain	Spain	·	Portugal
·	Germany	Germany	·	United Kingdom
·	Brazil	·	·	Germany
·	·	·	·	Brazil

- When imposing 100% certified raw materials (scenario 2), the best strategy is to purchase, as much as possible, certified wood and to resort to pulp supply to fulfil the demand. That is why only 25% of the production is non-integrated (Table 17.2). The environmental impact of transportation increases by 65% and the combined impact of pulp purchasing and paper production increases by 14%. Since raw-materials are all certificated, carbon credits increase by 36% (Fig. 17.4).
- When removing the intermodal transportation option (scenario 3), Brazilian suppliers are not chosen (results not shown) and the certified wood supplied decreases to 45.5% (Table 17.2). Both certified and non-certified wood is bought mostly from Portuguese and Spanish suppliers (to their maximum capacity). The remaining needs are met through non-integrated production. The nearness of the suppliers leads to a significant reduction of the environmental impact concerning transportation activities (by 43%, as shown in Fig. 17.5).
- Finally, when maximizing the economic performance of the supply chain (scenario 4) non-integrated production increases by 44.8%, which corresponds to the maximum allowed under the imposed constraints (Table 17.2). The solution is then to source certified wood only if the distance compensates the extra cost of certified wood (which only happens if the certified wood is bought locally) and then meet the remaining needs using non-certified wood (59.3%). Even though the combined impact of pulp production at suppliers and paper production increases by 25%, the environmental impact of transportation reduces by 75% (Fig. 17.5). Carbon credits, which are not included in the economic objective function, also decrease in consequence of the reduction of certified wood supply. Overall scenario 4 solution leads to an increase of 111% increase in net present value and 18.4% in the total environmental impact (Fig. 17.2).

All in all, results show that by widening the options available through increasing decision integration in this type of models one may uncover solutions that result in an overall lower environmental impact, without imposing such a burden on the economic performance of the company. Also, from a different perspective, it may also be that only focusing on the environmental impact of a specific part of the complex system, that is a supply chain, may actually result in both higher costs and environmental impact.

17.5 Conclusions

In this work an analysis of the potential impact of a wider or narrower decision integration level in the design and planning decisions of green supply chains was performed. More precisely, the impact that including more or less decisions (for example, including supplier selection) as well as options within those decisions (for instance, including intermodal transportation options versus simply including road transportation) has on the environmental and economic performances of the system. Different options in terms of the number of integrated decisions were investigated with ToBLoOM, a mixed integer linear programming model for the design and planning of sustainable supply chains.

The results of a case-study based on the UWF paper company corroborated that the impact of the level of decision integration is significant both on the design and planning decisions and on the environmental and economic performance of supply chains. This particularly sheds light on the implications that certain measures traditionally considered environmentally friendly in sustainability reports and sustainability policies can actually have.

This work shows how important it is to carefully design the studies that support supply chain design and planning decision making as well as sustainability policy making in order to reduce environmental collateral damages that can take place on other supply chain activities.

Future work should further look into the trade-offs among the three pillars of sustainability: economic, environmental and social. Additional case-studies of different sectors as well as uncertainty analysis on the main parameters affecting these decisions should be included in future studies.

Acknowledgements This work was partially supported by the Fundação para a Ciência e a Tecnologia (FCT) through the projects PTDC/EMS-SIS/1982/2012, MITP-TB/PFM/0005/2013, UID/MAT/00297/2013, and grants SFRH/BD/51947/2012 and SFRH/BSAB/128453/2017.

References

1. T. Boukherroub, A. Ruiz, A. Guinet, J. Fondrevelle, An integrated approach for sustainable supply chain planning. Comput. Oper. Res. **54**, 180–194 (2015)
2. M. Brandenburg, K. Govindan, J. Sarkis, S. Seuring, Quantitative models for sustainable supply chain management: developments and directions. Eur. J. Oper. Res. **233**(2), 299–312 (2014)
3. M. Eskandarpour, P. Dejax, J. Miemczyk, O. Péton, Sustainable supply chain network design: an optimization-oriented review. Omega **54**, 11–32 (2015)
4. M. Goedkoop, R. Heijungs, M. Huijbregts, A. De Schryver, J. Struijs, R. Van Zelm, Recipe 2008. A life cycle impact assessment method which comprises harmonised category indicators at the midpoint and the endpoint level, 1, 2009
5. G. Mavrotas, Effective implementation of the ε-constraint method in multi-objective mathematical programming problems. Appl. Math. Comput. **213**(2), 455–465 (2009)
6. B. Mota, M.I. Gomes, A. Carvalho, A.P. Barbosa-Povoa, Sustainable supply chains: an integrated modeling approach under uncertainty. Omega (2017)
7. P. Taticchi, P. Garengo, S.S. Nudurupati, F. Tonelli, R. Pasqualino, A review of decision-support tools and performance measurement and sustainable supply chain management. Int. J. Prod. Res. **53**(21), 6473–6494 (2015)

Chapter 18
Scheduling of Nonconforming Devices: The Case of a Company in the Automotive Sector

Mariana Araújo Nogueira, Maria Sameiro Carvalho and José António Vasconcelos

Abstract This article presents a project developed in a company's quality department aiming at scheduling the nonconforming devices analysis' process. The company faced a problem of low compliance with pre-established time requests, resulting in large fines paid to its customers of the automotive sector. In order to overcome this problem, scheduling tools were developed and tested, with the goal of minimizing the number of tardy tasks in identical parallel machines. The simulation of different scheduling rules allowed confirmation that the current prioritization rule is not the most effective one. Preliminary simulations were carried out using Lekin software [18], showing that other criteria promote better results. The use of a newly developed algorithm, combining two different criteria, resulted in a reduction of tardy tasks, thus decreasing tardiness fines paid to customers. Despite the preliminary status of present results, it is possible to foresee some improvements in the analysis process performance, by using decision making support tools based on scheduling algorithms. This way, a significant improvement on the number of analysis which fulfills the defined pre-requirements will be achieved.

Keywords Quality · Complaints · Priorization · Scheduling · Lekin

18.1 Introduction

With market globalization and the increase of competition in the industrial sector, product quality has gradually become a key factor in a company's success [9]. High consumer demand and global competition has been forcing companies to

---

M. A. Nogueira (✉)
Universidade do Minho, Braga, Portugal
e-mail: araujomariana@hotmail.com

M. S. Carvalho · J. A. Vasconcelos
Centro ALGORITMI, Universidade do Minho, Guimarães, Portugal
e-mail: sameiro@dps.uminho.pt

J. A. Vasconcelos
e-mail: zan@dps.uminho.pt

A. I. F. Vaz et al. (eds.), *Operational Research*, Springer Proceedings in Mathematics & Statistics 223, https://doi.org/10.1007/978-3-319-71583-4_18

continuously improve the quality level in its products and services [20], so they can provide effective and efficient support services to their clients according to their needs [17].

In order to improve the customers' support service, it is essential that companies possess a complete vision of all the customer's processes and are therefore able to identify and analyze all the relevant information to aid in their management and planning [12]. Since planning is a rather complex problem that directly influences the service's performance, it should be supported by an efficient task scheduling process. While planning concerns about the tasks to be done, scheduling concerns when and who carries the tasks out [10].

Scheduling is an important decision making process and widely used in the production systems of many companies. Scheduling consists in the allocation of resources throughout determined time periods and its goal is to optimize one or more objectives. Tasks in the scheduling process are characterized by: (1) a certain level of priority; (2) an earliest possibly starting date; (3) an effective date for beginning; (4) a processing time; (5) a due date and (6) a deadline. The goal of using a scheduling process is to minimize or maximize the objective function, so that the company is able to obtain good solutions and reach its goals [21].

The project presented in this article was developed in an industrial environment, in a company's Warranty Laboratory. It arose from the necessity of improving the complaints' process for nonconforming devices, since the analysis times of nonconforming devices was not in compliance with the pre-established temporal requirements. The noncompliance of the mentioned requests, particularly the established timeline for the identification of the origin of the malfunction, involves the payment of heavy financial fines.

This project proposes the definition of sequencing rules and prioritization of the devices to be analyzed, with the objective of improving the complaints process and consequently minimize the number of analysis done in the established timeline, diminishing the fines paid to the clients.

The remainder part of the paper is organized as follows: Sect. 18.2 presents a literature review on scheduling problems, Sect. 18.3 presents the main problem and its characterization, Sect. 18.4 illustrates the applicability of the scheduling model and discussion of the main results of the project and, finally, Sect. 18.5 reports the main conclusions, limitations and final considerations regarding future developments.

18.2 Literature Review

A customer complaints service plays a key role in organizations in order to obtain and recover customer satisfaction [16]. Therefore offering adequate responses to the complaints services is essential in order to correct problems and keep the customers as service providers [22]. Supplying an adequate customer support service is the key "ingredient" in product quality and competitive success [8]. Due to the current competitive environment among companies, scheduling has been becoming a need for their survival, once companies have to increasingly comply with the customers' established response times [21].

Scheduling is a process optimization in which limited resources are allocated through time between parallel and sequential activities. Such situations occur in factories, editors, transportations, universities, hospitals, airports, warranty services, among others [2]. Scheduling deals with the allocation of scarce resources and activities with the objective of optimizing one or more performance indicators [15].

Due to the varied possible combinations between different machine environments, different objectives and different side constraints, hundreds of useful scheduling problems can be obtained. As a result of this abundance of problems, as well as the constant attention of researchers from different optimization areas, throughout the last decades, scheduling literature has become rather extensive [11]. Scheduling currently represents a knowledge basis about models, techniques and knowledge on real systems. Considering scheduling as if it includes pure allocation problems, the formal development of techniques and optimization models for modern scheduling theories probably started in the years previous to World War II. In 1950 the first formal articles began to be recognized about characteristics of scheduling problems and only in 1960 did the first books arise [3]. During the 1970 s and 1980 s a greater abundance of articles and books about scheduling appeared, examples being [7, 13] that focus on deterministic scheduling aspects. Since then different and diverse scheduling problems articles have been developed [3, 21].

According to [21], a scheduling problem presents a taxonomy described according to three classes, $\alpha/\beta/\gamma$: α describes the machine environment and contains a single entry, β describes details about characteristics and processing restrictions, containing a single or multiple entries and γ describes the optimization criteria or performance measures.

Dispatching rules have always received great attention from the scientific community because of its simplicity of implementation and ease of understanding [5, 19]. Priority rules are used to select the following task to be processed from a set of tasks that await the availability of a machine.

Some examples of priority rules are Shortest Process Time (SPT), that assigns priority to the tasks that have the shortest processing time in that machine, Earliest Due Date (EDD) that assigns priority to the task with the earliest due date, Longest Processing Time (LPT) that assigns priority to the task with the longest processing time in that machine, First-In-First-Out (FIFO), where the task that arrived first will be processed first, Minimum Slack (MS) assigns the task that possesses the shortest slack time until the delivery date, Earliest Release Date (ERD) assigns the latest task to be entered in the system [4, 6].

Scheduling problems in parallel machines environments have been studied throughout several years due to its importance in research and industrial environments, since the delay of tasks relating to their limit conclusion time is considered to be an essential indicator in measuring its performance [1]. Even after recognizing and studying problems with the objective of minimizing tardiness, such as minimizing the sum of tardy tasks, there has been less progress in problems regarding minimizing the number of tardy tasks in a parallel machine environment [14, 23]. In scheduling problems for parallel machines there usually are two decisions that need to be taken. One consists in allocating tasks to specific machines and the other regards determining the sequence of tasks in each machine [23].

18.3 The Problem

In the case study, the complaint process of a nonconforming device is divided in 3 stages: customer complaint, where the client detects the device failure and claims the device to the company; the analysis process, where the device is analyzed in the Warranty Laboratory that looks for the root cause of the failure, and finally the response to the client, communicating about the reported problem. This process has a response time for the complaint based on company standards and/or customer requests.

This project will focus on the stage of the analysis process of nonconforming devices in the Warranty Laboratory defined in Fig. 18.1.

There are many cases of nonconforming devices not fulfilling the company's and the clients' standards in its analysis in the Warranty Laboratory. This can be related to several factors such as: (i) priorities of analysis are not adequate to the temporal request of clients and simply follow FIFO; (ii) technicians that perform the analysis do not comply with the priority orders indicated by the system currently used in the company and; (iii) the high Backlog time (waiting time before analysis) of the nonconforming devices.

In 2015 only 1099 out of 2736 (40%) of the clients' complaints complied with the response times according to the company's standards. Therefore, penalties paid to customers in 2015 were in the order of millions of euros.

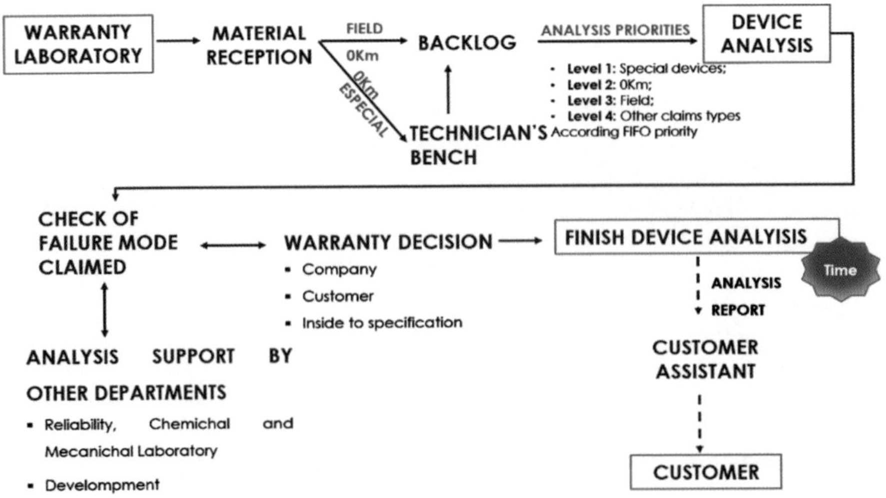

Fig. 18.1 Analysis process of nonconforming devices

Using scheduling notation, this problem can be defined as:

- Nature of variables: stochastic scheduling, since the analysis time can be uncertain/variable. However, in this first model mean values were assumed and a deterministic scheduling model was developed.
- α – Machine environment: identical parallel machine problem, $\alpha = Pm$, since the technicians responsible for each group possess the same expertise and similar speed executing the analysis;
- β – Characteristics and processing restrictions:
  - Release Date, $\beta = rj$, the analysis of a nonconforming device can only be initiated the moment a device enters the system;
  - Interruptions, $\beta = prmp$, a device's analysis can be interrupted to proceed to a second level analysis from other departments;
  - Machine eligibility restrictions, $\beta = Mj$, when a result is sent to the Warranty Laboratory after a support in the analysis from another department, the resuming of the analysis will be attributed to the technician that started the analysis.
- γ – Performance measures or optimization criteria:
  - Minimize the number of tardy jobs, $\gamma = \min(\sum wjUj)$ where wj is the task weighting and Uj is equal to 1 if task j is tardy.

Scheduling tools can play an important role in order to understand the main issues of a scheduling problem. Therefore, Lekin software [18] was used to explore some scenarios relative to the established priorities currently used in the Warranty Laboratory. Lekin software was developed by the Stern School of Business (1998) and contains a series of scheduling algorithms and heuristics that the user can use. Data regarding the Warranty Laboratory was tested to verify which rule or priority criteria obtained the best result relative to the objective function: the minimum weighted number of tardy tasks. However, Lekin software has some limitations, so it was necessary to simplify the characteristics of the analysis process of nonconforming devices. According to the $\alpha/\beta/\gamma$ notation, the problem simulated on Lekin presents the following characteristics: α = Parallel Identical Machines, Pm; β = Release Date, rj; γ = minimize the weighted number of tardy tasks, $\min(\sum wjUj)$ and deterministic nature.

The following conditions were taken into account:

- Four Warrant Laboratory technicians;
- Seven daily hours shift per technician;
- Average Analysis time for each task pre-defined;
- Only analysis that depend entirely on the Warranty Laboratory have been selected – no preemption allowed;
- Different weights associated to the tasks according to the relevance the company attributes to the type of customer complaint;
- Simulation ran on a determined time period.

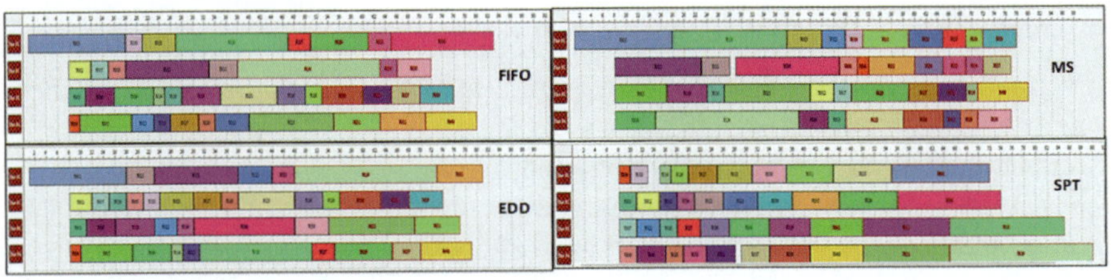

Fig. 18.2 Results of simulations on Lekin software

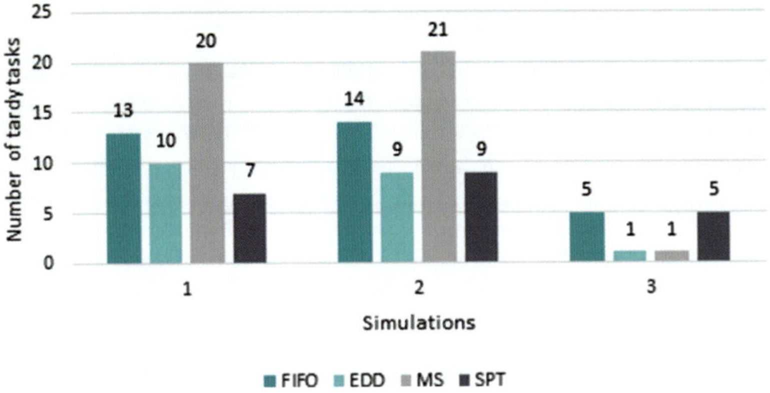

Fig. 18.3 Number of tardy tasks by priority criteria

Different scenarios were ran with the objective of exploring different priority criteria: First-In-First-Out (FIFO), Earliest Due Date (EDD), Minimum Slack (MS) and Shortest Processing Time (SPT). Figure 18.2 shows a Gantt diagram with results produced by the software:

Figure 18.3 shows that the priority criteria currently used in the Laboratory is not the one that obtains the best results for the selected objective function (minimize the number of tardy tasks). The following data supports this conclusion:

- In the first simulation, the FIFO criteria results in 32.5% tardy tasks, with EDD resulting in 25% and SPT 17.5%;
- The second simulation shows the percentage of tardy tasks goes from 58% to 37.5% if the EDD or SPT criteria is used instead of FIFO;
- The third simulation shows that FIFO causes 25% of tardy tasks while EDD only causes 5%;
- In the total of the 3 simulations, EDD caused 23.8% tardy tasks, SPT 25%, FIFO 38% and MS 50%.

We can then conclude that the criteria that achieve the best results are EDD and SPT. These results happen when no new tasks arrive until all the tasks in analysis are concluded.

Despite the software limitations, the interesting results obtained, motivated the development of a new tool that allowed adding more inherent characteristics to the process under analysis.

A scheduling algorithm has been proposed with the objective of finding a better solution for the prioritization in the analysis of nonconforming devices and consequently obtaining better results in the amount of tardy tasks.

18.4 Solution Method and Results

The definition of prioritization and process scheduling rules used in the algorithm are based in the Moore and Hodgson Algorithm [13] and in the results obtained in the simulations ran in Lekin, where EDD and SPT criteria proved to obtain the best results.

The scheduling algorithm takes into account the combination of two priority criteria: EDD orders the device list to allocate the technicians and later assign priorities to the devices with larger weight. SPT influences the weight attributed to the devices. Besides the SPT criteria, the weighing of devices is also influenced by the type of customer complaint and the state of the analysis.

The solution approach includes four main stages.

The first stage is the characterization of the problem according to the scheduling classes $\alpha/\beta/\gamma$: α = Parallel Identical Machines, Pm; β = Release Date, rj; β = preemption, $prmp$ e β = machine eligibility restrictions, Mj; γ = minimize the weighted number of tardy tasks, $\min(\sum wjUj)$.

The second stage is the description of each j task's data according the notation of scheduling problems.

The third stage defines the instant the scheduling algorithm should be run for analysis in the Warranty Laboratory.

The fourth stage is to mathematically define the algorithm rules:

1. Check for unallocated j tasks;
2. Check technicians' availability;
3. Calculate priorities according to the EDD criteria;
4. Select the highest priority j task;
5. Select the technician with earliest release date;
6. Allocate task to technician;
7. Calculate the moment the technician becomes available again by adding the tasks' start instant with the task total time;
8. Remove j task from the priority list;
9. Check for technicians with availability instant in less than seven hours then (1); if none, wait for new action.
10. Calculation of the objective function.

Besides the rules defined previously, new rules were defined for different steps in which the algorithm assigns a new task for the technician in a dynamic context:

Table 18.1 Average results of numerical experiments of the new algorithm and existing warranty laboratory rules

	Tardy tasks	Tasks done on time	Total tasks
New algorithm	3	28	31
Warranty laboratory rules	5	26	31

- The first time that the algorithm runs;
- Whenever a technician finishes the task he has been allocated or when he is free;
- When an analysis is assigned to a second level process (preemption can occur whenever further analysis are required);
- Whenever a technician finishes the task outside the expected date.

Lastly, the algorithm was tested using JAVA language.
Features of the algorithm include:

- Schedules according to EDD and SPT criteria;
- Schedules tasks submitted to further analysis in other departments allowing preemption;
- Incorporated real time information if a technician finishes the analysis earlier or later than the scheduled date it readjusts his tasks to the moment he becomes free; reschedule in case the real analysis time differs from the estimated time.

Preliminary numerical experiments were carried out to compare scheduling results of the new algorithm and the current practice in the Warranty Laboratory (Table 18.1).

Table 18.1 shows that the number of tardy tasks, using the new algorithm, is 3 (9.67%), therefore lower than according the Warranty Laboratory priorities, which was 5 (16.12%). A reduction of 40% in the number of tardy task (from 5 to 3) can be seen as very promising but still requires further developments to provide the company with a significant decrease in the total penalties paid.

18.5 Discussion and Conclusion

The process of analysis of nonconforming devices has shown to be very complex and to have great variability in the analysis times, as well as in the operations made during the analysis. Such process variability increases complexity and influences the results obtained in this project.

The simulation of several priorities rules in Lekin software allowed concluding that EDD and SPT criteria produced the best results. The criteria currently used in the Warranty Laboratory, FIFO, obtained the worst results. The EDD criteria resulted in 23.8% of tardy tasks, SPT in 25%, FIFO in 38% and MS 50%. Using the EDD criteria instead of FIFO results in a 14% reduction in the number of tardy tasks.

A new algorithm has been developed to overcome Lekin software limitations, namely to include real and estimated times and to combine EDD criteria with SPT;

additionally, in the new approach a new task will be assigned to a technician that finishes the previous task earlier. Although the limited number of numerical experiments carried out, the scheduling of tasks using EDD criteria combined with SPT is promising towards an increase performance of the Laboratory.

The implementation of this algorithm for the prioritization of analysis in the Warranty Laboratory can be said to bring significant improvements relative to the number of tardy tasks and therefore a reduction on penalties paid to the clients of about 40% relatively to the current prioritization. However, it is necessary to take into account that the obtained results are preliminary and need further validation by widening the number of tests and performing a data sensibility analysis.

Throughout this project some limitations were felt that directly or indirectly affected the obtained results. One of the larger difficulties was the fact that the complaints process has some activities that are not completely defined. Besides, the variability and uncertainty in the analysis time, the root causes associated with the complained defect, the description of the complained defect, the different tests done during the analysis, the type of support provided by other departments and the data in the systems were severe limitations that prevented the project from obtaining more specific and relevant results. Another limitation was the fact that the company did not provide access to the external costs of the clients' indemnities, making it impossible to measure the impact of this project in cost reduction.

Future work will start by increasing the number of algorithm tests, a sensitivity data analysis and the use of improvement heuristics.

Acknowledgements This work has been supported by COMPETE: POCI-01-0145-FEDER-007043 and FCT– Fundação para a Ciência e Tecnologia within the Project Scope: UID/CEC/00319/2013.

References

1. V. Armentano, D. Yamashita, Tabu search for scheduling on identical parallel machines to minimize mean tardiness. J. Intell. Manuf. **11**, 453–460 (2000)
2. T. 'Bagchi, *Multiobjective Scheduling by Genetic Algorithms* (Kluwer Aademic Pubishers, Dordrecht, 1999)
3. K. 'Baker, D. Trietsch, *Principles of Sequencing and Scheduling* (Wiley, London, 2009)
4. H. Barksdale, T. Powell, E. Hargrove, Complaint voicing by industrial buyers. Ind. Mark. Manag. **13**, 93–99 (1984)
5. J. Blackstone, D. Philliphs, G. Hogg, A state-of-the-art survey of dispatching rules for manufacturing job shop operations. Int. J. Prod. Res. **20**, 27–45 (1982)
6. D. Carvalho. Programação da Produção. Universidade do Minho (2000)
7. E. Coffman, *Computer and Job-Scheduling Theory* (Wiley, London, 1976)
8. M.A. Cohen, H.L. Lee, Out of touch with customer needs? Spare parts and after sales service. Sloan Manag. Rev. **31**, 55 (1990)
9. M. Colledani, T. Tolio, Impact of quality control on production system performance. CIRP Ann. Manuf. Technol. **55**, 453–456 (2006)
10. T. 'Dean, S. Kambhampati, Planning and Scheduling (1997)
11. B.C. Dean. Approximation Algorithms for Stochastic Scheduling Problems. Massachusetts Institute of Techonology, 2005

12. U. Dombrowski, C. Malorny, Process identification for customer service in the field of the after sales service as a basis for lean after sales service. Procedia CIRP **47**, 246–251 (2016)
13. S. 'French, *Sequencing and Scheduling: And Introduction to the Mathematics of the Job-Shop* (Ellis Horwood, Chichester, 1982)
14. J.C. Ho, Y.-L. Chang, Minimizing the number of tardy jobs. Eur. J. Oper. Res. **84**, 343–355 (1995)
15. J. 'Leung, *Handbook of scheduling: Algorithms, Models and Performance Analysis* (Chapman and Hall/CRC, Boca Raton, 2004)
16. C. 'Lovelock, J. Wirtz, H. Keh, X. Lu, *Services Marketing in Asia, Managing People, Techonology ans Strategy* (Prentice Hall International Incorporation, New Jersey, 2002)
17. S. Murali, S. Pugazhendhi, C. Muralidharan, Modelling and investigation the relationship of after sales service quality with customer satisfaction, retention and loyalty - a casa study of home appliances business. J. Retail. Consum. Serv. **30**, 67–83 (2016)
18. NYU Stern School of Business. LEKIN–flexible job-shop scheduling system (1998), http://web-static.stern.nyu.edu/om/software/lekin/
19. H. Oliver, C. Rajendran, Efficient dispatching rules for scheduling in a job shop. Int. J. Prod. Econ. **48**, 87–105 (1997)
20. M.A. Pešić, V.J. Milić, J. Stanković, Significance of business quality management for increasing competitiveness of serbian economy. Serb. J. Manag. **7**, 149–170 (2012)
21. M.L. Pinedo, *Scheduling: Theory, Algorithms, and Systems* (Prentice Hall, New Jersey, 2002)
22. V.T. Quy, The impact of organizational responses to complaints on posto purchase behavioral intentions via recovery satisfaction - the case of saigon commercial bank. Strateg. Manag. Q. **2**, 44–79 (2014)
23. S.-O. Shim, Y.-D. Kim, Scheduling on parallel identical machines to minimize total tardiness. Eur. J. Oper. Res. **177**, 135–146 (2007)

Chapter 19
Understanding Complexity in a Practical Combinatorial Problem Using Mathematical Programming and Constraint Programming

Beatriz B. Oliveira and Maria Antónia Carravilla

Abstract Optimization problems that are motivated by real-world settings are often complex to solve. Bridging the gap between theory and practice in this field starts by understanding the causes of complexity of each problem and measuring its impact in order to make better decisions on approaches and methods. The Job-Shop Scheduling Problem (*JSSP*) is a well-known complex combinatorial problem with several industrial applications. This problem is used to analyse what makes some instances difficult to solve for a commonly used solution approach – Mathematical Integer Programming (MIP) – and to compare the power of an alternative approach: Constraint Programming (CP). The causes of complexity are analysed and compared for both approaches and a measure of MIP complexity is proposed, based on the concept of load per machine. Also, the impact of problem-specific global constraints in CP modelling is analysed, making proof of the industrial practical interest of commercially available CP models for the *JSSP*.

Keywords Job-shop scheduling problem · Mathematical programming
Constraint programming · Global constraints · Complexity

19.1 Introduction

The Job-Shop Scheduling Problem (*JSSP*) and its many extensions and variations have been thoroughly studied on the field of Optimization, due to its computational difficulty. This problem is NP-hard [8] and was even qualified as "one of the most computationally stubborn combinatorial problems" [3]. For these reasons, it will be analysed in this paper as a starting point to understand the meaning of *complexity* for both approaches: Mathematical Programming and Constraint Programming.

B. B. Oliveira (✉) · M. A. Carravilla
INESC TEC and Faculty of Engineering, University of Porto, Porto, Portugal
e-mail: beatriz.oliveira@fe.up.pt

M. A. Carravilla
e-mail: mac@fe.up.pt

A. I. F. Vaz et al. (eds.), *Operational Research*, Springer Proceedings
in Mathematics & Statistics 223, https://doi.org/10.1007/978-3-319-71583-4_19

The *JSSP* summarily consists of scheduling *J* jobs to be processed in *M* machines with the goal of minimizing the time it takes for all the operations to end (makespan). Each job takes a fixed processing time in each machine and there is a certain sequence of machines that it should follow – each job has its own specific sequence. The operations must be completed – once a job starts on a specific machine, it must take the full processing time. Also, a machine can only handle one job at the time.

The *JSSP* has numerous applications in the industry, namely on production planning and scheduling. In the past years, several extensions and adaptations are arising in the literature [9, 12], many concerning real applications.

This paper has two main objectives. Firstly, it aims to gain some insights regarding the factors that induce complexity on the Mathematical Programming approach to this problem. As this approach is commonly used to solve real-world problems, understanding these factors can help practitioners identify beforehand the challenges and/or opportunities that arise from the characteristics of the problem and therefore make better decisions regarding e.g. solution methods.

Secondly, it aims to understand whether the alternative approach of Constraint Programming (CP) is useful to tackle to the "difficult to solve" instances. Above all, there is the goal to understand the power of the commercially available CP models for the *JSSP* and, consequently, whether they are a realistic solid alternative for practitioners.

The paper is thus divided into two main sections. The first section is related to the Mathematical Programming approach. It presents a mixed-integer formulation for this combinatorial problem that is commonly used in the literature and assesses its computational power with eighty-two literature instances. The main objective is to understand which factors are more relevant to explain the complexity and "stubbornness" of the *JSSP*, namely to understand the impact of size. Therefore, on Sect. 19.2.1, the MIP model is presented; the computational results are discussed on Sect. 19.2.2. In this section, a potential measure of instance *complexity* is also presented and tested. The second part is devoted to the Constraint Programming approach to the *JSSP* (Sect. 19.3). Here, three different CP formulations are presented (Sect. 19.3.1) and tested against the performance of the MIP model (Sect. 19.3.2). One of the CP formulations is a model commercially available, including global constraints specifically developed for this problem. The main goal is herein to understand the power and limitations of CP programs, namely the commercially available one. Finally, on Sect. 19.4 the main conclusions are drawn.

The overall purpose of this paper is to draw attention to the impact of understanding and measuring impact when tackling specially complex problems, such as real-world applications of the *JSSP*. This purpose is materialised in the following contributions:

- A quantitative-based discussion on factors that induce complexity for a MIP approach to the *JSSP*, namely the total size of the instance and relationship between number of jobs and machines;
- A proposed measure of MIP-complexity of *JSSP* instances based on load per machine;

- A comparison of the computational power of different CP formulations for the *JSSP*, with insights related with commercially available CP models, which show significant advantages when tackling complex instances;
- A comparison between CP and MIP approaches to the *JSSP*, establishing their advantages and shortcomings;
- A quantitative evidence of the benefits of using commercially available CP models to tackle complex *JSSP* instances and settings.

19.2 Mathematical Programming Approach

19.2.1 MIP Model

This model follows the one proposed by Applegate and Cook [3], based on the *disjunctive programming problem* presented in Balas [5].

Sets and indexes

$\mathscr{J}$ set of jobs ($|\mathscr{J}| = J$)
$\mathscr{M}$ set of machines $|\mathscr{M}| = M$)
$i, j \in \mathscr{J}$ index of jobs ($i, j = \{1, \ldots, J\}$)
$k \in \mathscr{M}$ index of machines ($k = \{1, \ldots, M\}$)

Parameters

p_{ik} the kth operation (machine) for job i
t_{ik} processing time of job i on machine k
L significantly large value

Decision Variables

x_{ik} starting time of job i on machine k
y_{ijk} (= 1) if job i is scheduled before job j in machine k
c makespan

Mathematical Program

$$\min \quad c \tag{19.1}$$

$$\text{s.t.} \quad x_{i\,p_{ik}} \geq x_{i\,p_{ik-1}} + t_{i\,p_{ik-1}}, \qquad \forall i \in \mathscr{J}, \forall k \in \mathscr{M}\setminus\{1\} \tag{19.2}$$

$$x_{ik} \geq x_{jk} + t_{jk} - L y_{ijk}, \qquad \forall k \in \mathscr{M}, \forall i, j > i \in \mathscr{J} \tag{19.3}$$

$$x_{jk} \geq x_{ik} + t_{ik} - L(1 - y_{ijk}), \qquad \forall k \in \mathscr{M}, \forall i, j > i \in \mathscr{J} \tag{19.4}$$

$$c \geq x_{i\,p_{iM}} + t_{i\,p_{iM}} \qquad \forall i \in \mathscr{J} \tag{19.5}$$

$$x_{ik}, c \geq 0, \qquad \forall i \in \mathscr{J}, k \in \mathscr{M} \tag{19.6}$$

$$y_{ijk} \in \{0, 1\}, \qquad \forall i, j > i \in \mathscr{J}, k \in \mathscr{M} \tag{19.7}$$

The goal of this model is to minimize the total makespan (Eq. 19.1). The first constraint (Eq. 19.2) ensures that a job can only start to be processed in a machine if it has finished its processing on the previous machine (clearly, this does not apply to the first operation of the job).

The second and third constraints are related to the order of the jobs in each machine (Eqs. 19.3 and 19.4). They state that if a job precedes another job in a certain machine, then its starting time is not limited by the latter. However, if a job comes after another job in a machine, it can only start when this one has finished. The following constraint (Eq. 19.5) states that the total makespan is given by the job that finishes the processing last (i.e., based on the last machine: the Mth machine the job will go through).

Finally, the decision variable that sets the starting time of each job on each machine as well as the makespan should be non-negative, and the precedence variable is set as a binary (Eqs. 19.6 and 19.7).

19.2.2 Computational Tests and Results

In order to understand the behaviour of this MIP model, 82 instances from the literature were run. The computational tests were run on a Toshiba Personal Computer with 16 Gigabyte of RAM memory, and with a processor Intel(R) Core(TM) i7-4600U CPU @ 2.10 GHz 2.70 GHz. The model was developed in a OPL Project in IBM ILOG CPLEX Optimization Studio 12.6.1.0 and the MIP Solver used was CPLEX. The time limit set for the resolution of each instance was 30 min. The instances' characteristics and solution can be found on the Appendix (Table 19.4) and the main results are summarised on Table 19.1.

Table 19.1 Summary of the average results of the MIP model

J	M	#instances	Avg. sol. time (s)	Sol. time std. dev. (s)	Avg. gap (%)	Gap std. dev.
6	6	1	0	0	0	0
10	5	5	13	18	0	0
10	10	18	142	286	0	0
15	5	5	*1.800*	0	29	5%
15	10	5	*1.800*	0	11	8%
15	15	5	*1.800*	0	12	3%
20	5	6	*1.800*	0	47	5%
20	10	10	*1.800*	0	39	9%
20	15	8	*1.800*	0	37	2%
20	20	4	*1.800*	0	22	4%
30	10	5	*1.800*	0	55	2%
50	10	10	*1.800*	0	76	3%

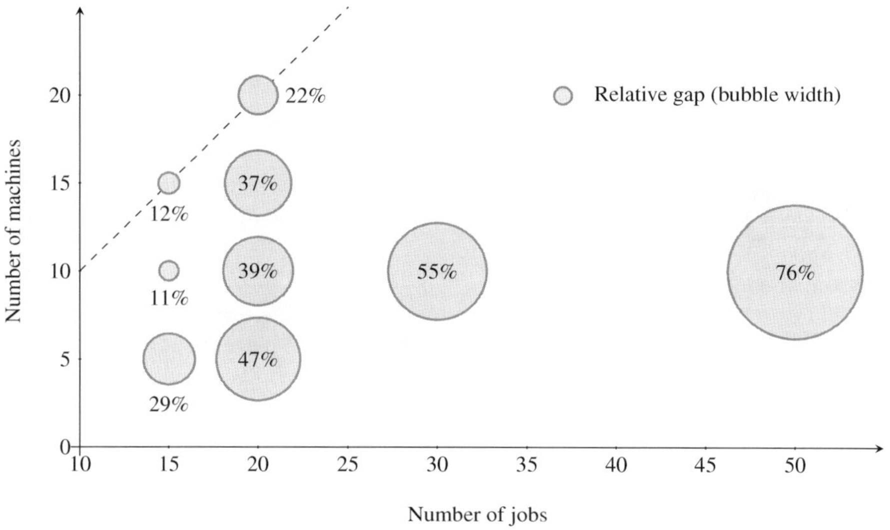

Fig. 19.1 Average MIP relative optimality gap attained per instance size

As it is possible to observe on Table 19.1, the solver was only able to prove optimality within the time limit (30 min) for the twenty-four smallest instances.[1] The results are herein analysed for both resulting groups:

Instances Solved to Optimality:

For the smallest instances, solved to optimality, the average solution time increases with the size of the instances. Nevertheless, the magnitude of the variability of solution times for instances of the same size (given by its standard deviation, in Table 19.1) suggests that the size is not the only factor that translates into complexity and, consequently, to longer run times.

Instances Not Solved to Optimality:

For the fifty-eight instances that were not solved to optimality in 30 min, the value of the optimality gap attained may shed some light on the matter of complexity. Firstly, it is interesting to compare the average results of the instances of type $(J, M) = (20, 5)$ and of type $(J, M) = (15, 15)$. The latter are bigger in terms of number of variables and constraints, yet the solver is able to achieve significantly lower gaps for them (47% vs. 12%). This is another evidence that the "easiness to solve" is not only linked with the size of the problem.

Figure 19.1 represents the average gap attained for instances of the same size. The horizontal axis represents the number of jobs in the instances, the vertical axis represents the number of machines, and the width of the bubbles depicts the size of the average optimality gap. On an horizontal reading, it is possible to see that the optimality gap increases as instances tackle an increasing number of jobs. This may

[1]Instances `ft06`, `ft10`, `abz5`, `abz6`, `la01–la05`, `la16–la20`, `orb01–orb10`.

be due to the fact that the number of jobs influences the most the size and thus, at some (limited) degree, the complexity of the model and its difficulty to solve to optimality. In fact, due to the structure of the binary variable $y_{ijk}, \forall i, j > i \in \mathscr{J}, k \in \mathscr{M}$, it can be seen that $J = |\mathscr{J}|$ has more impact on the number of variables that $M = |\mathscr{M}|$. Due to the structure of Constraints 19.3 and 19.4, it also strongly influences the number of constraints. In conclusion, this horizontal increase seems a natural consequence of the size of the instances, which is heavily dependent on the number of jobs (J).

However, if one considers a fixed number of jobs (for example, $j = 20$ on the horizontal axis) and analyses the impact of the number of machines in the optimality gap, the effect is reversed. It seems that, for a fixed number of jobs, the optimality gap achieved in the same run time is better for more balanced instances, with the same number of jobs and machines (even if that means a bigger number of variables and constraints).

Regarding the observable "easiness" to solve balanced instances (in terms of number of jobs and machines), one may consider several hypotheses. Please consider the following definition of optimality gap for a minimization problem: $Gap = (UB - LB)/LB$, where LB stands for the *lower bound*, given by the best relaxed solution found on the Branch-and-Bound nodes not pruned at a time; and UB stands for the *upper bound*, which is given by the best integer solution found so far. Due to the complexity of the methods and heuristics utilized by CPLEX to speed up the Branch-and-Bound procedure, it is difficult to understand how the *lower bound* evolves in different instances. Therefore, the main hypothesis presented to explain this behaviour is related with the update velocity of the *upper bound*. Analysing the data herein presented, it may be possible to say that the balanced structure (i.e., a similar number of jobs and machines) makes it easier for the solver to find admissible (integer) solutions and thus update the *upper bound*.

One could also argue that the effect of the increasing number of machines (that reduces the optimality gap for the same number of jobs) is not a matter of balance but of the existence of alternative admissible solutions. That is to say that it is not the *balance between the two dimensions* of the problem that is at stake but *only the number of machines*. In fact, if a certain problem is characterized by a big number of machines (and if the precedence of operations is not too restraining) it is possibly easier to find alternative optima, as well as admissible integer solutions that update the *upper bound*.

Other factors may influence the complexity of the instance, namely the precedences and processing times that may hinder or help the procedure to find admissible integer solutions, which may explain the high standard deviation of the solution times observed for the twenty-four smallest instances.

A Measure of Complexity

In order to further understand the concept of instance complexity, a measure of the load of solicitations applied to each machine throughout the scheduling horizon was designed. Dividing the horizon in partitions of the same size, the goal was to understand whether the load applied in the partitions (the need or demand of jobs to be scheduled there) was constant, and what was the maximum value it achieved throughout the horizon. So, for some partition p of the horizon, the *needs-to-capacity ratio* (r) is given by:

$$r_p = \frac{\text{count}(ES_{ik} \in p)}{M} \quad , \forall i \in \mathcal{J}, \forall k \in \mathcal{M}$$

Here, M is, as before, the number of machines, and ES_{ik} stands for the *earliest start* of job i in machine k. The earliest start is defined by the precedence rules of each job:

$$ES_{ik} = \sum_m t_{im} \quad , \forall m \in \mathcal{M} : p_{io} = m,\ p_{io'} = k,\ o < o'$$

Here, as before, t_{im} stands for the processing time of job i on machine m and p_{io} represents the oth machine where job i is processed. Therefore, the earliest start of a job on a machine is the result of the sum of the processing times of that same job on the preceding machines. For example, if job 1 can only be processed in machine C after being processed in machines A and B then the *earliest start* of job 1 on machine C is equal to the time it takes to be processed in machines A and B. Therefore, the *needs-to-capacity ratio* is able to represent, at a certain extent, what are the needs of jobs to be scheduled in each machine, taking into consideration the precedences.

This measure was calculated for the eighty-two instances run. In order to adapt the partitions of the scheduling horizon to each instance, the length of the partitions was based on the instances' minimum value of the processing times:

$$\text{length} = 2 \cdot \min_{ik}(t_{ik}), \quad \forall i \in \mathcal{J}, \forall k \in \mathcal{M}$$

Figure 19.2 represents the evolution of this ratio for instances `la10` and `la37`. Both instances consider 15 jobs yet in `la10` there are 5 machines whilst in `la37` there are 15 machines. Despite the smaller size, the MIP solver achieved an optimality gap of 35% for the former (vs. 15% for the latter).

Fig. 19.2 Evolution of the *needs-to-capacity ratio* for the first 40 partitions of the scheduling horizon of instances `la10` and `la37`

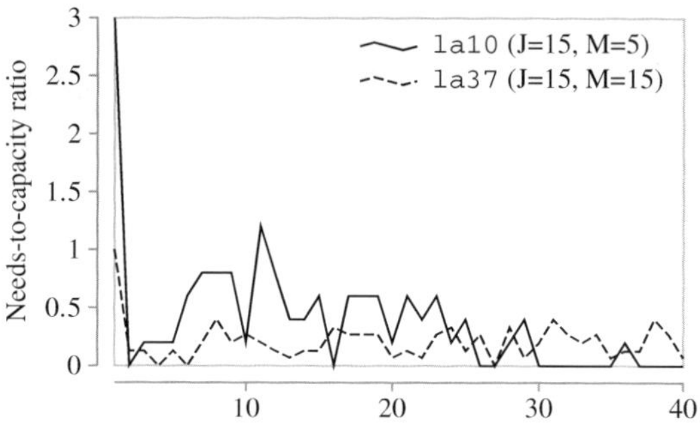

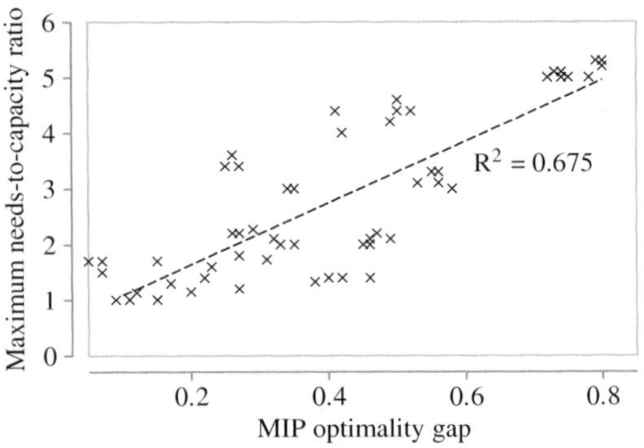

Fig. 19.3 Correlation between the optimality gap attained by the MIP solver and maximum *needs-to-capacity ratio* of the instances in Footnote 2

As explained before, the *needs-to-capacity ratio* represents, for each partition of time, the number of jobs whose *earliest start* in some machine falls within the partition, divided over the total number of machines. It is thus an average measure of the load per machine. Figure 19.2 is hence able to display the additional complexity of `la10`: the maximum and overall level of solicitations (needs) per machine is higher than in `la37` and has more variation throughout time. This supports the statement that the precedences and processing times in the instance may hinder the solver to find admissible integer solutions and influence the solution time/quality more than the number of variables and constraints.

In order to compare instances in a easier way, the *maximum* ratio attained throughout the partitions of the scheduling horizon was calculated for each instance. Figure 19.3 represents the correlation between the maximum ratio and the optimality gap attained by the MIP solver (for those instances in which the solver was unable to prove optimality within the time limit[2]). Table 19.5 in the Appendix also presents these metrics, together with the quartile in which each maximum ratio falls when comparing the maximum needs-to-capacity ratio of the eighty-two instances. This table also contains information regarding the number of partitions, their length, and the standard deviation of the ratio throughout the horizon.

Figure 19.3 represents each instance by its optimality gap and maximum *needs-to-capacity ratio*. It is possible to observe that the ratio increases as the gap increases, i.e. as the MIP solver finds it more and more difficult to reach (or prove) optimality. The R^2 statistics with a value of 67% shows a good correlation between these two metrics, which supports the hypothesis that the ratio is a good indicator of an instance's complexity for the MIP solver. That is to say, the relation between the load of jobs that, at a certain point in time, are soliciting the machines and the capacity the system has to respond to these solicitations (i.e. the number of machines) has a direct influence on the difficulty the solver has to reach and prove optimality for a given instance.

[2]Instances in which the MIP solver was unable to prove optimality within the 30 min time limit: `abz7–abz9`, `ft20`, `la06–la15`, `la21–la40`, `swv01–swv20`, `yn1–yn4`.

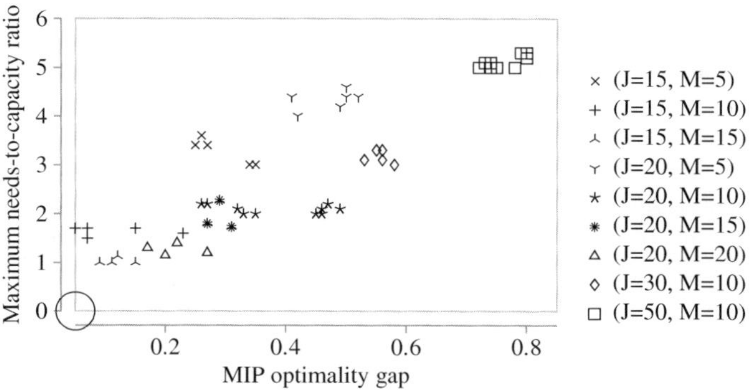

Fig. 19.4 Correlation between the optimality gap attained by the MIP solver and maximum *needs-to-capacity ratio* of the previous instances, categorized by size (number of jobs J and number of machines M)

Figure 19.4 shows the size characteristics (number of jobs and machines) of the instances in Fig. 19.3. It is interesting to see from a different perspective the conclusion drawn before: an increase of the size does not imply an increase of complexity. At the same time, if one considers a specific group of instances of the same size, videlicet the instances with $(J, M) = (20, 20)$ (triangle-shaped points), the maximum ratio attained is very similar amongst the instances while the optimality gap still shows some variation, not fully explained by the ratio (the dots representing the instances are spread horizontally on Fig. 19.4). This shows that are still other factors that may contribute to the complexity of the instance besides size and load of solicitations on the machines.

19.3 Constraint Programming Approach

"CP solving is based on intelligently enumerating all possible variable-value combinations" [13].

Constraint Programming is based on a flexible and intuitive modelling language, allowing for logic constraints, non-linear constraints, and almost every expression possible over the variables. It also requires that the domain of the variables (finite set of possible values) is set beforehand [13].

A CP program assigns values to variables without violating the constraints [4]. A search tree is built with several nodes, where each node is a solution, with some variables with a value already assigned to them. Each node then branches to several other nodes, each representing a similar solution but with other variable (from the remaining without assignment) allocated to some possible value. There are as many *child nodes* as there are possible values in the domain of the variable to assign [1]. The mechanism of constraint propagation consists on decreasing the domain of the variables not assigned yet due to the value assigned to a certain value in a iteration, and thus decreasing the number combinations to test [13].

Global constraints are a powerful element of CP programs. A global constraint is, by definition, a constraint that involves more than two variables. When designed to seize to the full potential the structure of the problem, these can be a powerful tool to facilitate constraint propagation and thus reduce significantly the search tree. It is important to recall that, although there are widely used global constraints (e.g. *alldiff*), some of the best performing global constraints that solve specially difficult problems are usually designed specifically for them. For the most studied problems in CP, global constraints have been designed and enhanced in the past years [10]. It is also important to note that in CP, unlike MIP problems, increasing the number of constraints may indeed be beneficial as it helps propagation.

Due to their characteristics, it is often the case that CP programs should perform well, by nature and theory, at finding admissible solutions, although only being able to prove optimality by explicitly searching all nodes and comparing their results. MIP solvers, on the other hand, although having more difficulty at finding admissible solutions, are able to prove optimality by setting upper and lower bounds and determining the optimality gap, i.e. it is not needed to *explicitly* explore the full solution space.

Scheduling problems are a classical application of CP due to their above-discussed "stubbornness". Therefore, specific well-performing global constraints have been designed and tested for it. Moreover, since it was previously inferred that the difficulty to find admissible integer solutions may be an important factor on the "stubbornness" of this problem (Sect. 19.2.2), Constraint Programming should be an appropriate approach to tackle this problem. So, in the next section, three different CP models are presented and tested. The first two are CP variations of the previously presented MIP model. The third one uses global constraints developed specifically for this problem and different decision variables.

19.3.1 CP Models

Three CP models are used in this paper to understand and compare the power of Constraint Programming models, namely and especially of global constraints.

The first model (CP1), is nearly a translation of the MIP model presented in Sect. 19.2.1. Since logic constraints can be used in CP, Constraints 19.3 and 19.4 (of the MIP model) can be translated into one disjunctive equation, and thus there is no need to use the auxiliary binary variables y_{ijk}. The second model (CP2) is a variation of CP1, where the *makespan* variable (c) is eliminated (as well as Constraint 19.5). Since CP does not require its model elements to be linear, the objective function can be stated as the minimization of the maximum value amongst several, which was not possible using Mathematical Linear Programming. Finally, the third model (CP3) is based on the IBM ILOG CPLEX Optimization Studio example *sched_jobshop* for

CP and seizes all the advantages of Constraint Programming: the decision variables' representation is more appropriate and two specifically developed global constraints are used.

19.3.1.1 CP1

Using the same sets, indexes and parameters of the MIP Model presented:

Decision Variables

$$x_{ik} \in \{0, \ldots, 10000\} \textit{ and integer} \text{ starting time of job } i \text{ on machine } k$$
$$c \in \{0, \ldots, 10000\} \textit{ and integer} \text{ makespan}$$

CP1 Model

$$\min \quad c \tag{19.8}$$
$$\text{s.t.} \quad x_{i\,p_{ik}} \geq x_{i\,p_{i\,k-1}} + t_{i\,p_{i\,k-1}}, \qquad \forall i \in \mathscr{J}, \forall k \in \mathscr{M}\backslash\{1\} \tag{19.9}$$
$$x_{ik} \geq x_{jk} + t_{jk} \quad \vee \quad x_{jk} \geq x_{ik} + t_{ik}, \qquad \forall k \in \mathscr{M}, \forall i, j > i \in \mathscr{J} \tag{19.10}$$
$$c \geq x_{i\,p_{iM}} + t_{i\,p_{iM}} \qquad \forall i \in \mathscr{J} \tag{19.11}$$

The main difference of this model (vs. the MIP model) is the absence of the y_{ijk} variables and the presence of the disjunctive Constraint 19.10. In fact, CP models allow for this kind of logic constraints, which state that either the inequality on the left or the inequality of the right must hold. That is to say that for each machine either job i waits for job j to end (if it is scheduled after it) or the other way around. This type of constraint allows for this disjunction "or-or" to be stated without the auxiliary binary variable that states whether or not job i precedes job j.

19.3.1.2 CP2

Using the same sets, indexes and parameters of the MIP Model presented:

Decision Variables

$$x_{ik} \in \{0, \ldots, 10000\} \textit{ and integer} \text{ starting time of job } i \text{ on machine } k$$

CP2 Model

$$\min \quad \max_i \{x_{i\,p_{iM}} + t_{i\,p_{iM}}\} \tag{19.12}$$
$$\text{s.t.} \quad x_{i\,p_{ik}} \geq x_{i\,p_{i\,k-1}} + t_{i\,p_{i\,k-1}}, \qquad \forall i \in \mathscr{J}, \forall k \in \mathscr{M}\backslash\{1\} \tag{19.13}$$
$$x_{ik} \geq x_{jk} + t_{jk} \vee x_{jk} \geq x_{ik} + t_{ik}, \qquad \forall k \in \mathscr{M}, \forall i, j > i \in \mathscr{J} \tag{19.14}$$

When comparing with CP1, this model has a different objective function and, consequently, the variable c and the Constraint 19.11 are no longer needed. This is possible because CP allows for the use of non-linear equations, such as Eq. 19.12.

19.3.1.3 CP3

Using the same sets, and indexes of the MIP Model presented:

Parameters

O_{ik} : (m, t) operation of job i in the kth machine, represented by a tuple of the machine in question (m) and the associated processing time (t)

Decision Variables

w_{ik}, sizeO_{ik} : t **interval** that represents the operation of job i on machine k, which must have the length (size) of the processing time t of operation O_{ik}

s_k **sequence** of the above-defined intervals (jobs) on machine k; built from the intervals for all i jobs and o machines such that $O_{io} : m = k$

CP3 Model

$$\min \quad \max_i \{\text{endOf}(w_{iM})\} \tag{19.15}$$

$$\text{s.t.} \quad \text{noOverlap}(s_k) \qquad \forall k \in \mathcal{M} \tag{19.16}$$

$$\text{endBeforeStart}(w_{ik}, w_{ik+1}) \qquad \forall i \in \mathcal{J}, \forall k \in \mathcal{M} \backslash \{M\} \tag{19.17}$$

This model is considerably different from the previous two (and, consequently, from the MIP Model) mainly because the decision variables were changed in order to enable the use of two global constraints (Constraints 19.16 and 19.17), which were designed specifically for the *JSSP* and are able to fulfil the Constraint Programming potential to its maximum in this type of problems. In fact, our variables are now *intervals*, representing the length of the job operations by its size (which is a parameter) and their positioning by its start time (which is the decision variable) and finish time, and *sequences of intervals*, which represent the job scheduling in each machine. In fact, although the second decision variable is somehow repeating what the first one already says, it is needed in order to build Constraint 19.16, which will bring significant computational advantages.

The objective function is similar to the one on CP2, and attempts to minimize the makespan, i.e. the last interval or operation to end. Constraint 19.16 uses the structure of the decision variable sequence s, and ensures that there is no overlap of jobs in every machine. It has the same function as Constraints 19.3 and 19.4 in the MIP Model. Constraint 19.17 ensures that a job only starts on a machine after it has finished on the previous. It has the same function as Constraint 19.2 in the MIP Model.

19.3.2 Computational Tests and Results

In order to understand the behaviour and performance of these CP models versus the MIP model, eleven instances were chosen from the eighty-two tested for the Mathematical Programming approach (Sect. 19.2.2). These instances were chosen based on their performance with the MIP Model. For each size of instance (number of jobs and number of machines), the hardest instance to solve with Mathematical Programming was chosen, so as to increase sample variability. The "difficulty" criterion was the solution time if the instance was solved within the 30 min time limit, or the optimality gap attained if the solver was stopped without proving optimality. It is also possible to observe that the instances chosen come from all four quartiles, when classified using the maximum needs-to-capacity ratio (Table 19.5, in the Appendix), so as to increase diversity.

The computational tests were run on the same computer and the models were also developed and run on IBM ILOG CPLEX Optimization Studio 12.6.1.0 using the CP Optimizer. The time limit set for the resolution of each instance was also 30 min. The eleven instances chosen and the summary of the results are presented on Table 19.2. Here, the values in bold are the best solutions found across models.

These results point to the initial hypothesis that Constraint Programming is most useful when it involves well-performing global constraints, which were able to capture and use the problem structure to propagate constraints and prune branches more

Table 19.2 Results of CP models, comparing with MIP model

	Objective function				Solution time (s)			
	MIP	CP1	CP2	CP3	MIP	CP1	CP2	CP3
la10	**958**	**958**	**958**	**958**	*1.800*	*1.800*	*1.800*	0
la14	**1.292**	**1.292**	**1.292**	**1.292**	*1.800*	*1.800*	*1.800*	0
orb01	**1.059**	1.096	1.070	**1.059**	1.013	*1.800*	*1.800*	28
la23	**1.032**	**1.032**	**1.032**	**1.032**	*1.800*	*1.800*	*1.800*	0
la30	**1.355**	**1.355**	**1.355**	**1.355**	*1.800*	*1.800*	*1.800*	2
la37	1.418	1.430	1.410	**1.397**	*1.800*	*1.800*	*1.800*	5
la34	1.749	1.721	1.740	**1.721**	*1.800*	*1.800*	*1.800*	1
abz7	702	691	691	**663**	*1.800*	*1.800*	*1.800*	*1.800*
swv10	2.024	1.915	1.961	**1.827**	*1.800*	*1.800*	*1.800*	*1.800*
yn4	1.054	1.044	1.061	**986**	*1.800*	*1.800*	*1.800*	*1.800*
swv13	4.307	3.829	3.948	**3.149**	*1.800*	*1.800*	*1.800*	*1.800*

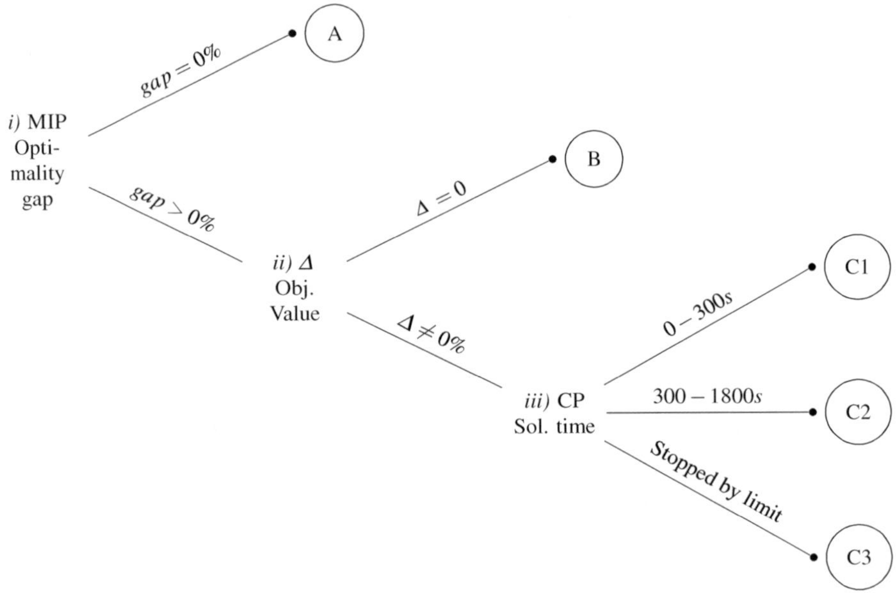

Fig. 19.5 Characteristics of the different types of instances

easily. In fact, when comparing CP1 or CP2 with the MIP model, one is able to conclude that these Constraint Programming models bring little or no advantage. As the instances get bigger, the combinations of the domains to test get exponentially large and hinder a good performance of the model. However, when the global constraints and a more appropriate representation are used (CP3), this complexity is more easily tackled and the results point to a significantly better performance of CP when compared with a MIP approach. Comparing the performance of CP1 and CP2 (which were very similar), it is also possible to conclude that reducing the number of constraints does not bring clear advantages, as is more usual in MIP programs.

In order to compare in more detail the Mathematical Programming and Constraint Programming approaches, the eighty-two instances were run with the full-performing CP model (CP3). The results are presented in the Appendix (Table 19.6[3]). From the results, the instances were categorized in six categories (A, B, C1, C2, C3), depending on three sequential criteria: *(i)* whether or not the MIP model was able to prove the optimality of the solution found (i.e. attaining 0% optimality gap), *(ii)* the difference on the objective function values found by both approaches, and *(iii)* the run time of the CP model. Figure 19.5 depicts the instance categorization and the respective characteristics.

Instances of type A are the instances that the MIP model solved within the time limit, i.e. the optimality of the solution was proven (with 0% gap). The instances where the MIP solution optimality was not proven were divided in two groups: the

[3]This table also repeats the MIP results in order to facilitate the comparison instance by instance.

Table 19.3 Summary of average results for CP model CP3, comparing with MIP model

Instance type	(#)	Avg size[5]	Avg MIP sol. time (s)	Avg CP sol. time (s)	Avg Δ obj. value (%)
A	(24)	87	109	6	0
B	(17)	154	*1.800*	0	0
C1	(18)	278	*1.800*	44	3
C2	(1)	225	*1.800*	313	−1
C3	(22)	341	*1.800*	*1.800*	−11

instances where the solution value found by the CP Optimizer (within the time limit, hence optimal) was equal to the one attained by the MIP model (type B) and the instances where the solution values were different[4] (type C). The instances of type C were also divided according to the solution time of the CP program. One should notice that the program was stopped by the time limit only for category C3; for C1 and C2, it stopped because it had explored all possible solutions. Table 19.3 presents the summary of the average results for each type of instance.

The average size[5] of the instances increases from A to C, yet there is no linear relation with size and the distribution of the instances of type C in C1, C2 and C3. In fact, C1 contains both smaller instances (15 jobs and 5 machines) and some of the biggest instances (50 jobs and 10 machines). At the same time, other instances of the same sizes are classified as C3. This is an indication that also in Constraint Programming size is not a clear indicator of complexity.

Instances of type A (the smallest) were solved to optimality by both approaches, and their optimality was proven also in both cases.[6] Nevertheless, the CP approach was considerably faster.

For instances of type B, since the CP Optimizer was able to explore all possible solutions under the time limit, it proved the optimality of the best solution found. As there is no difference between the solution value found by CP and MIP, it is possible to state that both approaches reached the optimum. However, although the MIP model was also able to reach the optimum for these instances, it was not able to prove so within the time limit. Considering that for these 17 instances the CP Optimizer took, in average, under a second to attain the optimal solution, this approach was able to almost instantaneously prove optimality whilst the MIP solver was not able to do so in 30 min. This conclusion does not contrast with the notion that Constraint Programming is a technique more prone to admissibility rather than optimality: in

[4] I.e., there existed a delta on the objective function value given by $\Delta = (\text{ObjValue}_{CP} - \text{ObjValue}_{MIP})/\text{ObjValue}_{MIP}$.

[5] Since the number of variables and constraints is different in the two models, size is herein considered as the number of jobs multiplied by the number of machines.

[6] MIP model: proven by optimality gap. CP Model: the only way for a CP program to prove optimality is by exploring all possible solutions; therefore, if the solver stops before the time limit, it means that optimality was proven.

fact, it was its power to efficiently check admissibility that allowed for the optimality to be proven in these cases.

As for instances of type C, it is important to notice that there are 18 instances categorized as C1, 22 as C3 yet only one as C2. Moreover, the instance categorized as C2 is close to the threshold of C1 (the run time is 313 s and the threshold is 300 s). In fact, either the CP Optimizer is swift to explore all possible solutions (under 5 min) or it reaches the time limit (30 min) without being able to do so. Moreover, as it as mentioned above, the size is not a good indicator of this behaviour: when analysing the biggest instances (`swv11-swv20`), it is possible to conclude that in half of the instances the optimal solution is found in under 3 s while in the other half the CP Optimizer is stopped after 30 min without being able to explore all possible solutions. Nevertheless, there are significant improvements on the objective function value for all sub-categories of C, especially on C3.

It is therefore interesting to study how swiftly the CP Optimizer finds its best solution in 30 min. As an example, Fig. 19.6 presents the value of the incumbent solutions found throughout the run time for instance `swv12`. The dotted line represents the best solution found by the MIP solver for this instance (with an optimality gap of 79%). It is possible to see that the steep improvement on the objective function value occurs in the beginning. In fact, approximately 80% of the improvement (vs. the MIP solution) occurs in 20 s, and 95% occurs in 2 min.

It is also interesting to analyse the swiftness to find a good solution for those instances in which both approaches were stopped by the 30 min time limit and where the CP presented only a slightly better performance in terms of the best objective function value found in this limit. Figure 19.7 shows the evolution of the incumbent value for both approaches for instance `yn1` (20 jobs, 20 machines) in which the improvement attained by the CP approach (vs. the MIP approach) was the smallest (1%).

As it is possible to see, the CP model was able to achieve a good solution before the MIP model. In fact, for the CP approach there was one major improvement in the

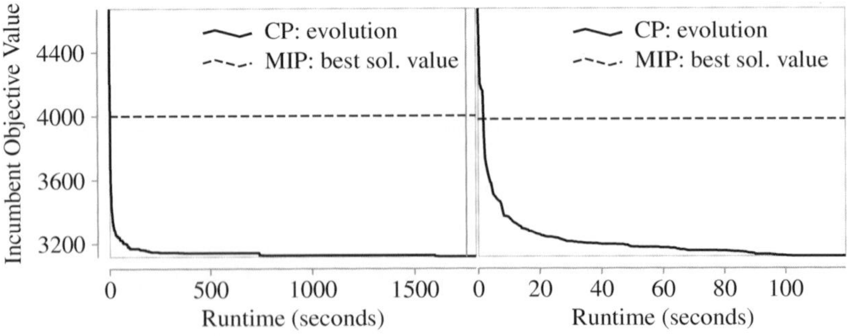

(a) Overview of the 30 minutes run time (b) Detail of the first 2 minutes

Fig. 19.6 Evolution of the incumbent objective function value for instance `swv12` throughout the 30 min run time with model CP3

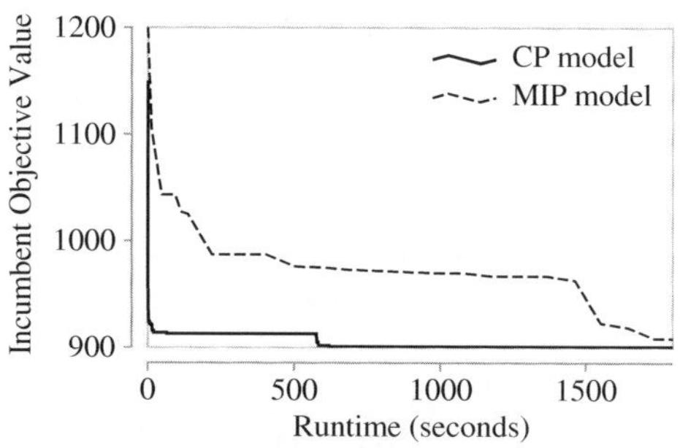

Fig. 19.7 Evolution of the incumbent objective function value for instance `yn1` throughout the 30 min run time with the MIP model and CP model (CP3)

first seconds and a smaller one after approximately 10 min. As for the MIP model, there were three major improvements: the first after a few seconds, the second at around 2–3 min, and the third in the last 10 min of the time horizon. This may lead to the conclusion that this full-performing CP model is quicker than the MIP model at finding good solutions even in the case where both approaches reach the same solution within the same time limit.

One final note to highlight the fact that the needs-to-capacity ratio was not a good predictor of complexity (or "difficulty to solve") for the CP Optimizer. In fact, there was not a clear correlation between the CP solution time and the needs-to-capacity ratio quartile of the instances (Table 19.5, in the Appendix).

19.4 Conclusions

The first objective of this paper was to understand the impact of the instances' size on the intrinsic complexity of the *JSSP* and on its consequent "difficulty to solve". The problem was described and a MIP Model from the literature was presented and used to solve eighty-two instances. The results enabled a detailed analysis of the instances' characteristics and responsiveness to the MIP model.

From the analysis of the performance of the Mathematical Programming approach, it was possible to conclude that an instance size has a significant impact on its easiness to solve, yet it is not the only significant factor. In fact, there was significant variability on solution time for same-size instances, and for bigger instances that were not solved to optimality there was not always visible a straightforward relationship between size and optimality gap achieved.

In this approach, the size of the problem is determined by two characteristics: the number of jobs and the number of machines. From the structure of the problem and its formulation, the number of jobs has a bigger impact on the number of variables and number of constraints, hence a bigger impact on the size. Moreover, the increase on the number of jobs seems to have a straightforward relationship with the increase of optimality gap achieved. As for the number of machines, it was possible to conclude that its impact is not direct. For a specific number of jobs, it seems that an increasing or more balanced number of machines makes the instances easier to solve despite being of a bigger size. An hypothesis proposed to explain this behaviour is related to the easiness of the solver to find admissible integer solutions in such a problem, hence being able to update the upper bound and the optimality gap more frequently.

In order to further understand the concept of *complexity*, a measure of the load applied to each machine throughout the scheduling horizon was designed. The *needs-to-capacity ratio* developed shows a significant correlation with the optimality gap attained by the MIP solver, thus being an adequate predictor of the "difficulty to solve" of an instance, for a Mathematical Programming approach.

The measure proposed, as well as the insights gained regarding the factors that induce complexity, can be useful e.g. when tackling an industrial scheduling application of the *JSSP*. Analysing the number of jobs and machines and other characteristics of the actual problem, it will be possible to understand whether the MIP model can easily solve the problem to optimality thus eliminating the need for more complex solution approaches or whether these are justifiably needed.

The second objective of this paper was to compare the impact of using Constraint Programming instead of Mathematical Programming to tackle the complexity of the *JSSP*. It was concluded that the main power of CP when tackling these hard, complex, combinatorial problems comes from the tailored-made global constraints that are able to swiftly improve propagation, find admissible solutions and thus significantly decrease the solution time. In fact, the CP models that used the same variables as the MIP model and similar constraints – which benefited only from part of the CP modelling flexibility by the use of logic and non-linear constraints – performed as well as, or worse than, the MIP model for the instances tested. The CP model that used a different representation of the variables and solid global constraints, however, was able to get significant improvements either in solution time and quality of the solutions found.

It was also possible to conclude that for some instances, although the MIP solver presented high optimality gap values at the end of the time limit, it was indeed able to reach the optimum. Nevertheless, it was not able to prove so. This conclusion is built on the fact that, within seconds, the (well-performing) CP model was able to test all possible admissible solutions and state that said solution was the best possible. Moreover, it was possible to conclude that the instance size has a significantly smaller impact in Constraint Programming than in Mathematical Programming. In fact, half of the largest instances tested were solved to optimality in a few seconds while the run of the remaining half was stopped by the 30 min time limit. There were also indicators that the full-performing CP approach may be swifter than the MIP approach at finding good solutions, even when it is not able to prove optimality in the specified time limit and it is not able to attain major improvements on the best solution value found.

Since the best performing CP model, which presents all the above-mentioned advantages, is commercially available, these conclusions may have a significant impact on practical application of optimization methods. In fact, it is pointed out that there are Constraint Programming models available for this type of problem that may be a better alternative to the "traditional" MIP approach for some problems.

Future work will be focused on comparing other characteristics of both approaches, such as the number of feasible solutions found in a specific time frame, which might be useful when using hybrid solution methods such as matheuristics to solve this problem. Additionally, it would be interesting to study the reasons why the needs-to-capacity ratio does not appear to be a good predictor of “difficulty to solve” for CP models.

Acknowledgements The first author was supported by grant SFRH/BD/103362/2014 from FCT - Fundação para a Ciência e Tecnologia (Portuguese Foundation for Science and Technology). This work was also partially financed by the ERDF - European Regional Development Fund through the Operational Programme for Competitiveness and Internationalisation - COMPETE 2020 Programme within project “POCI-01-0145-FEDER-006961”, and by National Funds through the FCT - Fundação para a Ciência e Tecnologia (Portuguese Foundation for Science and Technology) as part of project UID/EEA/50014/2013.

Appendix

See Tables 19.4, 19.5 and 19.6.

Table 19.4 Size and characteristics of the instances run and corresponding results with the MIP model

Instance name	Jobs	Machines	Obj. function value	Solution time (s)	Optimality gap (%)
ft06	6	6	55	0	0
abz5	10	10	1.234	19	0
abz6	10	10	943	4	0
ft10	10	10	930	122	0
la01	10	5	666	7	0
la02	10	5	655	6	0
la03	10	5	597	3	0
la04	10	5	590	2	0
la05	10	5	593	45	0
la16	10	10	945	7	0
la17	10	10	784	13	0
la18	10	10	848	3	0
la19	10	10	842	6	0
la20	10	10	902	4	0
orb01	10	10	1.059	1.013	0
orb02	10	10	888	9	0

(continued)

Table 19.4 (continued)

Instance name	Jobs	Machines	Obj. function value	Solution time (s)	Optimality gap (%)
orb03	10	10	1.005	590	0
orb04	10	10	1.005	38	0
orb05	10	10	887	17	0
orb06	10	10	1.010	596	0
orb07	10	10	397	14	0
orb08	10	10	899	36	0
orb09	10	10	934	56	0
orb10	10	10	944	14	0
la06	15	5	926	1.800	33.91
la07	15	5	890	1.800	26.07
la08	15	5	863	1.800	26.66
la09	15	5	951	1.800	25.00
la10	15	5	958	1.800	34.66
la21	15	10	1.071	1.800	15.07
la22	15	10	932	1.800	7.41
la23	15	10	1.032	1.800	23.42
la24	15	10	940	1.800	4.68
la25	15	10	979	1.800	6.54
la36	15	15	1.281	1.800	10.54
la37	15	15	1.418	1.800	15.31
la38	15	15	1.207	1.800	11.76
la39	15	15	1.240	1.800	8.55
la40	15	15	1.245	1.800	14.78
abz7	20	15	702	1.800	31.45
abz8	20	15	715	1.800	28.95
abz9	20	15	721	1.800	27.04
ft20	20	5	1.198	1.800	48.51
la11	20	5	1.222	1.800	50.31
la12	20	5	1.039	1.800	40.69
la13	20	5	1.150	1.800	41.91
la14	20	5	1.292	1.800	52.46
la15	20	5	1.207	1.800	49.59
la26	20	10	1.218	1.800	26.93
la27	20	10	1.288	1.800	33.39
la28	20	10	1.224	1.800	31.74
la29	20	10	1.220	1.800	26.42
la30	20	10	1.355	1.800	34.64

(continued)

Table 19.4 (continued)

Instance name	Jobs	Machines	Obj. function value	Solution time (s)	Optimality gap (%)
swv01	20	10	1.563	1.800	46.45
swv02	20	10	1.579	1.800	45.40
swv03	20	10	1.618	1.800	47.42
swv04	20	10	1.625	1.800	48.51
swv05	20	10	1.633	1.800	45.81
swv06	20	15	1.868	1.800	40.26
swv07	20	15	1.736	1.800	38.28
swv08	20	15	2.046	1.800	41.88
swv09	20	15	1.907	1.800	41.85
swv10	20	15	2.024	1.800	45.65
yn1	20	20	913	1.800	17.50
yn2	20	20	954	1.800	21.91
yn3	20	20	945	1.800	20.33
yn4	20	20	1.054	1.800	27.42
la31	30	10	1.784	1.800	55.77
la32	30	10	1.850	1.800	52.68
la33	30	10	1.719	1.800	54.53
la34	30	10	1.749	1.800	57.72
la35	30	10	1.888	1.800	55.58
swv11	50	10	3.767	1.800	77.84
swv12	50	10	4.002	1.800	78.68
swv13	50	10	4.307	1.800	80.37
swv14	50	10	4.081	1.800	79.66
swv15	50	10	3.722	1.800	78.02
swv16	50	10	2.952	1.800	73.41
swv17	50	10	2.915	1.800	72.10
swv18	50	10	2.950	1.800	74.16
0swv19	50	10	3.107	1.800	74.67
swv20	50	10	3.009	1.800	74.46
abz5 – abz9 in Adams et al. [2]			ft06, ft10, ft20 in Fisher and Thompson [6]		
la01 – la40 in Lawrence [7]			orb01 – orb10 in Applegate and Cook [3]		
swv01 – swv20 in Storer et al. [11]			yn1 – yn4 in Yamada and Nakano [14]		

Table 19.5 Needs-to-capacity ratio calculation for each instance

Needs-to-capacity ratio						
Instance name	Optimality gap (%)	Maximum value	Quartile (of max)	Std. deviation	# partitions	Partition length
ft06	0	1.17	1st	0.30	27	2
abz5	0	2.00	3rd	0.55	9	100
abz6	0	1.20	2nd	0.34	18	40
ft10	0	1.00	1st	0.11	159	4
la01	0	2.60	3rd	0.63	17	24
la02	0	2.40	3rd	0.64	15	24
la03	0	2.00	3rd	0.43	31	14
la04	0	2.00	3rd	0.35	42	10
la05	0	2.20	3rd	0.39	39	10
la16	0	1.00	1st	0.19	51	14
la17	0	1.00	1st	0.17	67	10
la18	0	1.00	1st	0.17	65	10
la19	0	1.00	1st	0.19	56	10
la20	0	1.20	2nd	0.20	60	12
orb01	0	1.10	1st	0.18	68	10
orb02	0	1.00	1st	0.16	50	12
orb03	0	1.00	1st	0.16	66	10
orb04	0	1.10	1st	0.17	76	10
orb05	0	1.10	1st	0.18	55	10
orb06	0	1.00	1st	0.19	61	12
orb07	0	1.20	2nd	0.25	28	10
orb08	0	1.20	2nd	0.18	66	10
orb09	0	1.00	1st	0.17	69	10
orb10	0	1.10	1st	0.17	78	10
la06	34	3.00	3rd	0.63	29	14
la07	26	3.60	4th	0.77	25	16
la08	27	3.40	4th	0.59	38	10
la09	25	3.40	4th	0.64	26	14
la10	35	3.00	3rd	0.51	44	10
la21	15	1.70	2nd	0.28	48	14
la22	7	1.50	2nd	0.23	59	10
la23	23	1.60	2nd	0.23	73	10
la24	5	1.70	2nd	0.24	75	10
la25	7	1.70	2nd	0.25	77	10
la36	11	1.00	1st	0.17	69	14
la37	15	1.00	1st	0.15	95	10
la38	12	1.13	1st	0.15	98	10

(continued)

Table 19.5 (continued)

Needs-to-capacity ratio						
Instance name	Optimality gap (%)	Maximum value	Quartile (of max)	Std. deviation	# partitions	Partition length
la39	9	1.00	1st	0.15	95	10
la40	15	1.00	1st	0.15	91	10
abz7	31	1.73	2nd	0.41	19	22
abz8	29	2.27	3rd	0.53	21	22
abz9	27	1.80	2nd	0.54	23	22
ft20	49	4.20	4th	0.44	111	4
la11	50	4.60	4th	0.90	31	14
la12	41	4.40	4th	0.76	41	10
la13	42	4.00	4th	0.71	42	10
la14	52	4.40	4th	0.72	45	10
la15	50	4.40	4th	0.80	32	12
la26	27	2.20	3rd	0.36	50	14
la27	33	2.00	3rd	0.27	64	10
la28	32	2.10	3rd	0.31	81	10
la29	26	2.20	3rd	0.31	76	10
la30	35	2.00	3rd	0.30	77	10
swv01	46	2.00	3rd	0.14	356	2
swv02	45	2.00	3rd	0.15	330	2
swv03	47	2.20	3rd	0.15	334	2
swv04	49	2.10	3rd	0.15	351	2
swv05	46	2.10	3rd	0.14	390	2
swv06	40	1.40	2nd	0.09	483	2
swv07	38	1.33	2nd	0.09	449	2
swv08	42	1.40	2nd	0.09	527	2
swv09	42	1.40	2nd	0.09	476	2
swv10	46	1.40	2nd	0.13	231	4
yn1	17	1.30	2nd	0.23	33	20
yn2	22	1.40	2nd	0.31	37	20
yn3	20	1.15	1st	0.28	35	20
yn4	27	1.20	2nd	0.27	37	20
la31	56	3.30	4th	0.46	73	10
la32	53	3.10	3rd	0.43	79	10
la33	55	3.30	4th	0.46	75	10
la34	58	3.00	3rd	0.43	60	10
la35	56	3.10	3rd	0.45	68	10
swv11	78	5.00	4th	0.30	397	2
swv12	79	5.30	4th	0.31	390	2

(continued)

Table 19.5 (continued)

Needs-to-capacity ratio						
Instance name	Optimality gap (%)	Maximum value	Quartile (of max)	Std. deviation	# partitions	Partition length
swv13	80	5.20	4th	0.31	382	2
swv14	80	5.30	4th	0.31	382	2
swv15	78	5.00	4th	0.32	339	2
swv16	73	5.10	4th	0.31	372	2
swv17	72	5.00	4th	0.33	308	2
swv18	74	5.00	4th	0.32	331	2
swv19	75	5.00	4th	0.31	339	2
swv20	74	5.10	4th	0.32	333	2

Table 19.6 Instances run (with J jobs and M machines) with MIP model and CP model CP3

Instance	MIP optimality gap (%)	MIP obj. function value	CP3 Obj. function value	MIP sol. time (s)	CP3 sol. time (s)
ft06	0.0	55	55	0	0
abz5	0.0	1.234	1.234	19	7
abz6	0.0	943	943	4	2
ft10	0.0	930	930	122	12
la01	0.0	666	666	7	0
la02	0.0	655	655	6	0
la03	0.0	597	597	3	0
la04	0.0	590	590	2	1
la05	0.0	593	593	45	0
la16	0.0	945	945	7	2
la17	0.0	784	784	13	2
la18	0.0	848	848	3	2
la19	0.0	842	842	6	10
la20	0.0	902	902	4	3
orb01	0.0	1.059	1.059	1.013	28
orb02	0.0	888	888	9	7
orb03	0.0	1.005	1.005	590	19
orb04	0.0	1.005	1.005	38	10
orb05	0.0	887	887	17	7
orb06	0.0	1.010	1.010	596	13
orb07	0.0	397	397	14	4
orb08	0.0	899	899	36	5
orb09	0.0	934	934	56	3
orb10	0.0	944	944	14	2

(continued)

Table 19.6 (continued)

Instance	MIP optimality gap (%)	MIP obj. function value	CP3 Obj. function value	MIP sol. time (s)	CP3 sol. time (s)
la06	33.9	926	926	1.800	0
la07	26.1	890	890	1.800	0
la08	26.7	863	863	1.800	0
la09	25.0	951	951	1.800	0
la10	34.7	958	958	1.800	0
la21	15.1	1.071	1.046	1.800	76
la22	7.4	932	927	1.800	9
la23	23.4	1.032	1.032	1.800	0
la24	4.7	940	935	1.800	35
la25	6.5	979	977	1.800	52
la36	10.5	1.281	1.268	1.800	13
la37	15.3	1.418	1.397	1.800	5
la38	11.8	1.207	1.196	1.800	313
la39	8.5	1.240	1.233	1.800	18
la40	14.8	1.245	1.222	1.800	82
abz7	31.5	702	663	1.800	1.800
abz8	29.0	715	680	1.800	1.800
abz9	27.0	721	686	1.800	1.800
ft20	48.5	1.198	1.165	1.800	1
la11	50.3	1.222	1.222	1.800	0
la12	40.7	1.039	1.039	1.800	0
la13	41.9	1.150	1.150	1.800	0
la14	52.5	1.292	1.292	1.800	0
la15	49.6	1.207	1.207	1.800	0
la26	26.9	1.218	1.218	1.800	1
la27	33.4	1.288	1.235	1.800	254
la28	31.7	1.224	1.216	1.800	31
la29	26.4	1.220	1.153	1.800	1.800
la30	34.6	1.355	1.355	1.800	2
swv01	46.4	1.563	1.413	1.800	1.800
swv02	45.4	1.579	1.475	1.800	203
swv03	47.4	1.618	1.406	1.800	1.800
swv04	48.5	1.625	1.488	1.800	1.800
swv05	45.8	1.633	1.438	1.800	1.800
swv06	40.3	1.868	1.710	1.800	1.800
swv07	38.3	1.736	1.667	1.800	1.800
swv08	41.9	2.046	1.810	1.800	1.800
swv09	41.8	1.907	1.698	1.800	1.800

(continued)

Table 19.6 (continued)

Instance	MIP optimality gap (%)	MIP obj. function value	CP3 Obj. function value	MIP sol. time (s)	CP3 sol. time (s)
swv10	45.7	2.024	1.827	1.800	1.800
yn1	17.5	913	901	1.800	1.800
yn2	21.9	954	910	1.800	1.800
yn3	20.3	945	910	1.800	1.800
yn4	27.4	1.054	986	1.800	1.800
la31	55.8	1.784	1.784	1.800	1
la32	52.7	1.850	1.850	1.800	0
la33	54.5	1.719	1.719	1.800	0
la34	57.7	1.749	1.721	1.800	1
la35	55.6	1.888	1.888	1.800	1
swv11	77.8	3.767	3.021	1.800	1.800
swv12	78.7	4.002	3.119	1.800	1.800
swv13	80.4	4.307	3.149	1.800	1.800
swv14	79.7	4.081	2.970	1.800	1.800
swv15	78.0	3.722	2.960	1.800	1.800
swv16	73.4	2.952	2.924	1.800	0
swv17	72.1	2.915	2.794	1.800	0
swv18	74.2	2.950	2.852	1.800	0
swv19	74.7	3.107	2.843	1.800	2
swv20	74.5	3.009	2.823	1.800	0

References

1. 4COutreachProgram. CSP tutorial (2005), http://4c.ucc.ie/web/outreach/tutorial.html
2. J. Adams, E. Balas, D. Zawack, The shifting bottleneck procedure for job shop scheduling. Manag. Sci. **34**, 391–401 (1988)
3. D. Applegate, W. Cook, A computational study of the job-shop scheduling instance. ORSA J. Comput. **3**, 149–156 (1991)
4. K.R. APT, *Priciples of Constraint Programming* (Cambridge University Press, Cambridge, 2003)
5. E. Balas, Disjunctive programming. Ann. Discret. Math. **5**, 3–51 (1979)
6. H. Fisher, G.L. Thompson, Probabilistic learning combinations of local job-shop scheduling rules, in *Industrial Scheduling*, ed. by J.F. Muth, G.L. Thompson (Prentice-Hall, Englewood Cliffs, 1963), pp. 225–251
7. S. Lawrence, Resource constrained project scheduling: an experimental investigation of heuristic scheduling techniques (supplement), in *Graduate School of Industrial Administration*, Carnegie-Mellon University, Pittsburgh, Pennsylvania (1984)
8. J.K. Lenstra, Computational complexity of discrete optimization problems. Ann. Discret. Math. **4**, 121–140 (1979)
9. J. Poppenborg, S. Knust, J. Hertzber, Online scheduling of flexible job-shops with blocking and transportation. Eur. J. Ind. Eng. **6**, 497–518 (2012)

10. F. Rossi, P. van Beek, T. Walsh (eds.), *Handbook of Constraint Programming* (Elsevier, Amsterdam, 2006)
11. R.H. Storer, S.D. Wu, R. Vaccari, New search spaces for sequencing instances with application to job shop scheduling. Manag. Sci. **38**, 1495–1509 (1992)
12. K. Thörnblad, A.B. Strömberg, M. Patriksson, T. Almgren, Scheduling optimisation of a real flexible job shop including fixture availability and preventive maintenance. Eur. J. Ind. Eng. **9**, 126–145 (2015)
13. W.J. van Hoeve, Introduction to constraint programming, in *ACP Summer School on Theory and Practice of Constraint Programming*, September 24–28, 2012, Wrocław, Poland (2012)
14. T. Yamada, R. Nakano, A genetic algorithm applicable to large-scale job-shop instances, in *PPSN'2 Proceedings of the 2nd International Workshop on Parallel Problem Solving from Nature*, ed. by R. Manner, B. Manderick (1992), pp. 281–290

Chapter 20
A Dynamic Programming Approach for Integrating Dynamic Pricing and Capacity Decisions in a Rental Context

Beatriz B. Oliveira, Maria Antónia Carravilla and José Fernando Oliveira

Abstract Car rental companies have the ability and potential to integrate their dynamic pricing decisions with their capacity decisions. Pricing has a significant impact on demand, while capacity, which translates fleet size, acquisition planning and fleet deployment throughout the network, can be used to meet this price-sensitive demand. Dynamic programming has been often used to tackle dynamic pricing problems and also to deal with similar integrated problems, yet with some significant differences as far as the inventory depletion and replenishment are considered. The goal of this work is to understand what makes the car rental problem different and hinders the application of more common methods. To do so, a discrete dynamic programming framework is proposed, with two different approaches to calculate the optimal-value function: one based on a Mixed Integer Non Linear Program (MINLP) and one based on a Constraint Programming (CP) model. These two approaches are analyzed and relevant insights are derived regarding the (in)ability of discrete dynamic programming to effectively tackle this problem within a rental context when realistically sized instances are considered.

Keywords Car rental · Dynamic programming · Dynamic pricing · Fleet deployment · Optimization model · Constraint programming

20.1 Introduction

This work deals with the integration of dynamic pricing decisions with resource capacity, deployment and consumption decisions within the car rental context. The goal is to decide, for the time horizon of a specific selling season:

B. B. Oliveira (✉) · M. A. Carravilla · J. F. Oliveira
INESC TEC and Faculty of Engineering, University of Porto, Porto, Portugal
e-mail: beatriz.oliveira@fe.up.pt

M. A. Carravilla
e-mail: mac@fe.up.pt

J. F. Oliveira
e-mail: jfo@fe.up.pt

A. I. F. Vaz et al. (eds.), *Operational Research*, Springer Proceedings
in Mathematics & Statistics 223, https://doi.org/10.1007/978-3-319-71583-4_20

- How many cars to have in the fleet,
- When to acquire them,
- How to deploy them among rental stations throughout the time horizon,
- How to assign them to rentals (that start and end throughout the time horizon and rental stations),
- How to price these rentals.

Car rental companies face a significantly price-sensitive demand. Since online sale channels have allowed companies to change their prices virtually instantaneously and with no cost, dynamic pricing is becoming a critical demand-management tool, not only in this sector but also in airlines, hotels and other businesses that rely on revenue management techniques (including pricing) to seize price-sensitivity and other demand segmentation characteristics.

In car rental, unlike the above-mentioned (more traditionally studied) sectors, the fleet is highly flexible and mobile, since the vehicles (resources) are easy(ier) to transfer, deploy and acquire. However, there is a myriad of products—the rentals—that share the same fleet capacity. The rentals are broadly characterized by their start and end date and location. Other elements (such as vehicle group required, for example) may characterize a type of rental. Nonetheless, for the sake of simplicity and clarity, throughout this work the fleet is assumed to be homogeneous and the pricing decisions, made for each rental type on its broader definition, can be considered as "reference prices" to which others are indexed (e.g. variations according to antecedence of purchase or extra conditions required). For a more detailed review on car rental fleet and revenue management works, the reader may refer to [5].

Recently, some interesting works have been dealing with the integration of dynamic pricing with capacity and inventory decisions [1, 7]. This integration has been becoming relevant for companies that can change and update their pricing policies and inventory and capacity decisions in an increasingly easier way, due to the improvement of the above-mentioned technological systems. The methodological approach applied often involves dynamic programming due to its affinity with the problem. Also, for the stochastic problem, other non-parametric approaches such as robust optimization have been developed. For a thorough and relevant review regarding dynamic pricing, especially when learning processes regarding demand are considered, the reader should refer to [2].

The work herein presented aims to tackle a similar problem, which differs on the type of capacity/inventory decisions made. In previously studied cases, the capacity/inventory was decided and considered to be available at the start of the horizon (or at multiple points throughout the horizon, through multiple capacity decisions) and then depleted by the demand until the end of the horizon. In the car rental (actually any rental) context, the capacity is not only affected by these decisions but also by "returning" (re-usable) resources. That is to say, the resource is not depleted by demand but only temporarily occupied and it will become available capacity again, possibly at a different location. This difference has a significant impact on the structure of the problem and motivated the research presented in this paper.

The goal of this work is to study the possibility to develop a solution method based on one of the most applied methodologies in the similar problem presented above—dynamic programming—and understand its advantages, limitations and drawbacks in this context. Besides its common application within the dynamic pricing and revenue management setting, dynamic programming has also been used on works that deal with car rental operational and fleet management problems, such as fleet size and deployment [4].

In this work, a general discrete dynamic programming framework is developed as well as two approaches to calculate the value of the decisions at each stage and state, which are presented in Sect. 20.2. Then, in Sect. 20.3, some illustrative numeric examples are used to analyze the limitations and advantages of the method. Finally, in Sect. 20.4, some conclusions are drawn.

20.2 Discrete Dynamic Programming Formulation

One important characteristic of this problem is that most decisions are intrinsically integer, such as the number of vehicles to acquire and deploy or the number of fulfilled rentals. Due to the detail considered for rental types, which aggregate rentals that start and end at specific locations and times, the order of magnitude of these decisions is relatively small and an approximate result obtained by relaxing the integrality constraints might be significantly impaired. Therefore, a discrete formulation was developed.

Dynamic programming provides a general framework to solve different problems, where a multistage structure is latent and can be used to decompose a complex problem into simpler sub-problems. Within this context, a *stage* represents a point where decisions are made. The goal is to formulate the problem so that at any stage the only information needed to make decisions is summarized on one or more *state* variables. The state at a specific stage is fully defined (on the deterministic case herein considered) by the state and decisions of the previous state, translated on a *state transition function*. At each stage, an *optimal-value function* can be defined, dependent on the current state and on the decisions made. Dynamic programming is then based on the recursive computation of the optimal-value function [3].

In this section, the stages, state variables and spaces, and transition functions will be defined. Also, two approaches to calculate the optimal-value function will be presented.

The notation for indexes and sets used throughout this paper is as follows:

n Index for stage;
e Index for state;
r Index for type of rental;
s, c Indexes for rental station;
p Index for price level;
$\mathscr{E}^n$ Set of states possible to achieve in stage n;

$\mathscr{S}$ Set of rental stations;
$\mathscr{R}$ Set of types of rental;
$\mathscr{R}_n^{\text{start}}$ Set of types of rentals that start at stage n;
$\mathscr{R}_{n,s}^{\text{start}}$ Set of types rentals that start at station s at stage n;
$\mathscr{P}$ Set of possible price levels.

Also, the following parameters will be considered:

T Number of time periods on the time horizon;
HC_n Holding cost for the fleet of vehicles existent at stage n (cost per vehicle);
TT_{sc} Empty transfer time between station $s \in \mathscr{S}$ and station $c \in \mathscr{S}$;
TC_{scn} Cost of initiating an empty transfer from station $s \in \mathscr{S}$ to station $c \in \mathscr{S}$ at stage n (cost per vehicle);
$DEM_r(q_r)$ Demand for type of rental $r \in \mathscr{R}$, dependent on the price q_r that is charged for this type of rental;
DEM_{rp} Demand for type of rental $r \in \mathscr{R}$ if price level $p \in \mathscr{P}$ is charged for this type of rental (alternative notation);
PRI_p Monetary value associated with price level $p \in \mathscr{P}$.

20.2.1 Stages

In the car rental context, the start and end dates that characterize the rental types can be aggregated in e.g. weeks. The same unit can be used for the capacity decisions due to the flexibility of some vehicle acquisition modes, such as leasing. These time units mark the decision points throughout the time horizon and are the most notorious element that contributes to the multistage structure of the problem.

The computation will follow the *backward induction* method, as it will start at the last time period and end at the first. That is to say, the calculation will start at $n = 0$, where n defines the number of stages missing to end the time period, and end at $n = T$.

The decisions made at each stage n are represented by the following variables:

u_r^n Number of rentals of type $r \in \mathscr{R}_n^{\text{start}}$ fulfilled at stage n;
q_r^n Price charged for rentals of type $r \in \mathscr{R}_n^{\text{start}}$;
w_s^n Number of vehicles acquired to be available in rental station $s \in \mathscr{S}$ at stage n;
y_{sc}^n Number of vehicles to deploy from station $s \in \mathscr{S}$ to station $c \in \mathscr{S}$ by an empty transfer that starts at stage n.

20.2.2 State Variables, Transition Function and State Spaces

At any stage, the state variables should provide all the information required to make the previously mentioned decisions. Dynamic formulations for inventory problems and integrated pricing and capacity problems use the stock or inventory available at each stage as the state variables.

In this case, in order to decide on number of rentals fulfilled (u-type decision), two types of information are required: the amount of demand for this type of rental, which is solely dependent on the pricing decision made at the same stage, and the stock of vehicles of each rental type available at the starting station, which depends on decisions from previous periods and should thus be summarized on the state variables.

At each station $s \in \mathcal{S}$ and stage n, this stock depends on the previous stock, the vehicles that leave the station (either occupied by rentals or transfers) and the vehicles that arrive (either at the end of rentals or transfers or by their acquisition) and can be computed by the following equation:

$$\begin{aligned}\text{stock}_s^n = \text{stock}_s^{n+1} &- \text{rentals that leave}_s^n - \text{transfers that leave}_s^n \\ &+ \text{rentals that arrive}_s^n + \text{transfers that arrive}_s^n + \text{vehicles aquired}_s^n \quad (20.1)\end{aligned}$$

As previously discussed, since the state variables fully define the state, the transition function should only depend on the previous state. However, since the rentals and empty transfers may last for more than one time period, Eq. 20.1 requires information from stages other than the immediately subsequent.

Therefore, an artifact was developed and a second type of state variable introduced to represent the capacity occupied by current rentals or transfers that will be available in a later period. Figures 20.1 and 20.2 present an example to better explain these variables. As exemplified with the solid arrow, if there is a rental type that starts in $t = 1$ (station B) and ends in $t = 3$ (station A), the arrival of these rentals will affect the stock of vehicles available at station A in $t = 3$ (Fig. 20.1). However, this decision occurs on a stage other than the immediately consequent. With an additional stock variable, it is possible to memorize, for any stage, how many vehicles are currently occupied and will be available in a certain station in a certain number of time periods. In the example presented, as shown by the dashed arrows in Fig. 20.2, in $t = 2$, rentals of this type will increase the stock of vehicles currently occupied that will be available in station A in the next period. Then, in $t = 3$, the stock in station A will be increased by these units.

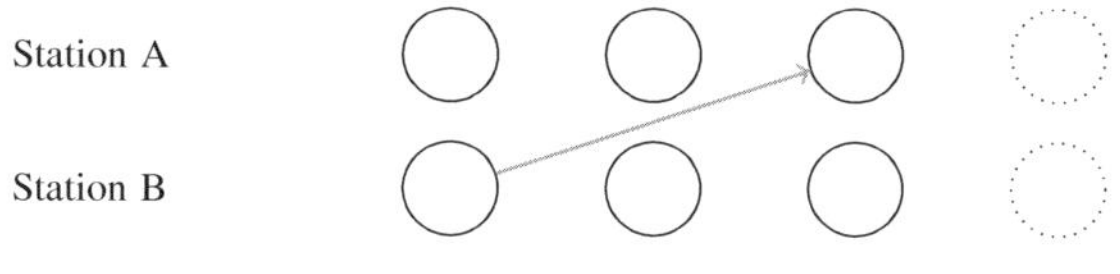

Fig. 20.1 Space-time network of an explanatory example, with 2 rental stations and 3 time periods. The arrow represents a rental type that starts in station B in $t = 1$ and ends in station A in $t = 3$

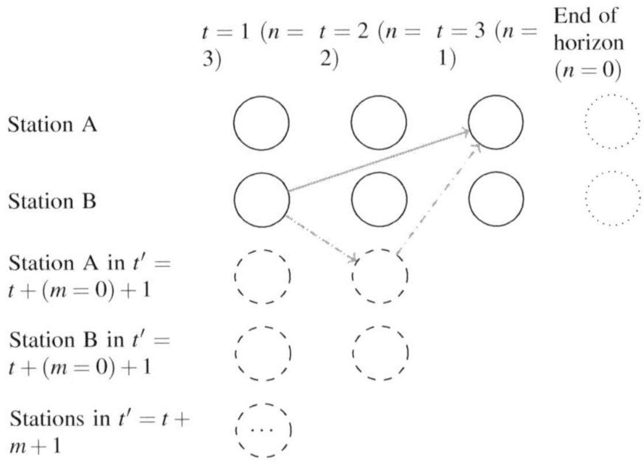

Fig. 20.2 Space-time network of Fig. 20.1, extended to include the representation of the additional state variables. The solid arrow is now decomposed in two dashed arrows using the additional state variable so that the state transition depends solely on the previous period

For each stage, the state variables can thus be defined as:

x_s^n Number of vehicles available in station $s \in \mathscr{S}$ (stock), at stage n;

o_{sm}^n Number of vehicles that are currently occupied and will be available in station $s \in \mathscr{S}$, at stage $n + m + 1$.

Thus, at each stage n, the state transition function t^n takes the following form:

$$\begin{aligned} \text{state}^{n-1} &= t^n(u_r^n, q_r^n, w_s^n, y_{sc}^n, \text{state}^n) \\ &= \begin{cases} x_s^{n-1} = x_s^n - \sum\limits_{r \in \mathscr{R}_{s,n}^{\text{start}}} u_r^n - \sum\limits_{c \in \mathscr{S}} y_{sc}^n + o_{s0}^n + w_s^n, \quad \forall s \in \mathscr{S} \\ o_{sm}^{n-1} = o_{s,m+1}^n + \sum\limits_{r \in \mathscr{R}_n^{\text{start}} \cap \mathscr{R}_{n+m+1,s}^{\text{end}}} u_r^n + \sum\limits_{c : c \in \mathscr{S}, TT_{cs} = m+2} y_{cs}^n, \\ \qquad \forall s \in \mathscr{S}, m = \{0, \dots, n-2\} \end{cases} \end{aligned} \tag{20.2}$$

State Space:
As for the state space, it was assumed that at the beginning of the time horizon ($n = T$), the state variables will be null (meaning that there are no vehicles occupied or in stock). For the remaining stages, an upper bound $XMAX$ must be defined for the x-type stock variables and another $OMAX$ for the o-type occupation variables. Each state is a combination of the possible values of these state variables. Therefore, the following equation defines the number of possible states NS, per stage $n < T$:

$$NS = \max[1, (OMAX + 1)^{|\mathscr{S}|(n-1)}] \times (XMAX + 1)^{|\mathscr{S}|} \tag{20.3}$$

Therefore, there are $NS \times |\mathscr{S}|$ x-type state variables per stage n and $NS \times |\mathscr{S}|(n-1)$ o-type state variables per stage $n > 1$.

20.2.3 Optimal-Value Calculation

At each stage n and state s_n, the maximum possible return over the n missing stages is given by the optimal-value function v_n. As previously discussed, this function v_n is defined recursively, based on the return f of this stage, which depends on the current decisions and state, and on the optimal-value function of the previous stage. Since the overall goal is to optimize the profit, the recursive optimization problem is broadly given by:

$$v^n(\text{state}^n) = \max \left\{ f^n(u_r^n, q_r^n, w_s^n, y_{sc}^n, \text{state}^n) + v^{n-1}\left(t^n(u_r^n, q_r^n, w_s^n, y_{sc}^n, \text{state}^n)\right)\right\}$$
$$\text{s.t. Constraints on decisions} \tag{20.4}$$

The return function f, in this case the profit obtained, is given by the difference between the revenue obtained from the price charged for each of the fulfilled rentals and the costs of holding the fleet of vehicles existent at this stage and the cost of initiating empty transfers:

$$f^n = \sum_{r \in \mathscr{R}_n^{\text{start}}} u_r^n \times q_r^n - z^n \times HC_n - \sum_{s \in \mathscr{S}} \sum_{c \in \mathscr{S}} y_{sc}^n \times TC_{scn} \tag{20.5}$$

The auxiliary decision variable z summarizes the total fleet and is defined, at each stage, by the sum of the vehicles acquired (decision variable), the vehicles in stock (state variable) and the vehicles occupied on rentals or transfers (state variable):

$$z^n = \sum_{s \in \mathscr{S}} \left(w_s^n + x_s^n + \sum_{m=0}^{n-1} o_{sm}^n \right) \tag{20.6}$$

The constraints on decisions are as follows:

- The price-dependent demand $(DEM_r(q_r))$ is an upper bound on the number of rentals that can be fulfilled:

$$u_r^n \leq DEM_r(q_r), \quad \forall r \in \mathscr{R}_n^{\text{start}} \tag{20.7}$$

- The overall capacity in a station limits the rentals fulfilled and the empty transfers leaving the station:

$$\sum_{r \in \mathscr{R}_{n,s}^{start}} u_r^n + \sum_{c \in \mathscr{S}} y_{sc}^n \leq x_s^n + w_s^n, \quad \forall s \in \mathscr{S} \tag{20.8}$$

- Also, an auxiliary constraint ensures that no empty transfers start and end in the same station:

$$y_{ss} = 0, \quad \forall s \in \mathscr{S} \tag{20.9}$$

- All decisions should lead to integer values.
- Additionally, the resulting state must be possible to achieve (i.e. be defined).

Two relevant characteristics of this optimization problem are the integrality of the decision variables and the non-linearity of the objective function. Therefore, two adequate optimization models and consequent resolution strategies were applied: a Mixed Integer Non Linear Program (MINLP) and a Constraint Programming (CP) model.

For each stage and state, the MINLP model proposed is a straightforward adaptation of the optimization problem summarized in (20.4) and (20.5). The main difference is related with the price decision variable q that is transformed into a binary variable q_{rp} that indicates whether or not a specific *price level* p from a previously defined set $\mathscr{P}$, which is associated with a monetary value PRI_p, is chosen for rental type r. This causes minor adaptations in the objective function and on the demand constraint (20.7). Also, binary variables st_e are added, to indicate whether or not the state achieved on the consequent stage from the decisions made will be state $e \in \mathscr{E}^{n-1}$ or not. This requires additional constraints to associate the binary variables with the consequent states and to ensure that at least one possible state is achieved. The model is thus as follows[1]:

$$
\begin{aligned}
\max f^n + v^{n-1} = & \left[\sum_{r \in \mathscr{R}_n^{\text{start}}} u_r^n \times \sum_{p \in \mathscr{P}} \left(q_{rp}^n \times PRI_p \right) - z^n \times HC_n - \sum_{s \in \mathscr{S}} \sum_{c \in \mathscr{S}} y_{sc}^n \times TC_{scn} \right] \\
& + \left[\sum_{e \in \mathscr{E}^{n-1}} st_e \times v^{n-1}(st_e) \right] \\
\text{s.t. } & (20.6), (20.8), (20.9) \\
& u_r^n \le DEM_{rp} + M(1 - q_{rp}), \quad \forall r \in \mathscr{R}_n^{\text{start}}, p \in \mathscr{P} \\
& [x_s^{n-1}]_{t^n} \le [x_s^{n-1}]_e + M(1 - st_e), \quad \forall e \in \mathscr{E}^{n-1}, s \in \mathscr{S} \\
& [x_s^{n-1}]_{t^n} \ge [x_s^{n-1}]_e - M(1 - st_e), \quad \forall e \in \mathscr{E}^{n-1}, s \in \mathscr{S} \\
& [o_{sm}^{n-1}]_{t^n} \le [o_{sm}^{n-1}]_e + M(1 - st_e), \quad \forall e \in \mathscr{E}^{n-1}, s \in \mathscr{S}, m \in \{0, ..., n-2\} \\
& [o_{sm}^{n-1}]_{t^n} \ge [o_{sm}^{n-1}]_e - M(1 - st_e), \quad \forall e \in \mathscr{E}^{n-1}, s \in \mathscr{S}, m \in \{0, ..., n-2\} \\
& \sum_{e \in \mathscr{E}^{n-1}} st_e \ge 1 \\
& u_r^n \in \mathscr{Z}_0^+, \forall r \in \mathscr{R}_n^{\text{start}}; \quad z^n \in \mathscr{Z}_0^+; \quad y_{sc}^n \in \mathscr{Z}_0^+, \forall s, c \in \mathscr{S} \\
& q_{rp}^n \in \{0, 1\}, \forall r \in \mathscr{R}_n^{\text{start}}, p \in \mathscr{P}; \quad st_e \in \{0, 1\}, \forall e \in \mathscr{E}^{n-1}
\end{aligned}
\tag{20.10}
$$

The second approach is based on Constraint Programming (CP), which aims to solve combinatorial problems by combining search and constraint solving, following

[1]The symbol $[\text{state}]_{t^n}$ indicates that the state expression was calculated based on the transition function and thus involves decision variables, while the symbol $[\text{state}]_e$ refers to an input/parameter: the state variables associated with state e.

the basis of logic programming [6]. This modeling and solving approach is suitable for integer, finite domain decision variables that are related by a set of constraints. Due to the characteristics of its associated search procedures, non-linearity presents no issue for CP models. Also, logic constraints such as "if-then-else" and implication statements can be implemented, which simplifies the model when compared with (20.10). For the sake of comparison between approaches, the price decision variable also refers to *price levels*, yet in this case it indicates directly the level, instead of having a binary variable per level. A similar reasoning is applied to the decision variable indicating the consequent state. The variable domains were considered as follows:

$$
\begin{aligned}
&u_r = \{0, \ldots, DUB_{s:\text{start}_r}\}, && \forall r \in \mathcal{R}_n^{\text{start}}\\
&q_r \in \mathcal{P}, && \forall r \in \mathcal{R}_n^{\text{start}}\\
&w_s = \{0, \ldots, DUB_s\}, && \forall s \in \mathcal{S}\\
&y_{sc} = \{0, \ldots, x_s^n\}, && \forall s, c \in \mathcal{S}\\
&z = \{0, \ldots, \textstyle\sum_{s \in \mathcal{S}} DUB_s\}\\
&st \in \mathcal{E}^{n-1}, && \forall e \in \mathcal{E}^{n-1}
\end{aligned}
$$

The demand upperbound per station DUB_s was calculated by:

$$
DUB_s = \sum_{r \in \mathcal{R}_{n,s}^{\text{start}}} \left(\max_{p \in \mathcal{P}} DEM_{rp} \right) \tag{20.11}
$$

The CP model is then similar to the previous one:

$$
\begin{aligned}
&\max f^n + v^{n-1} = \left[\sum_{r \in \mathcal{R}_n^{\text{start}}} u_r^n \times q_r^n \times PRI_p - z^n \times HC_n - \sum_{s \in \mathcal{S}} \sum_{c \in \mathcal{S}} y_{sc}^n \times TC_{scn} \right] + \left[v^{n-1}(st) \right]\\
&\text{s.t. } (20.6), (20.8), (20.9)\\
&\quad u_r^n \leq DEM_{rq_r}, \quad \forall r \in \mathcal{R}_n^{\text{start}}\\
&\quad [x_s^{n-1}]_{t^n} = [x_s^{n-1}]_e \Rightarrow st = e, \quad \forall e \in \mathcal{E}^{n-1}, s \in \mathcal{S}\\
&\quad [o_{sm}^{n-1}]_{t^n} = [o_{sm}^{n-1}]_e \Rightarrow st = e, \quad \forall e \in \mathcal{E}^{n-1}, s \in \mathcal{S}, m \in \{0, \ldots, n-2\}
\end{aligned} \tag{20.12}
$$

Base Case:
To start the recursive calculation, it is important to define the base case for $n = 0$. Since in this problem it represents the end of the horizon, when no more rentals or vehicles are considered, it was assumed that $v^0 = 0$.

20.3 Illustrative Numeric Examples

Scope

The goal of this section is to provide some numerical examples that illustrate the drawbacks and limitations of this method in order to support the discussion on its adequacy. The main characteristics of the problem that influence will be identified to understand the potential and limits of its application.

From the discussion on the number of states and state variables, it was possible to verify that four main characteristics of the problem could significantly influence the effectiveness of the method proposed: the upper bound on the number of vehicles in stock in each station $XMAX$, the upper bound on the number of vehicles currently occupied to be available in a specific future period of time and station $OMAX$, the number of stations S and the number of time periods (i.e. stages).

From Eq. (20.3) it is possible to observe that the number of states in a stage easily explodes. Therefore, as an example, considering two rental stations and three periods of time: for $XMAX = 10$ and $OMAX = 5$,[2] the maximum number of states in a stage goes above 4.300, which leads to more than 17.000 state variables. If these numbers are doubled ($XMAX = 20$, $OMAX = 10$), the number of states becomes bigger than 53.000, with over 213.000 state variables.

The main issue is that this makes the effectiveness of the model highly dependent on two characteristics that are not intrinsic to the problem (although the maximum stock could have a clear parallel with the number of parking spaces available), and indirectly on the scale of the problem.

Data

Instances:
These numeric experiments are based on three cases that were adapted from instances provided for the Capacity-Pricing Problem in car rental,[3] which present a "photograph" of the rental system at a certain time, showing the demand for each type of rental, as well as the remaining parameters. The instances chosen were the ones where the number of rental types was (i) the smallest, (ii) the biggest and (iii) the median value. It is important to analyze how the approach performs varying this indicator since the number of rentals is one of the most relevant drivers of complexity to solve each sub-problem and, at the same time, it has virtually no impact on the number of states and stages, i.e. on the number of sub-problems to solve.

Experiment Environment:
The algorithms and MINLP and CP models were developed in C++/IBM ILOG Concert Technology and were run on a workstation computer with 48 Gigabyte of RAM memory, with 2 CPUs (Intel(R) Xeon(R) X5690 @ 3.46 GHz), with a 64-bit

---

[2]It is reasonable to assume that $OMAX \leq XMAX$.

[3]*Capacity-Pricing Model: car rental instances*, 2017, available at doi:http://dx.doi.org/10.17632/g49smv7nh8.1.

Operating System. The MINLP Solver used was CPLEX 12.6.1 and the CP solver used was CPLEX CP Optimizer 12.6.1.

Due to the number of stages and states, a time limit was set for calculating the optimal-value function. This makes it possible that the value obtained is not the optimum, yet it was considered as a mandatory control of the overall run time, since the MINLP or CP solver could experience difficulties in finding the optimum value or proving its optimality. The time limit chosen was 3 s. Preliminary experiments indicated that within this limit both solvers would often reach and prove the optimal result and that increasing it to 5 or 10 s would not significantly impact the results obtained. Nevertheless, it was considered that the possibility of no solution being found for a few specific stages and states was significant and impactful and therefore a "back-up" mechanism was developed so that in this case the time limit was slightly increased. Moreover, in the last stage (corresponding to the first time period), since only one state is possible, the time limit for its calculation was increased to 60 s in order to improve the probability of achieving and proving the optimum.

Results and Discussion

Figure 20.3 presents the best value obtained for each instance, with different combinations of the parameters $XMAX$ and $OMAX$. These numeric examples illustrate the potential and limitations of the approach proposed since they represent small configurations of the problem and already embody significant insights.

Firstly, if one compares the overall values obtained by both approaches, the one that uses the Constraint Programming model to calculate the optimal-value function (henceforward referred to as "CP approach" for the sake of brevity) obtains better

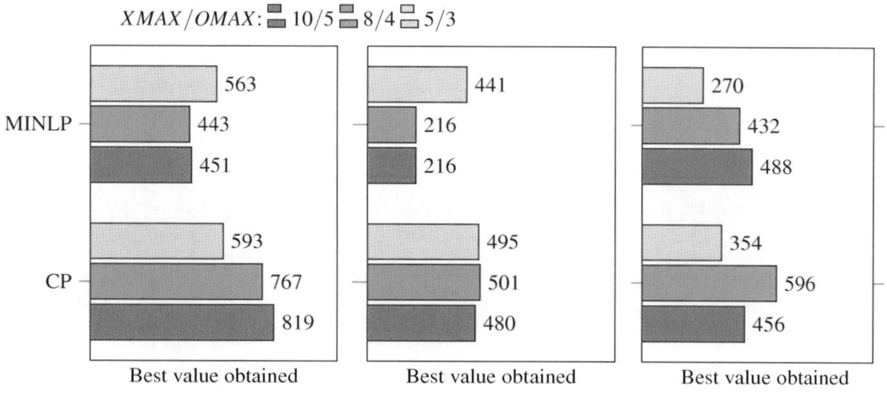

(a) Smallest instance ($|\mathscr{R}| = 428$) (b) Median instance ($|\mathscr{R}| = 865$) (c) Largest instance ($|\mathscr{R}| = 2369$)

Fig. 20.3 Best profit values obtained by each approach, for each instance, with different combinations of $XMAX/OMAX$ parameters results than the one that uses the Mixed Integer Non Linear Program ("MINLP approach"), especially for smaller instances.[4]

An interesting analysis can be made regarding the effect of the parameters $XMAX$ and $OMAX$ (directly connected with the number of states). It could be expected that an increase of these parameters would lead to higher values being obtained, since they represent a constraint on the "original problem" and their increase can be compared to a gradual relaxation of this constraint. Nevertheless, for the MINLP approach, this only happens for the biggest instance. In the remaining instances, increasing these parameters leads to a lower value. This might be explained by the effect of the time limitation imposed to each sub-problem. Due to this limit, the solver may not reach the optimum solution. Increasing the parameters makes the problem more complex to solve and thus makes the optimum more difficult to achieve. In fact, when the parameters are increased, the number of states increases and the sub-problems (which have decision variables and constraints dependent on the number of states) get more complex to solve.

As for the CP approach, a similar tendency is not as visible and it becomes difficult to draw conclusions regarding the relative strength of each of the two contradictory effects of increasing the parameters $XMAX$ and $OMAX$: (1) the "original problem" is more relaxed and thus better solutions could be achieved, and (2) the problems become more complex and, due to the time limit, the optimum value is more difficult to achieve.

From this analysis rises an interest in observing the actual run times to understand if the hypothesis related with the optimum being reached more easily is supported. Two bounds on expected time to solve can be drawn. The first is based on the number of optimization problems to solve and the time limit to solve them, not considering the previously mentioned extra-time rules. The second is the actual upper bound on time that considers all the extra-time rules. In order to reach the latter, the extra time would have to be used to its maximum for all optimization problems. Figure 20.4 presents the time to solve these numeric examples and reach the values presented in Fig. 20.3.

As expected, with an increase in instance size, there is a trend to increase the time to solve. Also, as it can be easily observed, the MINLP approach is consistently faster than the CP approach. In fact, the former is consistently below the expected bound (not considering extra time) while for the latter this only happens for the smallest instance. This means that the MINLP approach was often able to prove optimality in less than the time limit imposed, while the CP approach often used the extra time allowed. This does not fit in a straightforward way with the results previously discussed when comparing the best values obtained by each approach. In fact, the integrated analysis of Figs. 20.3 and 20.4 supports the claim that the CP approach quickly finds good solutions yet takes longer to reach the optimum (or prove optimality)

[4]Throughout this discussion, the notion of instance *size* will be associated with the intrinsic parameter being analyzed: the number of rental types. It thus especially related with the complexity of the optimization problems solved for each stage and state (not the number of stages and states *per se*).

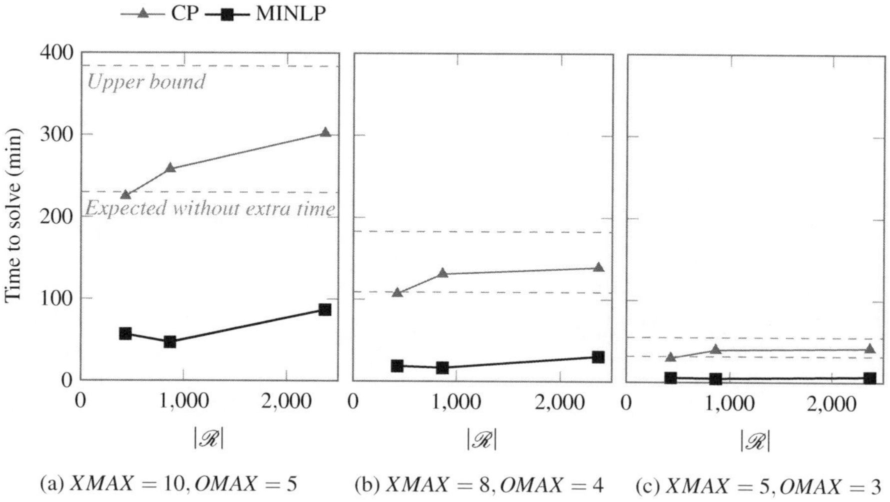

Fig. 20.4 Time to solve the numeric examples using the two approaches to calculate the optimal-value function, plotting the instances by the number of rental types $|\mathscr{R}|$, for three possible combinations of the parameters $XMAX$ and $OMAX$

Table 20.1 Comparison of key results and differences

	CP approach	MINLP approach
Overall best profit values	Generally higher	Generally lower
Effect of increasing size-influencing parameters	No significant effect on profit	Lower profit values
Time to solve	Significantly slower	Significantly faster
Conclusions	Quickly finds good solutions yet has difficulty proving optimality, increasing significantly the time to solve	Achieving/proving optimality ranges from extremely fast to impossible within time limit, making it difficult to obtain better solution values

and that the ability and speed to prove optimality varies significantly more in the MINLP approach: from extremely fast to prove optimality to returning a feasible solution with a high gap. This seems reasonable considering the characteristics of each solution method and the fact that the complexity of the optimization problems varies significantly among stages (within the same instance).

Table 20.1 summarizes and compares the key differences and results from the two approaches. Overall, it is possible to conclude that the time limit imposed has a significant impact. Nevertheless, although it can lead to poorer overall results, if a time limit is not imposed the time to solve could make both approaches nonviable.

20.4 Conclusions

In this work, a dynamic programming approach was developed to deal with the integrated dynamic pricing and capacity problem in the car rental business. This methodology has been successfully applied to similar problems and from the multi-stage structure of the problem (and the consequent "stock available" type of control) can be seen as an adequate method. Nevertheless, the fact that the capacity is "re-usable" in the rental context raises significant applicability issues that were analyzed.

The first drawback of applying dynamic programming to this context is that the number of states and state variables easily explodes. Already with these small numeric examples (in terms of number of time periods and rental stations, and considering deterministic demand) this method shows computational limitations. This is mainly due to the fact that the problem is related with a *rental* context—and this is why car rental is not like any other pricing and capacity/inventory model: the number of states explodes because stock can "return" after being depleted and that makes it necessary to keep track of occupied vehicles, which relates with decisions from time periods other than the immediately previous one.

An additional limitation is that the number of states is based on parameters that are not derived from the original problem, although they may have a close parallel to actual operational constraints, such as the stock of vehicles in a station being limited by the available parking spots. Although it was possible to observe that increasing the maximum number of vehicles in stock and occupied (and thus increasing the number of states) may hinder getting a better solution due to time limitations, not increasing these parameters for a real-world application of the methodology is not a viable option. In fact, the values herein proposed fail to fully reflect the reality of the problem. Ideally, these parameters should have no impact on the optimum value. Nevertheless, from a quick analysis of the order of magnitude of the demand values, it is easily established that in these numeric examples they have had impact.

These conclusions do not support the claim that dynamic programming is an adequate method to tackle this problem. Nevertheless, this discussion was able to bring some insights related with the problem structure as well as the potential and limitations of CP and MINLP when embedded in a discrete dynamic programming approach.

As future work, other methodologies will be applied to this rental context, especially considering the case of uncertain demand and realistically sized problems.

Acknowledgements The first author was supported by grant SFRH/BD/103362/2014 from FCT—Fundação para a Ciência e Tecnologia (Portuguese Foundation for Science and Technology). This work was also partially financed by the ERDF—European Regional Development Fund through the Operational Programme for Competitiveness and Internationalisation—COMPETE 2020 Programme within project "POCI-01-0145-FEDER-006961", and by National Funds through the FCT—Fundação para a Ciência e Tecnologia (Portuguese Foundation for Science and Technology) as part of project UID/EEA/50014/2013.

References

1. E. Adida, G. Perakis, Dynamic pricing and inventory control: robust vs. stochastic uncertainty models-a computational study. Ann. Op. Res. **181**(1), 125–157 (2010). ISSN 02545330, https://doi.org/10.1007/s10479-010-0706-1
2. A.V. den Boer, Dynamic pricing and learning: historical origins, current research, and new directions. Surveys in Ope. Res. Manag. Sci. **20** (1), 1–18 (2015). ISSN 18767354, https://doi.org/10.1016/j.sorms.2015.03.001
3. A. Hax, S. Bradley, Dynamic programming (chapter 11), in *Applied Mathematical Programming*, vol. 26 (1977), pp. 320 – 362, https://doi.org/10.2307/2550876
4. Z. Li, F. Tao, On determining optimal fleet size and vehicle transfer policy for a car rental company. Comput. Op. Res. **37**(2), 341–350 (2010). ISSN 03050548, https://doi.org/10.1016/j.cor.2009.05.010
5. B.B. Oliveira, M.A. Carravilla, J.F. Oliveira, Fleet and revenue management in car rental companies: a literature review and an integrated conceptual framework. Omega **71**, 11–26 (2017). ISSN 03050483, https://doi.org/10.1016/j.omega.2016.08.011
6. F. Rossi, P. Van Beek, T. Walsh, *Handbook of Constraint Programming* (Elsevier, 2006). ISBN 0444527265
7. D. Simchi-Levi, X. Chen, J. Bramel, Integration of inventory and pricing (chapter 10), in *The logic of Logistics: Theory, Algorithms, and Applications for Logistics Management*, 3rd edn. (Springer Science+Business Media, Berlin, 2014), pp. 177–209

Chapter 21
Models and Advanced Optimization Algorithms for the Integrated Management of Logistics Operations

Telmo Pinto, Cláudio Alves and José Valério de Carvalho

Abstract In this paper, we describe a set of algorithms regarding real combinatorial optimization problems in the context of transportation of goods. These problems consist in the combination of the vehicle routing problem with the two-dimensional bin-packing problem, which is also known as the vehicle routing problem with two-dimensional loading constraints. We also analyzed two related problems, namely the elementary shortest path problem and the vehicle routing problem with mixed linehaul and backhaul customers. In both problems, two-dimensional loading constraints are explicitly considered. Two column generation based approaches are proposed for the vehicle routing problem with two-dimensional constraints. The elementary shortest path problem with two-dimensional constraints is addressed due to its importance in solving the subproblem of the column generation algorithms. To the best of our knowledge, we contribute with the first approach for this problem, through different constructive strategies to achieve feasible solutions, and a variable neighborhood search algorithm in order to search for improved solutions. In what concerns the vehicle routing problem with mixed linehaul and backhaul customers and two-dimensional loading constraints, different variable neighborhood search algorithms are proposed. All the proposed methods were implemented and experimentally tested. An exhaustive set of computational tests was conducted, using, for this purpose, a large group of benchmark instances. In some cases, a large set of benchmark instances was adapted in order to assess the quality of the proposed models.

Keywords Vehicle routing · Loading constraints · Computational study

T. Pinto (✉) · C. Alves · J. Valério de Carvalho
Departamento de Produção e Sistemas, Escola de Engenharia, Universidade do Minho, 4710-057 Braga, Portugal
e-mail: telmo@dps.uminho.pt

A. I. F. Vaz et al. (eds.), *Operational Research*, Springer Proceedings in Mathematics & Statistics 223, https://doi.org/10.1007/978-3-319-71583-4_21

21.1 Introduction

Vehicle routing plays a major role in the transportation and operations research field. The proposed approaches for this problem are not only relevant for the scientific community, but also for real-world situations since they impact significantly on the final cost of goods and commodities. It has been observed that the part corresponding to transportation represents from 10 to 20% of these final costs [20]. An efficient planning of routes seems very hard to achieve without the adequate analytical and technological tools [19]. The results of applying such tools are very much commented through the literature. In some cases, they may rise up to 20% as referred to in [20].

The economical waste that results from unnecessary or excess travel is also well documented. King and Mast [9] defined the excess travel as the difference between the real travel time and the potential travel time if all the travelled routes are optimal. They conclude that only in United States of America, this excess travel amounts to 7% of all the travel time. From all these observations, it seems clear that there is a true potential for improvements with both economic and environmental benefits [1]. The optimization problem that is behind this thematic is called the vehicle routing problem. It was proposed first in [2] as a generalization of the Travelling Salesman Problem (TSP). Applications of the vehicle routing problem are not restricted to logistics. The conceptual model that underlies this problem finds applications in diverse fields including computer networking and robotics.

Broadly speaking, the vehicle routing problem consists of finding the best set of routes of a fleet of vehicles that must visit a set of customers, taking into account operational constraints and human limitations resulting from the maximum time for the drivers' work, or the compulsory breaks for example. Additional constraints can be added to this general problem. One of the most usual constraints that are related in the literature is concerned to the capacity of the vehicles. However, as will be stated below, this type of constraints are not sufficient to reflect the complexity of some real systems.

In this paper, we describe a set of optimization tools for a family of combinatorial optimization problems that apply to the field of transportation and supply chain management in general. The focus is on routing problems with a concern on the real constraints that apply to these problems. One of the issues that is typically neglected by those that addressed the general vehicle routing problems is related to the loading component of these problems, concerning the shapes of the loads. When the loads are small compared to the size of the vehicles, one may characterize them through a single measure such as weight or volume. Most of the contributions in the literature follow this approach. In these cases, the capacities of the vehicles are represented by a single value. On another hand, when the loads are large compared to the sizes of the vehicles, deciding how to place them in the vehicle becomes a part of the problem. The former approaches may produce infeasible solutions for these problems. Indeed, it may happen that a given load whose weight (for example) does not exceed the capacity of a vehicle does not fit in the vehicle because of its dimensions. The related optimization problem is known in the literature as the vehicle routing problem with loading constraints. This problem integrates two hard combinatorial optimization problems, resulting in a very challenging optimization problem. Indeed, the methods applied to this variant have to consider both the definition of the routes and the packing problem in either two- (2L-CVRP) or three-dimensions (3L-CVRP).

Since the 2L-CVRP is an NP-hard problem, the vast majority of contributions presented in the literature is based in heuristics [3, 5, 21]. Few works tackled the 2L-CVRP by using exact methods. The first 2L-CVRP approach is due to Iori et al. [8]. The authors proposed an integer programming model formulation. In a first phase, the capacity-cut and loading constraints are removed and the linear programming (LP)-relaxation is strengthened using multi-star inequalities. Then, a branching scheme starts by applying separation procedures for the removed constraints. The procedure to check the loading feasibility of each route relies on a branch-and-bound approach. At the root, lower bounds for the two-dimensional bin packing problem are computed [10]. If the greatest lower bound is greater than one, clearly the route is infeasible. Otherwise, a bottom-left heuristic is applied taking into account all the loading constraints. If it is not possible to derive a valid packing, a search tree is used. At the root node an empty loading area is considered. For each item a descendent node is created by placing the item at the bottom-most and leftmost position. For each of these nodes, a child node is created for each item to be placed in each feasible position. Recently, based in [8], another exact method was presented in [6] for the 2L-CVRP and 3L-CVRP. The authors analyzed different routing and packing separation procedures. After deriving an integer solution provided by branch-and-bound, each route is verified concerning its feasibility. Furthermore, the authors suggested a strategy to find infeasible routes from a non-integer solution. To prove the feasibility of a given route, the authors resort to several heuristic methods. If these fail to prove its feasibility, exact methods can be used based in both branch-and-bound and constraint programming. Another exact approach for a variant of the 2L-CVRP was addressed by Pollaris et al. [16]. In this problem, pallets have to be sequenced in order to satisfy axle weight constraints. The pallets are alternately loaded into two horizontal stacks. The authors stressed that axle weight constraints are important since its violation causes not only an infringement of traffic regulations but also a risk to the road safety.

This paper is organized as follows. In Sect. 21.2, we define all the problems that are tackled in the described approaches. In Sect. 21.3 we describe the variable neighborhood search for the elementary shortest path problem with loading constraints. In Sect. 21.4, we describe the column generation based approaches for the vehicle routing problem with two-dimensional loading constraints, while in Sect. 21.5 we describe the approach for this problem using column generation based heuristics. In Sect. 21.6 we describe the metaheuristic algorithm for pickup and delivery problems with loading constraints. Finally, in Sect. 21.7 we draw some conclusions.

21.2 Vehicle Routing Problems with Loading Constraints

21.2.1 The Vehicle Routing Problem with 2-Dimensional Loading Constraints

The Vehicle Routing Problem with 2-Dimensional Loading Constraints (2L-CVRP) addressed in this paper can be defined in a complete directed graph whose nodes represent a single depot and the customers to be visited. There is a virtually unlimited and homogeneous fleet with a two-dimensional rectangular loading surface to serve the customers demand. The demand of each customer is composed by a finite number of two-dimensional rectangular items. The 2L-CVRP consists in finding a set which minimizes the total travel cost, satisfying the following constraints:

(C1) Each route must start and end at the depot;
(C2) All customers must be served in a single visit, i.e., the demand of each customer cannot be split;
(C3) A feasible orthogonal loading is required for each used vehicle, i.e., the items must be completely within the surface, must not overlap, and the edges of the items must be parallel to the surface edges. Additionally, the items cannot be rotated;
(C4) Unloading an item at a given customer must be performed in a straight movement, through a free passage towards the door, and whose width is greater than or equal to the width of the item.

The constraints (C4) are known as sequential constraints. They impose that, when unloading one item at a given customer, no movements are allowed for items belonging to other customers, and so these items cannot block the passage of the item that is being unloaded. During the unloading operation, each unloaded item must preserve its edges parallel to the edges of the surface. Note that there are no sequential constraints between items for the same customer. Figure 21.1 depicts this situation, where $I_{i,j}$ represents the item j of customer i: unloading items for customer 1 can be done by a straight movement for each item. However, in (a), when unloading items for customer 2, item $I_{3,1}$ blocks the passage of item $I_{2,1}$. Therefore, in (a), the loading is sequential infeasible. In (b), the sequential constraints are satisfied, since items belonging to the same customer can be unloaded in a straight movement without rearranging items to be delivered.

21.2.2 Elementary Shortest Path Problem with Loading Constraints

Iori and Martello [7] considered the use of column generation techniques as an open perspective to tackle routing problem with loading constraints. Several works

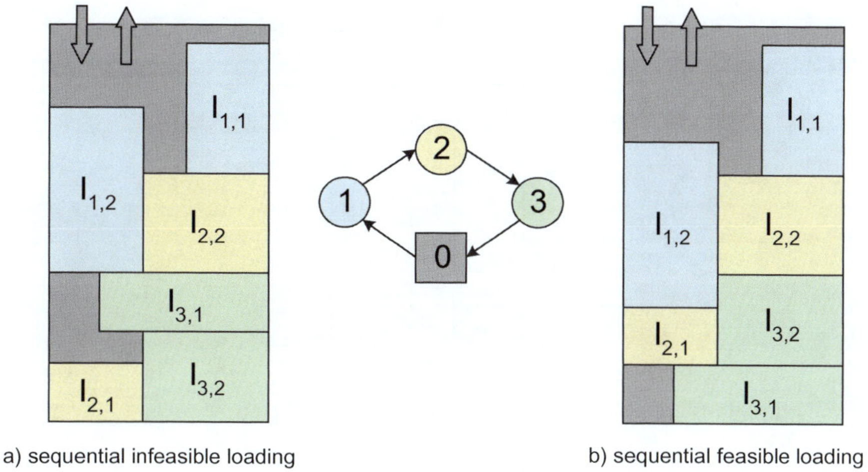

Fig. 21.1 Sequential constraints in the 2L-CVRP context

considered the application of column generation to the CVRP, and presented the application of the Dantzig-Wolfe decomposition to a CVRP. As a result, one can obtain a Restricted Master Problem (RMP) and a pricing subproblem which is a shortest path problem with resource constraints. These resource constraints are directly related to the capacity of vehicle for the CVRP. However, in the classical CVRP, the customer must be visited once. Since the solution of the pricing subproblem must be a valid solution of the CVRP, the obtained shortest path must be elementary.

The Elementary Shortest Path Problem with Resource Constraints (ESPPRC) is NP-hard and it can be effective in the resolution of the pricing problem of a column generation algorithm for VRP [4, 17, 18]. In this sense, it is possible to apply a Dantzig-Wolfe decomposition to the 2L-CVRP and then obtain an Elementary Shortest Path Problem with 2-dimensional Loading Constraints (2L-ESPP). To the best of our knowledge, there are no approaches for the 2L-ESPP.

The objective of the 2L-ESPP is to find the shortest path in a graph, starting and ending at the depot, visiting each customer at most once while satisfying (C3) and (C4). The cost of the edges may be negative and the demand of each node is composed by two-dimensional items.

21.2.3 2L-CVRP with Mixed Linehaul and Backhaul Customers

The Capacitated Vehicle Routing Problem with Mixed linehaul and Backhaul customers with 2-dimensional Loading constraints (2L-CVRPMB) is an extension of the 2L-CVRP. The main difference between this problem and the one addressed

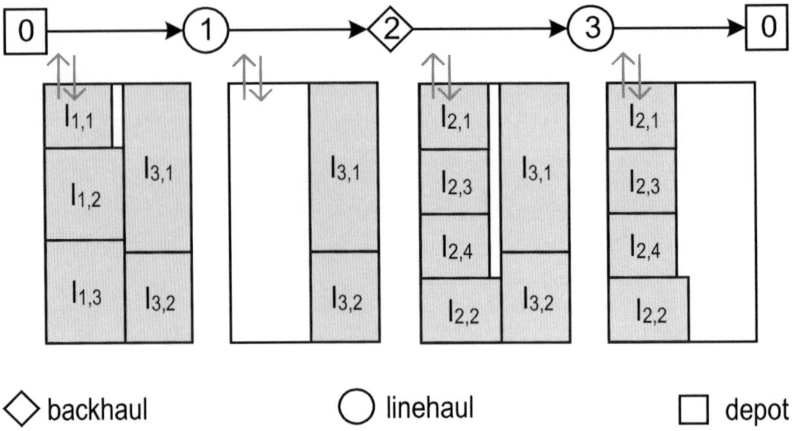

Fig. 21.2 A feasible solution of the 2L-CVRPMB

in Sect. 21.2.1 is that customers are divided into two different sets: the ones which require a given demand (linehaul customers), and the ones which provide a given supply (backhaul customers). The demands can only be provided by the depot, which is also the only destination of items provided by backhaul customers. Both the demand and the supply are composed by weighted and two-dimensional rectangular items. There is a limited and homogeneous fleet which is responsible to visit all customers exactly once. Each vehicle has a limited capacity in terms of weight, and it has a 2-dimensional rectangular loading area for placing items.

The 2L-CVRPMB calls for the minimization of travel costs while satisfying the following constraints:

(C5) The demand of each linehaul customer must be satisfied in a single visit. Analogously, the set of items provided by each customer must be picked up in a single visit;
(C6) The weight capacity of vehicles cannot be violated;
(C7) The items have a fixed orientation during loading and unloading operations;
(C8) The loading area of each vehicle must consist in a feasible orthogonal loading, and thus, items must be completely within the surface and must not overlap;
(C9) Each route is assigned to one vehicle, without exceeding the fleet size; each vehicle is only available to perform one route;
(C10) Unloading each item must be performed without moving other items or rearranging the layout. Therefore, each picked item cannot block items to be delivered.

In Fig. 21.2, a feasible solution of the 2L-CVRPMB is presented. For the sake of simplicity, only one route is considered. The layout of each arc is presented below the corresponding arc, and $I_{i,j}$ represents the item j of customer i.

21.3 Variable Neighborhood Search for the Elementary Shortest Path Problem with Loading Constraints[1]

In this section, we review the variable neighborhood search (VNS) approach for the elementary shortest path problem with two-dimensional loading constraints, as presented in Sect. 21.2.2. Different constructive heuristics are applied in order to generate feasible solutions. The customers are iteratively evaluated to be added to the path. There are two possible ways to select the first customer to be evaluated:

- the one that is nearest to the depot;
- random selection among all customers.

The following customers of the route are added to the route according to a nearest neighbor procedure, providing that their items can fit in the vehicle. In order to achieve a feasible layout (i.e., a feasible arrangement satisfying (C3) and (C4)), three placement strategies are considered:

- standard bottom-left strategy;
- adapted bottom-left strategy;
- level packing strategy.

When considering the level packing strategy, it is also required to select the level to place a given item, which is selected according to the following two strategies:

- the first level where the item fits;
- the level where the item best fits.

The order that items belonging to the same customer are placed in the vehicle relies on the two following criteria:

- non-increasing order of height of the items;
- non-increasing order of area of the items.

According to the strategies referred to above, 16 combinations for generating feasible solutions can be derived. Each one is used as a start solution of a VNS algorithm in order to improve the former solutions.

The neighborhood structures are divided into routing and packing neighborhoods structures. The routing neighborhood structures include swapping two customers or shifting one customer in the route, removing a customer from the route and possibly all its successors, and exchanging a customer by another in the route. The packing neighborhoods structures include shifting one item or swapping two items of a given customer in the sequence by which items are loaded to the vehicle.

To the best of our knowledge, we contributed with the first approach for the elementary shortest path with two-dimensional loading constraints. All variants were tested using benchmark instances for the 2L-CVRP [5, 8]. The obtained results show that the VNS algorithm improves significantly the solutions of the constructive heuristics within a small amount of computing time (only 3 s). Furthermore, it was

[1]The contribution presented in this section has been published in [12].

possible to compare all the strategies referred to above. To some extent, the percentage of improvement seems to be dependent on the quality of the initial solution, reaching larger values when the level packing strategy is used. The percentage of used space in the vehicle tends to be larger when bottom-left strategies are used.

The contribution of this approach is not confined to the problem itself. Indeed, the addressed problem may correspond to the subproblem of the vehicle routing problem with two-dimensional constraints, when Dantzig-Wolfe decomposition is applied. In this sense, the proposed methods were also used throughout the next two sections.

21.4 Column Generation Based Approaches for the Vehicle Routing Problem with Two-Dimensional Loading Constraints[2]

In this section, we review the branch-and-price algorithm for Capacitated Vehicle Routing Problem with Two-dimensional Loading constraints (2L-CVRP), as defined in Sect. 21.2.1. There are several column generation based approaches for many variants of the vehicle routing problem. In contrast, approaches for the variant with loading constraints through column generation are not quite explored.

We apply the Dantzig-Wolfe decomposition to the original problem, resulting in a reformulated model and in a pricing subproblem. The former corresponds to the master problem which is an integer programming model composed by partitioning constraints, while remaining constraints are considered in the subproblem. Each decision variable of the master problem corresponds to a column and each column represents a feasible route. Since the subproblem is limited, a given column corresponds to an extreme point of the valid space.

The subproblem corresponds to the Elementary Shortest Path Problem with 2-dimensional Loading Constraints (2L-ESPP), addressed in Sect. 21.2.2 which is solved heuristically through a variable neighborhood search algorithm described in Sect. 21.3. A family of dual inequalities is used, aiming to accelerate the convergence of the branch-and-price approach.

Branch-and-price is applied when it is not possible to add more attractive columns and the solution remains fractional. Our branching strategies rely on six different strategies to select the arc to branch on (single arc branching), and in four strategies to branch on the flow of a subset of arcs. In the former case, branching is performed in one variable of the original problem which corresponds to an arc. In the latter case, more than one original variable may be selected.

The branching tree is explored using a depth-first search strategy. We suggested three strategies for partial enumeration. According to the first strategy, the branching tree is explored by using only one partition rule while the second strategy applies a randomly selected partition rule each time branching is performed. In the third

---

[2]The contribution presented in this section has been published in [13].

strategy, each partition rule is applied along ten iterations without any improvement of the incumbent solution.

The computational experiments were conducted using a subset of benchmark instances for the 2L-CVRP [5, 8]. The computational results showed that it is possible to outperform significantly the value of the initial incumbent for instances with a small number of customers. However, the algorithm has some limitations for instances with a large number of customers.

21.5 Column Generation Based Heuristics for the Vehicle Routing Problem with Loading Constraints[3]

The analysis of the computational results of the approaches described in Sect. 21.4 (presented in [13]) demonstrates that the column generation algorithm for the 2L-CVRP has some limitations, particularly when dealing with a large number of customers. To overcome this limitation, we presented a set of new heuristic column generation based heuristics aiming to solve the 2L-CVRP. We make use of the reformulated model obtained in the previous section in order to develop this set of heuristic approaches.

The four heuristic approaches work in the formulation of the master problem. The first strategy relies in a restricted master heuristic, by solving the LP-relaxation of the master problem and enforcing all the variables to be integer. Therefore, this strategy uses all generated columns to solve a static integer programming model. The second strategy includes the first one: after solving the static integer programming model, the solution is composed by a set of columns corresponding to the set of routes. The column associated to the route with the best usage of the loading area is selected. This route is added to a partial solution. Then, the size of the reformulated model is reduced by eliminating all customers that are visited in the selected route, and the process is repeated starting from the LP-relaxation of the master problem. The third strategy consists in solving the LP-relaxation and in selecting the customer that is assigned to more columns. Then, the variables associated to that routes visiting that customer are enforced to be integer, and the resulting mixed integer programming model is solved. As a result, only one column will be associated to that customer. This column represents a route that will be added to a partial solution and the process is repeated, starting from the LP-relaxation. The fourth strategy is analogous to the third one, but the customer that is assigned to less columns is selected instead.

As in the previous section, all strategies were experimentally tested using a subset of benchmark instances for the 2L-CVRP [5, 8]. Among all strategies, the second one tends to present best average results, with some exceptions. On the contrary, and excluding the first strategy, the third one tends to present worse average results, leading to longer processing time in many cases.

---

[3]The contribution presented in this section has been published in [11].

21.6 A Metaheuristic Algorithm for Pickup and Delivery Problems with Loading Constraints[4]

In this section, we consider a pickup and delivery problem with two-dimensional loading constraints. More precisely, we address the capacitated vehicle routing problem with loading constraints and mixed linehaul and backhaul customers, as presented in Sect. 21.2.3. An insertion heuristic for generating feasible solutions is used. This heuristic starts by generating a solution by considering only linehaul customers. The backhaul customers are then inserted aiming to minimize the insertion cost, while satisfying all the constraints. The feasibility of routes is heuristically verified by an adapted bottom-left heuristic. We also suggested three different variants of the variable neighborhood search (VNS) algorithm in order to search for improved solutions: (VNS1) Variable Neighborhood Search; (VNS2) General Variable Neighborhood Search; (VNS3) Skewed General Variable Neighborhood Search.

The neighborhood structures are used both in shaking and local search phases and some of them are specific developed taking into account the context of this problem. All variants of the VNS use nine routing neighborhood structures and one packing and routing neighborhood structure, as follows:
Routing neighborhoods

- Swapping two linehaul customers within the route;
- Swapping two backhaul customers within the route;
- Shifting one customer within the route;
- Shifting one customer for another route;
- Swapping two consecutive customers of different sets;
- Swapping three consecutive customers of different sets;
- Swapping two consecutive customers of any set;
- Shifting two consecutive customers;
- Shifting three consecutive customers.

Packing and routing neighborhood

- Swapping two items and inserting one customer

The neighborhood structures are explored by the sequence they were presented. At each iteration of (VNS1), a neighbor solution is randomly obtained using one neighborhood structure. Then, local search is applied using the same neighborhood structure. Whenever a better solution is obtained, the process is restarted from the first neighborhood structure, in a first improvement strategy. The (VNS2) algorithm consists in an adaptation of the (VNS1) in which local search is performed using variable neighborhood descent method. This means that the neighborhood is completely explored, and then the best neighborhood is achieved. Finally, (VNS3) algorithm consists in an adaptation of (VNS2) in which worse solutions can be accepted, if they are sufficiently different from the incumbent solution. This procedure can drive the search to solutions far from the incumbent solution. A difference function is

[4]The contributions presented in this section have been published in [14, 15].

used aiming to measure the difference among two solutions. Additionally, the best obtained solution is always kept during the execution of the algorithm.

All the VNS approaches were computationally tested using instances adapted from the ones proposed for the 2L-CVRP [5, 8]. These instances were also replicated in order to diversify the number of backhaul customers in each instance. The values of initial solutions were clearly outperformed in all approaches. However, more significant improvements were reached using (VNS2) and (VNS3). These improvements are often related to a small number of backhaul customers.

21.7 Conclusion

In this paper, we described a set of optimization tools for the vehicle routing problem with loading constraints. We also explored different variants of the problem that are characterized by their set of constraints on the loading and routing part of the problem. The developed methods are based on column generation models and on metaheuristics. With these approaches, we aim to contribute with new algorithms to increase the efficiency in transportation and supply chain management, since it could improve the competitiveness companies which intend to differentiate its service level, taking into account complex constraints that can be found in real-world situations, while minimizing costs. In this sense, the contributions of the described methods are expected to have both scientific and practical relevance.

Acknowledgements This work has been supported by COMPETE: POCI-01-0145-FEDER-007043 and FCT—Fundação para a Ciência e Tecnologia within the Project Scope: UID/ CEC/00319/2013, and through the grant SFRH/BD/73584/2010, funded by QREN - POPH - Typology 4.1 - co-funded by the European Social Fund.

References

1. C. Alves, Planeamento de rotas em sistemas de recolha/distribuição. Master's thesis, University of Minho (2000)
2. G. Dantzig, J. Ramser, The truck dispatching problem. Manag. Sci. **6**(1), 80–91 (1959)
3. C. Duhamel, P. Lacomme, A. Quilliot, H. Toussaint, A multi-start evolutionary local search for the two-dimensional loading capacitated vehicle routing problem. Comput. Oper. Res. **38**(3), 617–640 (2011). ISSN 0305-0548
4. D. Feillet, P. Dejax, M. Gendreau, C. Gueguen, An exact algorithm for the elementary shortest path problem with resource constraints: application to some vehicle routing problems. Networks **44**(3), 216–229 (2004). ISSN 1097-0037, http://dx.doi.org/10.1002/net.20033
5. M. Gendreau, M. Iori, G. Laporte, S. Martello, A Tabu search heuristic for the vehicle routing problem with two-dimensional loading constraints. Networks **51**(1), 4–18 (2008). ISSN 00283045, http://doi.wiley.com/10.1002/net.20192
6. P. Hokama, F. Miyazawa, E. Xavier, A branch-and-cut approach for the vehicle routing problem with loading constraints. Expert Syst. Appl. **47**, 1–13 (2016).

ISSN 0957-4174, https://doi.org/10.1016/j.eswa.2015.10.013, http://www.sciencedirect.com/science/article/pii/S0957417415007009
7. M. Iori, S. Martello, Routing problems with loading constraints. TOP **18**(1), 4–27 (2010). ISSN 1134-5764, https://doi.org/10.1007/s11750-010-0144-x
8. M. Iori, J.-J. Salazar-Gonzalez, D. Vigo, An exact approach for the vehicle routing problem with two-dimensional loading constraints. Trans. Sci. **41**(2), 253–264 (2007). ISSN 0041-1655
9. G. King, T. Mast, Excess travel: causes, extent, and consequences. Trans. Res. Record **1111**, 126–134 (1987)
10. S. Martello, D. Vigo, Exact solution of the two-dimensional finite bin packing problem. Manag. Sci. **44**(3), 388–399 (1998)
11. T. Pinto, C. Alves, J.V. de Carvalho, Column generation based heuristics for the vehicle routing problem with loading constraints. Eletron. Notes Discret. Math. (accepted)
12. T. Pinto, C. Alves, J.V. de Carvalho, *Variable neighborhood search for the elementary shortest path problem with loading constraints*, vol. 9156, Lecture Notes in Computer Science, book section 34 (Springer International Publishing, Berlin, 2015a), pp. 474–489. ISBN 978-3-319-21406-1, https://doi.org/10.1007/978-3-319-21407-8_34
13. T. Pinto, C. Alves, J.V. de Carvalho, A branch-and-price algorithm for the vehicle routing problem with 2-dimensional loading constraints, vol. 9855, Lecture Notes in Computer Science (Springer International Publishing, Berlin, 2016), pp. 321–336, https://doi.org/10.1007/978-3-319-21407-8_34
14. T. Pinto, C. Alves, J.V. de Carvalho, Variable neighborhood search algorithms for pickup and delivery problems with loading constraints. Electron. Notes Discret. Math. **58**, 111–118 (2017). ISSN 1571-0653, https://doi.org/10.1016/j.endm.2017.03.015, http://www.sciencedirect.com/science/article/pii/S1571065317300513
15. T. Pinto, C. Alves, J.V. de Carvalho, A. Moura, An insertion heuristic for the capacitated vehicle routing problem with loading constraints and mixed linehauls and backhauls. FME Trans. **43**(4), 311–318 (2015)
16. H. Pollaris, K. Braekers, A. Caris, G. Janssens, S. Limbourg, Capacitated vehicle routing problem with sequence-based pallet loading and axle weight constraints. EURO J. Transp. Logist. 1–25 (2014). ISSN 2192-4384, https://doi.org/10.1007/s13676-014-0064-2
17. G. Righini, M. Salani, Symmetry helps: bounded bi-directional dynamic programming for the elementary shortest path problem with resource constraints. Discret. Opt. **3**(3), 255–273 (2006). ISSN 1572-5286, https://doi.org/10.1016/j.disopt.2006.05.007, http://www.sciencedirect.com/science/article/pii/S1572528606000417
18. G. Righini, M. Salani, New dynamic programming algorithms for the resource constrained elementary shortest path problem. *Networks*, 51 (3): 155–170, 2008. ISSN 1097-0037, https://doi.org/10.1002/net.20212
19. M. Savelsbergh, M. Sol, Drive: dynamic routing of independent vehicles. Oper. Res. **46**(4), 474–490 (1998)
20. P. Toth, D. Vigo, An overview of vehicle routing problems, in *The Vehicle Routing Problem*, Society for Industrial and Applied Mathematics (2002), pp. 1–26
21. E. Zachariadis, C. Tarantilis, C. Kiranoudis, A guided tabu search for the vehicle routing problem with two-dimensional loading constraints. Eur. J. Oper. Res. **195**(3), 729–743 (2009)

Chapter 22
Waste Collection Planning Based on Real-Time Information

Tânia Rodrigues Pereira Ramos, Carolina Soares de Morais and Ana Paula Barbosa-Póvoa

Abstract This paper studies the definition of dynamic routes regarding the waste collection problem. Based on access to real-time information, provided by sensors located at waste bin containers, a Vehicle Routing Problem with Profits (VRPP) solution approach is developed. This aims for the maximization of waste collected while minimizing the total distance travelled, resulting in a maximization of profit. Different scenarios are studied, based on real data. The conclusions clearly show that the usage of real-time information on containers fill-levels, coupled with an optimization approach to define dynamic routes potentially increases the profit of waste management companies.

Keywords Real-time information · Sensors · Waste collection · Vehicle routing problem with profits

22.1 Introduction

Waste collection companies deal every day with high transportation costs and, more importantly, with a highly inefficient usage of their resources as their trucks must often visit bins that are only partially full. A preliminary study conducted by [10] showed that 59% of containers which were visited by a recyclable waste collection company had a fill rate lower than 75%, resulting in several kilometers traversed only to collect small amounts of waste. This is a consequence, as stated by [14], of a highly variable accumulation rate of waste, which depends on several factors. Additionally, visiting bins that are only partially full, besides being an unnecessary use of resources, also results in avoidable polluting emissions [9].

This inefficiency could be mitigated if real-time information on the containers' fill-levels was available, which would lead to a general collection route optimization. Recent advances in Information and Communication Technologies (ICT) already make possible to access real-time information by placing volumetric sensors to measure the volume of waste (in m^3) inside waste bins, thus allowing for a smart

T. R. P. Ramos (✉) · C. S. de Morais · A. P. Barbosa-Póvoa
CEG-IST, Instituto Superior Técnico, Universidade de Lisboa, Av. Rovisco Pais, 1049-001 Lisboa, Portugal
e-mail: tania.p.ramos@tecnico.ulisboa.pt

A. I. F. Vaz et al. (eds.), *Operational Research*, Springer Proceedings in Mathematics & Statistics 223, https://doi.org/10.1007/978-3-319-71583-4_22

management of the containers' fill-level, avoiding the collection of empty or nearly-empty bins, which would lead to a better management of operations and a minimization of its associated costs.

In this context, this work aims to assess the impact on some key performance indicators (like distance travelled, amount of waste collected, profit, kg per km) of a dynamic waste collection system, accounting for the availability of sensors that provide real-time information on each bin's fill-level every morning, before the definition of the collection routes. To assess the impact of this new system, real data from a recyclable waste collection company will be used to compare two situations: (1) actual solution where no real-time information is used and static routes are defined a "blind collection"; (2) a future situation, where real-time information on containers fill-levels is transmitted every morning and dynamic routes are defined, i.e. different routes are created every day. Here, the term "blind collection" is used in the current situation to emphasize that bins are attended to without knowledge of their actual fill-level, in contrast with the future situation in study, where the routes are defined when the information of the actual fill-levels are transmitted by the sensors.

The definition of dynamic collection routes cannot be carried out through classical vehicle routing problem models, as each set of containers to visit is not a given, but rather these sets must be defined by the model which accounts for the containers "attractiveness", meaning their fill-levels. Vehicle Routing Problem with Profits (VRPP) must be explored in this case, pursuing the objectives of maximizing the quantity of waste collected (collecting the bins with higher fill-levels) while minimizing the total distance travelled. As recyclable waste has a price market, such objectives are translated into the maximization of the difference between revenues obtained from waste collected and the transportation cost of collecting that waste. As VRPP models are often difficult to solve, we propose a hybrid solution method based on the combination of a mathematical formulation and heuristic procedures to deal with the problem.

The paper is structured as follows: in Sect. 22.2, a literature review on the VRPP is presented. In Sect. 22.3 we describe the problem and Sect. 22.4 presents the solution method devised. The results from the application of the solution method to a real case study are provided and discussed in Sect. 22.5. Lastly, Sect. 22.6 is dedicated to the presentation of conclusions and directions for future works.

22.2 Literature Review

In the VRPP, each customer has an associated profit which defines his attractiveness to be selected for a visit. In contrast to the classical Vehicle Routing Problem (VRP), the set of customers is not a given but rather a decision that must be taken. The VRPP class has been explored in the literature with several different applications, namely, design of tourist trips to maximize the value of the visited attractions within a limited period [22]; identification of suppliers to visit so as to maximize the recovered claims with a limited number of auditors [11]; athlete recruiting from high schools for a college team [7]; customers selection in less-than-truckload transportation [5] or a mobile blood donation system [17]. Depending on the number of vehicles, type of objective function and constraints, several VRPP variants appear (see [5] for a comprehensive survey on VRPP and its variants). When the objective is to maximize profit and the main constraint is route duration, the Orienteering Problem (OP) is at stake when considering a single vehicle case (also known as Selective TSP or Maximum Collection Problem); and the Team Orienteering Problem (TOP) when considering multiple vehicles (also known as Selective VRP or Multiple Tour Maximum Collection Problem).

The TOP has not yet been much explored in literature. Butt and Cavalier [7] studied this problem for the first time calling it a Multiple Tour Maximum Collection Problem. Two years later, Chao and co-authors [8] formalized the TOP problem and defined it as an extension of the orienteering problem, in which given a fixed amount of time for each of M members of a team, the goal is to determine M paths from the start point to the end point through a subset of locations in order to maximize the total score. Chao and co-authors [8] explain TOP as a multi-level optimization problem, formed by three levels: the first is to select a subset of points for the team to visit; the second, to assign points to each member of the team; and the third, to construct a path through the points assigned to each member of the team. A heuristic was also presented in their work, which was applied to more than 350 generated test problems proving it computationally efficient.

Over the years, several TOP models and solution approaches have been developed. Vansteeswegen and co-authors [21] compile some best-performing TOP algorithms, classifying the one developed by Boussier and co-authors [6] as a very fast option for problems with 100 vertices, where the authors develop an exact column generation algorithm to solve the problem - a Branch-and-Price. Poggi and co-authors [15] also presented three integer programming formulations for the TOP and develop a Branch-Cut-and-Price algorithm to solve it, using column generation. Regarding heuristic approaches for the TOP, the more relevant work is done by Archetti and co-authors [4] where two versions of Tabu Search and two meta-heuristics implementations based on Variable Neighborhood Search are developed and presented as the best known solutions for the TOP benchmark instances; Ke and co-authors [12] develop two ant colonies variations and is able to get competitive results, reducing the overall computational time; while Souffriau and co-authors [19] develop a greedy randomized adaptive search with path relinking.

The TOP is also known in literature as the Selective Vehicle Routing Problem (Selective VRP). Allahviranloo and co-authors [2] considered uncertainty regarding the utility obtained from visiting nodes, using Selective VRP models. The authors propose three new formulations to account for different optimization strategies under an uncertain demand or utility level: reliable, robust and fuzzy selective VRPs; and develop three parallel genetic algorithms and a classic genetic algorithm. Valle and co-authors [20] develop two exact solution approaches for solving a min-max selective VRP in which not all clients need to be visited and the goal is to minimize the longest vehicle route based on a Branch-and-Cut algorithm, and also a meta-heuristic that relies on GRASP. The latter was capable of improving the best solution provided by the algorithm in a shorter computational time.

Another variant of the VRPP is the Profitable Tour Problem (PTP). Its objective is to maximize the difference between total collected profits and travelling costs, with no route constraints [5]. Also, a variant on this is the Capacitated PTP, where each customer has a demand and the vehicle has a prefixed capacity, which must not be exceeded by the route.

When considering the TOP applications to collection systems, only two works were identified in literature, to the best of the author's knowledge, and neither of these is related to the case of municipal solid waste collection. Aras and co-authors [3] studied the reverse logistics problems of a firm that aims to collect used cores from its dealers. The firm is not obliged to visit all dealers: cores are collected from a dealer only if it is profitable to do so. In this model, each visit to a dealer is associated with a gross profit and an acquisition price, to be paid in order to take the cores back. The authors formulate two mixed-integer linear programming models based on a Selective Multi-Depot VRP with pricing and develop a Tabu search based heuristic method to solve medium and large-sized instances which was proved to be promising. Aksen and co-authors [1] considered a biodiesel production company that collects waste vegetable oil from source points and attempts to decide which of them to include in the collection program. The problem addressed should define which source points to visit on each day and which periodic routing schedule to set over a time horizon so as to minimize the total collection cost, considering inventory and purchasing. Not all the source points should be visited if the required quantity for satisfying the input requirements for production is attended. In this paper, the authors develop two different formulations for the problem they call Selective and Periodic Inventory Routing Problem (Selective PIRP); they compare those formulations and apply the better one to a real-world problem with 36 scenarios.

In face of this literature review, we can conclude that there is space to explore the problem that we have defined above, which is the development of a TOP model aimed at the collection of recyclable waste. If real-time information is considered, the TOP model will define smart routes in order to maximize the recyclable waste collected and minimize the transportation cost.

22.3 Problem Description

The problem in study can be defined as follows.

Given a set of n nodes (waste bins), a set of v vehicles and a depot (where all vehicles start and end their routes), a complete directed graph is defined on $n + 1$ nodes, with distances d_{ij} for each arc (i, j) in the graph. Each node i has an Initial Stock S_i that corresponds to the data read and transmitted by the sensors in the morning about the volume of waste in m^3 and, additionally, an expected daily accumulation rate a_{ij} that is computed taken into account historical data read by the sensors. Each waste bin has a maximum capacity E_i. A travelling cost c per unit distance travelled and a revenue value r per kg of waste collected are considered. Each vehicle has a fixed capacity Q and each route has a maximum length M. At each day t, in the morning, the sensors transmit information about the fill-levels for each bin.

The problem at stake is to find the set of nodes to be visited in each day t (if any) and which visit sequence would maximize profit, while satisfying the vehicle capacity, route length and bin's capacity. Profit is given by the difference between the revenues from selling the waste collected to a recycling entity minus the transportation cost of collecting the waste bins.

22.4 Solution Method

In this section, we propose a solution method to define collection routes under a dynamic scenario, where bins have sensors installed within them and transmit the bin's fill-level every day. Routes are defined taking into account real-time information, selecting the bins that are worth to be collect given the waste amount inside and the transportation cost incurred to collect them.

A MILP is proposed to solve the problem for each time period t separately, considering only one vehicle at a time (see Fig. 22.1). The solution for each iteration is taken into account for the following iteration, i.e., the input data for day $t + 1$ takes into account if the waste bins were visited or not on day t, and the quantity of waste at each container is updated according with the information transmitted by the sensors. Regarding the vehicles, the input data for vehicle $k + 1$ takes into account the waste bins that were visited by vehicle k, and for those visited initial stock turns into zero. If the number of vehicles available for each day reaches the maximum ($k = K$), it means that all routes for day t are defined. Then, the time is updated and the next iteration is set to be carried out on the next day ($t = t + 1$).

The MILP model developed is described next.

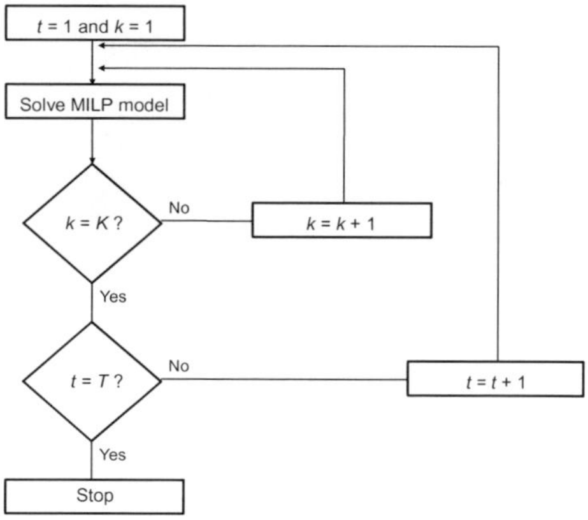

Fig. 22.1 Solution method to solve the waste collection problem under a dynamic scenario

Index sets
$I = 0, 1, \ldots, n$: set of waste bins and the depot 0
Parameters
c: travelling cost per distance unit
r: selling price per kg of a recyclable material
Q: vehicle capacity in kg
M: maximum route length in km
d_{ij}: distance between node i and node j
S_i: amount of waste in kg at bin i (calculated using the information given by the sensor (in m^3) and the material density (in kg/m^3))
a_i: expected daily accumulation rate of bin i in kg
E_i: bin capacity in kg
Decision variables
x_{ij}: binary variable indicating if arc (i, j) is traversed by a vehicle $(i, j \in I)$
y_i: binary variable indicating if node i is visited $(i \in I \setminus \{0\})$
u_i: auxiliary variable to prevent subtour $(i \in I \setminus \{0\})$
Model

$$\max \mathrm{P} = r \sum_{i \in I \setminus \{0\}} \sum_{j \in I \setminus \{0\}} y_i S_i - c \sum_{i \in I} \sum_{j \in I, (j \neq i)} x_{ij} d_{ij} \tag{22.1}$$

s.t.

$$\sum_{i \in I} \sum_{j \in I} x_{ij} d_{ij} \leq M \tag{22.2}$$

$$\sum_{i \in I \setminus \{0\}} y_i S_i \leq Q \tag{22.3}$$

$$y_i = 1, \forall i \in I \setminus \{0\} \wedge S_i \geq E_i - a_i \tag{22.4}$$

$$\sum_{j \in I, (j \neq i)} x_{ij} = y_i, \forall i \in I \setminus \{0\} \tag{22.5}$$

$$\sum_{i \in I} x_{ij} - \sum_{i \in I} x_{ji} = 0, \forall j \in I \tag{22.6}$$

$$u_i - u_j + n \times x_{ij} \leq n - 1 \quad 1 \leq i \neq j \leq n \tag{22.7}$$

$$x_{ij}, y_i \in \{0, 1\}, \forall i, j \in I, i \neq j \tag{22.8}$$

The objective function (22.1) considered is the profit P, defined as the difference between the revenues from selling the waste collected and the transportation cost (which is considered as a linear function of the distance travelled).

Constraint (22.2) ensures that the route length does not exceed a maximum amount of kilometers (M). Constraint (22.3) ensures that the waste collected does not exceed vehicle capacity. Constraint (22.4) guarantees that the capacity of a bin is not surpassed by imposing a visit to a bin when it fills up on that day. With this constraint, the service level is set to 100%, where all bins are collected before they overflow. Constraint (22.5) links variables x with variables y; if node i is visited, the outgoing degree of node i must be equal to 1. Route continuity is ensured by constraint (22.6). Constraint (22.7) is the [13] constraint to prevent subtours. Variables' domain is given in constraints (22.8).

22.5 Application to a Real Case Study

Valorsul (Valorização e Tratamento de Resíduos Sólidos, S.A.) is responsible for the collection of waste at 14 municipalities in Portugal, and has a homogeneous vehicle fleet based only at one depot, located at the Western Waste Treatment Center (Centro de Tratamento de Resíduos do Oeste, Cadaval).

The Valorsul system performs three different types of collection: undifferentiated, selective of recyclable materials (glass, paper/cardboard and plastic/metal), and organic. The recyclable waste collection process is planned with the help of software which estimates the bins' filling level and, based on the forecasts, decides whether to perform a specific route in that specific day. The routes are pre-defined and their definition does not take into consideration the bin's estimated fill-level. The available information regarding these levels comes directly from the collection team. During the collection, the team registers the bins fill-level, classifying them as empty (0%), less than half (25%), half (50%), more than half (75%) or full (100%) – this data allows the software to forecast an approximated fill-level of a specific bin.

Regarding the recyclable material paper/cardboard, 26 different routes are defined and performed periodically by Valorsul. In this work, route number 3 was selected to be analyzed as it is representative of the global operation. According to Valorsul's data, this route was performed 12 times within a six-month period. A time horizon of 30 days was established to allow for a comparative analysis between the current situation ("blind collection") and the future proposal (installation of sensors to get real-time information about bin's fill-levels). Within 30 days, all 68 bins in this route were visited three times; a total of 400 km was travelled and a total of almost 6400 kg of waste was collected. As there was no real-time information on the bin's fill-levels, more than 75% of the collected bins had fill-levels less or equal to 50%, leading to alow average monthly waste collection ratio of 16 kg/km. Table 22.1 presents the key performance indicators values for the current situation at Valorsul, for the time horizon under analysis (30 days).

The weight collected was calculated using the bins' fill rates registered by the collection team, and that filling rate was translated into kilos, considering that each bin has a capacity of 2.5 m^3 and the paper/cardboard density at the bins is 30 kg/m^3. The density value was obtained given the filling rate registered by the team and the total weight collected for each route (all vehicles are weighed when they arrive at the depot). It is noticeable that the maximum option to register is 100% (full), and there is no option to register if the bin is overflowed. Therefore, a projection was made for the fill-level given the expected daily accumulation rate (calculated dividing the

Table 22.1 Current situation

KPI	Day 1	Day 13	Day 23	Total
Profit (€)	63.69	65.47	77.13	206.29
Weight (kg)	2100.00	2118.75	2175.00	6393.75
Distance (km)	135.81	135.81	129.50	401.12
Attended bins	68	68	68	–
Ratio (kg/km)	15.46	15.60	16.80	15.94

Table 22.2 Projected current situation

KPI	Day 1	Day 13	Day 23	Total
Profit (€)	63.69	103.27	109.58	276.55
Weight (kg)	2100.00	2516.68	2097.20	6713.91
Distance (km)	135.81	135.81	129.50	401.12
Attended bins	68	68	68	–
Ratio (kg/km)	15.46	18.53	19.43	16.74

total filling levels registered for each bin for the six-month period by the number of days of the period - 149 days) multiplied by the time interval between routes. This projection shows greater values for the total of weight collected and for the kg per km ratio (see Table 22.2). Such values enable us to compare Valorsul's current situation with the results obtained when applying the proposed solution approach, since the expected daily accumulation rate is going to be used. From the analysis of this projection, we notice that only one bin overflows at day 13 and day 23, which represents a service level offered by Valorsul of 99% (67 out of 68 bins were collected without overflowing).

To simulate a future scenario where sensors are installed inside the waste bins and the actual bin's fill-level is known by the operations manager, we apply the proposed solution method to Valorsul's data, considering the time horizon of 30 days. Two scenarios were studied: (1) service level equal to 100% - all bins have to be collected before overflow; and (2) service level equals to 99% - 1 out of 68 bins is allowed to overflow without being collected immediately.

The values for the parameters used in the solution method are presented at Table 22.3. Transportation cost c was given by Valorsul and includes fuel consumption, maintenance of the vehicle and drivers' wages. Regarding the revenues, the value paid by Sociedade Ponto Verde for each tonne of paper/cardboard collected and sorted by a waste collection company is 136€/ton. This value intends to cover the costs with collection and sorting operations; it is estimated that the collection operation represents about 70% of the total cost [16]. Therefore, since in this work only the collection activity has been considered, the selling price (parameter r) is adjusted to 95.2€/ton (70% × 136€/ton). The vehicle capacity Q in kg was determined considering the vehicle's volume (20 m^3, according to Valorsul data) and the

Table 22.3 Parameters values and sources

Parameters	Values	Source
c	1 €/km	Valorsul
r	0.0952 €/kg	SPV
Q	5000 kg	Valorsul
d_{ij}	Adjusted Euclidian distance	–
S_i	Team records for day 1	Valorsul
E_i	75 kg	Valorsul
M	150 km	Valorsul

density of the material inside the vehicle, after being compacted (250 kg/m^3, again according to Valorsul data). The distance d_{ij} was calculated using the Euclidian distance, adjusted by a correction factor (to adjust the straight-line distance to the actual road distance). The correction factor was set to 1.58 [18]. The amount of waste at bin i, S_i, simulates the values that the sensors will read and transmit every day. To apply the solution method, we use the bin's fill-levels registered by the team for day 1 and the accumulation rate a_i for the following days. The bin capacity in kilos, E_i, was calculated multiplying the volume of the bin by the density of the paper/cardboard inside the bins. Finally, the maximum length of a route, parameter M, is an estimated value given by Valorsul that corresponds to the maximum possible duration of a route (7 h).

The solution method was applied considering a single available vehicle each day, and information about the fill-levels for the 68 bins. For Scenario 1, Eq. (22.4) is activated to obtain a solution with a service level of 100% (all bins are collected before they overflow). For scenario 2, Eq. (22.4) is relaxed to obtain a solution with 99% of service level (only one bin is allowed to overflow). The solution method was implemented in GAMS 24.5 and solved with CPLEX Optimizer 12.3.0, on an Intel Xeon CPU X5680 @ 3.33 GHz. The CPU time was limited to one hour for each day.

Tables 22.4 and 22.5 show the results for the two scenarios. To guarantee a service level of 100%, a collection route must be performed every 6 days during the time horizon (on day 1, 7, 13, 19 and 25). However, for a level of 99% only three routes are required during 30 days (on day 1, 13 and 25 for scenario 2). Comparing to the current situation, on day 1, only 50 bins are collected instead of all 68. This implies an increase of 73% in profit (110.39 € vs. 63.69 €) and a reduction of 37% in the distance travelled (85.55 km vs. 135.81 km).

Moreover, as the current situation corresponds to a service level of 99%, it can be seen that the method proposed for Scenario 2 found a solution that globally increases significantly both the kg per km ratio (from 16.74 to 24.84 kg/km), as well as the profit (from 276.55 € to 390.54 €), meaning that the vehicles travel less and collect more waste. Looking specifically at day 13, where a route is performed on both the current and the scenario 2 cases, in the current situation all bins are collected while in scenario 2 only 65 bins are collected, which results in a profit increase of

Table 22.4 Scenario 1 – service level of 100%

KPI	Day 1	Day 7	Day 13	Day 19	Day 25	Total
Profit (€)	110.39	24.84	24.34	24.09	25.21	208.85
Weight (kg)	2062.50	1227.78	1218.21	1309.91	1213.68	7032.10
Distance (km)	85.55	91.82	91.39	100.36	90.09	459.19
Attended bins	50	61	61	63	59	–
Ratio (kg/km)	24.11	13.37	13.33	13.05	13.47	15.31

Table 22.5 Scenario 2 – service level of 99%

KPI	Day 1	Day 13	Day 25	Total
Profit (€)	110.39	139.90	140.26	390.54
Weight (kg)	2062.50	2493.55	2577.30	7133.36
Distance (km)	85.55	96.99	104.58	287.12
Attended bins	50	65	68	–
Ratio (kg/km)	24.11	25.71	24.64	24.84

35% while reducing the distance travelled by 29%. Besides this, the last collection day for both cases is different: day 23, on the Valorsul solution, and day 25, on the proposed approach. Delaying the route just for two days leads to the collection of the same number of bins (68), but collecting a higher amount of waste, consequently increasing the profit and the kg per km ratio.

One interesting situation that arises in both scenarios is that the model, when defining the routes, decides to collect bins that are not full (e.g. Fig. 22.2, bin 5 with a fill-level of 20%). This is because the route passes nearby such bins and the revenues of collecting them surpasses the costs of doing so. Note that from Fig. 22.2 it can also be seen that bin 15, in red, is an example of a bin that is almost half full but was not collected as it was not near enough to be worth to visit.

Figure 22.3 shows the values for the kg per km ratio for all scenarios analyzed. As mentioned, Valorsul provides a service level of 99%. Maintaining that service level, it is possible to improve the kg per km ratio by 48% if sensors are implemented and our solution method applied to decide which bins should be visited in each day and what should the visit sequence be (Projected Current Solution vs. Scenario 2). To increase the service level to 100%, Valorsul will have to perform more routes in the time horizon, meaning that more kilometers will be travelled. This situation implies a reduction in the kg per km ratio of 9% (Scenario 1 vs. Projected Current Solution).

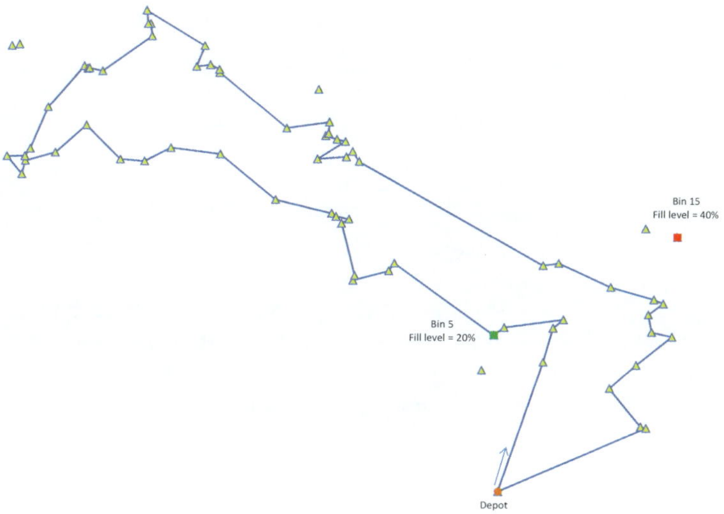

Fig. 22.2 Collection route of day 13 (scenario 2)

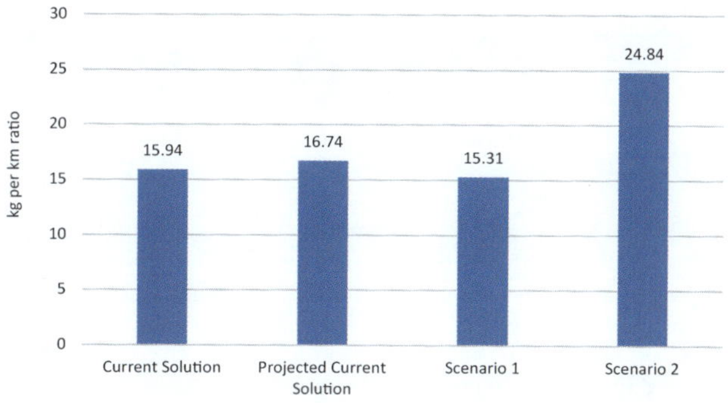

Fig. 22.3 Comparison of the kg per km ratio for the current situation and the scenarios studied

22.6 Conclusions

In this paper, a new paradigm for waste management is explored. Taking into consideration the advances on ICT, the uncertainty regarding the amount of waste available at each waste bin can be reduced by installing sensors capable of reading and transmitting the fill-level of each bin. This scenario will support dynamic routes definition, as opposed to a current situation where static routes are defined. In a dynamic context, the bins to be visited and the visit sequence are calculated for each day based on the information transmitted by the sensors. To calculate such routes and sequence, a solution method is developed based on real-time information scenario. This was applied to a real case-study and it's impact on profit, distance travelled and kg/km ratio was assessed. The results are promising, as within the service level actually provided, the use of sensors combined with the proposed solution method can improve the profits of the company in 48%.

As for future work, the solution method should be applied to a wider area (only 68 containers were analyzed) and also consider the other recyclable materials (glass and plastic/metal). The solution method is generic enough to be applied to the other recyclable materials; only the values of the parameters will change; for example, material density, selling price per kg of material and vehicle capacity have different values depending on the recyclable material.

Moreover, this study should be complemented with a cost-benefit analysis, taking into account the investment needed to install the sensors at the waste bins. Regarding the solution method, the maximization of profit for each day was considered as a main objective. However, maximizing profit for each day individually does not mean that profit will be maximized for a given time horizon. This issue also requires further study, and thus the improvement of the solution approach proposed should be pursued.

Finally, an important aspect related to the success of the analyzed routing definition system concerns the uncertainty associated with the fill-levels that may still exist on the sensors' information. This aspect opens a new research line on the treatment of uncertainty in these models that should also be explored.

References

1. D. Aksen, O. Kaya, F.S. Salman, Y. Akça, Selective and periodic inventory routing problem for waste vegetable oil collection. Opt. Lett. **6**(6), 1063–1080 (2012)
2. M. Allahviranloo, J.Y.J. Chow, W.W. Recker, Selective vehicle routing problems under uncertainty without recourse. Transp. Res. Part E: Logist. Transp. Rev. **62**, 68–88 (2014)
3. N. Aras, D. Aksen, M.T. Tekin, Selective multi-depot vehicle routing problem with pricing. Transp. Res. Part C: Emerg. Technol. **19**(5), 866–884 (2011)
4. C. Archetti, A. Hertz, M.G. Speranza, Metaheuristics for the team orienteering problem. J. Heuristics **13**(1), 49–76 (2007)
5. C. Archetti, M.G. Speranza, D. Vigo, Vehicle routing problems with profits **18**, 273 (2014)
6. S. Boussier, D. Feillet, M. Gendreau, An exact algorithm for team orienteering problems. 4OR Q. J. Oper. Res. **5**(3), 211–230 (2007)
7. S.E. Butt, T.M. Cavalier, A heuristic for the multiple tour maximum collection problem. Comput. Oper. Res. **21**(1), 101–111 (1994)
8. I.-M. Chao, B.L. Golden, E.A. Wasil, The team orienteering problem. J. Oper. Res. **88**(3), 464–474 (1996)
9. M. Faccio, A. Persona, G. Zanin, Waste collection multi objective model with real time traceability data. Waste Manag. **31**(12), 2391–2405 (2011)
10. D.S. Gonçalves, Tecnologias de informação e comunicação para otimização da recolha de resíduos recicláveis. Master's thesis, Lisboa: ISCTE (2014)
11. T. Ilhan, S.M.R. Iravani, M.S. Daskin, The orienteering problem with stochastic profits. IIE Trans. **40**(4), 406–421 (2008)
12. L. Ke, C. Archetti, Z. Feng, Ants can solve the team orienteering problem. Comput. Ind. Eng. **54**(3), 648–665 (2008)
13. C.E. Miller, A.W. Tucker, R.A. Zemlin, Integer programming formulation of traveling salesman problems. J. ACM (JACM) **7**(4), 326–329 (1960)

14. T. Nuortio, J. Kytöjoki, H. Niska, O. Bräysy, Improved route planning and scheduling of waste collection and transport. Expert Syst. Appl. **30**(2), 223–232 (2006)
15. M. Poggi, H. Viana, E. Uchoa, The team orienteering problem: Formulations and branch-cut and price. in *OASIcs-OpenAccess Series in Informatics*, vol. 14. Schloss Dagstuhl-Leibniz-Zentrum fuer Informatik (2010)
16. T.R.P. Ramos, M.I. Gomes, A.P. Barbosa-Póvoa, Assessing and improving management practices when planning packaging waste collection systems. Resour. Conserv. Recycl. **85**, 116–129 (2014)
17. F.G. Şahinyazan, B.Y. Kara, M.R. Taner, Selective vehicle routing for a mobile blood donation system. Eur. J. Oper. Res. **245**(1), 22–34 (2015)
18. R.F.R. Silva, The multi-compartment vehicle routing problem in the collection of recyclable municipal solid waste. Master's thesis, Lisboa: ISCTE (2016)
19. W. Souffriau, P. Vansteenwegen, G.V. Berghe, D. Van Oudheusden, A path relinking approach for the team orienteering problem. Comput. Oper. Res. **37**(11), 1853–1859 (2010)
20. C.A. Valle, L.C. Martinez, A.S. Da Cunha, G.R. Mateus, Heuristic and exact algorithms for a min-max selective vehicle routing problem. Comput. Oper. Res. **38**(7), 1054–1065 (2011)
21. P. Vansteenwegen, W. Souffriau, D. Van Oudheusden, The orienteering problem: a survey. Eur. J. Oper. Res. **209**(1), 1–10 (2011)
22. P. Vansteenwegen, D. Van Oudheusden, The mobile tourist guide: an or opportunity. OR Insight **20**(3), 21–27 (2007)

Chapter 23
Cargo Stability in the Container Loading Problem - State-of-the-Art and Future Research Directions

António G. Ramos and José Fernando Oliveira

Abstract The purpose of this paper is to present the current understanding and conceptualization of the cargo stability constraint within the context of the Container Loading Problem. This problem is highly relevant in the transportation industry due to the increasing pressure for a more economically, environmentally and socially efficient and sustainable cargo transportation. Stability is one the most important practical relevant constraints in the Container Loading Problem due to its strong influence on the cargo arrangement. Stability is usually divided into stability during loading operations (static) and stability during transportation (dynamic). Two main contributions are made. Firstly, an overview of recent developments in the literature on the two types of stability, static and dynamic, is provided. Secondly, of opportunities for future research are identified.

Keywords Container loading · Static stability · Dynamic stability

23.1 Introduction

The Container Loading Problem (CLP) is a combinatorial optimisation problem, which belongs to the wider combinatorial optimisation class of Cutting and Packing problems. In the Cutting and Packing problems, given a set of large objects and a set of small items (both defined in a number of geometric dimensions), the small items must be assigned to the large objects and a dimensional objective function is optimised satisfying two geometric conditions: all small items lie entirely within the large objects and the small items do not overlap [29].

The general definition of the CLP can be particularised for the three dimensional (3D) Cutting and Packing problem where a set of parallelepiped shaped boxes

A. G. Ramos (✉)
INESC TEC and CIDEM, School of Engineering, Polytechnic of Porto, Porto, Portugal
e-mail: agr@isep.ipp.pt

J. F. Oliveira
INESC TEC and Faculty of Engineering, University of Porto, Porto, Portugal
e-mail: jfo@fe.up.pt

A. I. F. Vaz et al. (eds.), *Operational Research*, Springer Proceedings
in Mathematics & Statistics 223, https://doi.org/10.1007/978-3-319-71583-4_23

(regular small items) must be packed orthogonally in a set of parallelepiped shaped containers (regular large objects), in a way that the boxes do not overlap and all the boxes of the subset lie entirely within the container. As an assignment problem it can have two basic objectives, the output value maximisation (knapsack problem) and the input value minimisation (bin packing problem). The first one refers to problems where the number of containers is not sufficient to accommodate all the boxes. The latter refers to problems where the number of containers is sufficient to accommodate all the boxes [29].

The CLP has received a lot of attention in literature due to the large spectrum of industrial applications that these problems address, in particular in the logistics and cargo transportation industry. However, the formulation of the CLP, in its essence a geometric assignment problem, is by itself of limited applicability for the industry, if a number of practical relevant constraints with a strong influence in the cargo arrangements are not taken into account when tackling the CLP.

In a review of the CLP literature, [7] presented a scheme to categorise practical constraints. The proposed categories are: container-related constraints (weight limit, weight distribution) box-related constraints (loading priorities, orientation, stacking) cargo-related constraints (complete shipments, allocation); positioning constraints; and load-related constraints (stability, complexity).

Of these practical relevant constraints stability is considered by some authors as the most important one [3, 21, 23], and received in recent years an increase of interest from researchers [7]. Even though it has been addressed in the literature, it as been done in an over-simplified way that does not actually translate real-world stability, limiting its applicability in the transportation industry.

This paper presents the current understanding and conceptualization of the cargo stability constraint within the context of the CLP, identifying research gaps and opportunities for future research.

The paper is organised as follows: in the Sect. 23.2 stability is framed within the CLP provides an over view of the literature concerning static and dynamic stability. Section 23.3 presents conclusions and opportunities for further research.

23.2 Cargo Stability in the Container Loading Problem

This article focuses on the cargo stability constraint in the CLP. It is therefore of the utmost importance to establish the meaning of what is considered as cargo stability and an evaluation of the approaches to stability in the CLP literature.

It is considered that cargo stability is the ability of each box to maintain the loading position without significant change during cargo loading and transportation operations. Stability during loading operations is usually addressed in the literature as static or vertical stability while stability during the transport operation is usually addressed in the literature as dynamic or horizontal stability [7].

The paper of [7] on CLP constraints was used as the starting point for the literature review on cargo stability in the CLP. Reference [7] reviewed CLP articles that focused on CLP and practical constraints, which were published or made available on-line in leading academic journals, edited volumes, and conference proceedings between 1980 and the end of 2011.

The same criteria were used and the time frame extended to the end of 2016, but the search was restricted to publications dealing with 3D Cutting and Packing problems that included additional cargo stability related constraints, considering the typology proposed by [29] for the Cutting and Packing Problems. This means that not only 3D Container Loading, but also 3D Pallet Loading Problems were taken into account.

In recent years cargo stability constraint has had an increase of interest from literature. Despite the growing interest, it has been treated in a rather over-simplified way by the majority of the authors. Some authors consider that, since the objective of the CLP is to obtain the maximum use of the container volume, a high volume occupation level would naturally translate into a stable cargo. Other authors consider that any solution is physically stable through the use of additional supports or filler materials. However, they do not take into account in their algorithms the extra cost and volume disposal that this approach implies.

The way static stability constraint has been addressed has not changed significantly since it was proposed by [8]: it is treated as a hard constraint which imposes that each base of the box to be fully supported by other boxes or by the container floor. Dynamic stability is addressed by less than 20% of the papers examined. As static stability, dynamic stability is approached in an over-simplified way: it is treated as a soft constraint and evaluated through the lateral support of each box. The most common condition for considering that a box has sufficient horizontal support is being surrounded by other boxes or the lateral walls of the container from at least three sides. The criteria used to enforce static stability or evaluate dynamic stability can be deconstructed with a few examples. A pyramid of boxes or a wall of boxes, can have full base support or three-sided support but are clearly unstable.

Stability is usually addressed on average with two other constraints, being the orientation constraint present in more than 90% of those cases.

23.2.1 *Static Stability*

The approaches to static stability found in the CLP literature can be classified according to the type of stability constraint to be enforced: full base support, partial base support or static mechanical equilibrium. Both full base support and static mechanical equilibrium guarantee static stability, while partial support does not.

- **Full base support** requires the entire base of a box to be in contact with the base of the container or with the top surface of other boxes. As a result, no overhanging boxes are allowed. Examples can be found in [4, 6, 15, 17, 31].
- **Partial base support** requires either the entire base of a box to be in contact with the base of the container, or a pre-specified percentage of the area of the base of the box base to be in contact with the top surface of other boxes, thereby allowing

overhanging. As an example, [8] requires the contact area to fall in the range of 95–75%, while [9] require a minimum of 80%, [12, 14, 20, 28], 75%, [13], 70% and [19], 55%.

- **Static mechanical equilibrium** requires the entire base of a box to be in contact with the base of the container or,
  - the sum of external forces acting on the box is zero and;
  - the sum of torque exerted by the external forces acting on the box is zero.

  Examples can be found in [11, 25].

Another element that should be taken into account when addressing static stability is the feasibility of the physical packing sequence. The physical packing sequence is the sequence by which each box is placed inside the container in a specific location determined by a CLP algorithm [22] and it is closely related with static stability.

It must be observed that the sequence by which solutions are generated by CLP algorithms, that is, the sequence by which the algorithm fills the space, does not necessarily correspond to the actual loading sequence. As such, only when the CLP problem incorporates additional requirements (e.g. multi-drop situations), or the loading sequence is directly related with the nature of the problem, as when the CLP is combined with the VRP, is there the need to address the loading or unloading sequence [7].

Therefore, the problem of determining the actual packing sequence of boxes has been ignored in the majority of the CLP approaches. However, there are authors that make a distinction between placing sequence and physical packing sequence [4], and address the generation of the physical packing sequence, by proposing packing in vertical and horizontal layers [22]. Nevertheless, they do not provide any physical packing sequence algorithm [4, 22]. In algorithms that use a back-bottom-left criterion to select the location for inserting a box, and place one box at a time [11], the placing sequence can be a viable physical packing sequence. Within the context of the CLP [26] developed of a Physical Packing Sequence Algorithm, that, given a container loading arrangement, generates the actual sequence by which each box is placed inside the container, considering static stability and loading operations efficiency constraints. The proposed algorithm introduces a Static Stability Algorithm based on static mechanical equilibrium conditions and includes ergonomic factors that influence manual floor loading operations in the loading sequence strategy.

Most recent approaches to stability in the literature consider the concept of static stability as equivalent to enforcing full base support [2, 15, 31]. Consequently these authors developed both algorithms that enforce full base support and algorithms that do not have such requirement (Unsupported). The performance measure used by these algorithms is the percentage of volume loaded with or without full base support. As a result, the goal is not to obtain a loading arrangement that is statically stable but a loading arrangement where all boxes have full support.

The use of full base support as a stability constraint is very costly for the efficiency of a CLP algorithm, particularly in the strongly heterogeneous instances. Figure 23.1 presents the difference in percentage points between the average results of the CLP

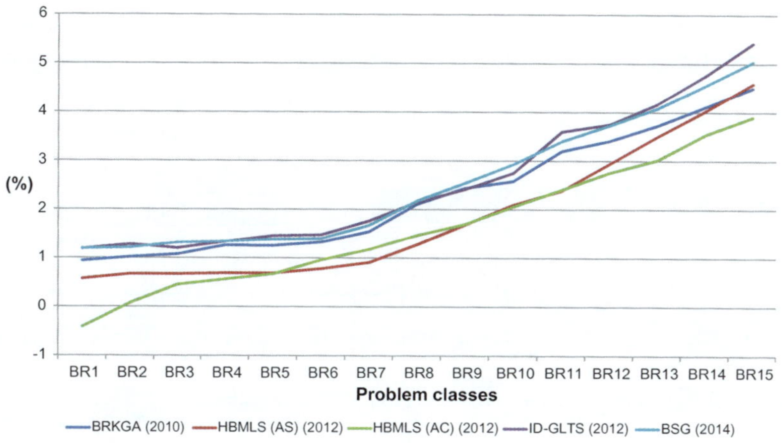

Fig. 23.1 Difference in percentage points between the results of the unsupported and static stability variants

algorithms solutions without considering static stability (Unsupported) and the CLP algorithms solutions considering static stability, for each of the 15 classes of problems proposed by [4, 10]. Classes BR1 to BR7 are considered to be weakly heterogeneous classes while BR8 to BR15 are considered to be strongly heterogeneous. In the Figure, BRKGA refers to the multi-population biased random-key genetic algorithm of [15], HBMLS refers to the block-loading heuristic based on multi-layer search of [30] (two versions are presented, the AS version that uses simple blocks and the AC version that uses composite bocks), ID-GLTS refers to the iterative-doubling greedy-lookahead algorithm of [31] and BSG refers to the Beam Search Approach of [2].

It can be observed that in each algorithm the difference follows a similar pattern, that is, for the weakly heterogeneous classes BR1 to BR7 the value of the difference is almost constant, while from classes BR8 to BR15, the strongly heterogeneous, the difference increases with the increase of the heterogeneity of the classes. The average difference of all algorithms for classes BR1 to BR7 is 1.03 *pp* and for classes BR8 to BR15 is 3.17 *pp*. In the most heterogeneous class BR15 the average difference is above 4 *pp*.

The first authors to actually mention the static mechanical equilibrium conditions in the context of the CLP are [11]. However they have only formulated the constraints and did not provide an algorithm to enforce them. Other authors, enforce stability conditions that are related with the static mechanical equilibrium conditions applied to rigid bodies. Reference [18] enforce the centre of gravity condition which requires the centre of gravity of a box be located above the contact surface of its supporting boxes. However, by itself, the centre of gravity condition does not guarantee static stability. Reference [19] stability condition requires the centre of gravity of a box to be located above the contact surface of its supporting boxes and the supported base of each box, calculated in relation to its supporting boxes and the container floor, to be greater than a predetermined value. Figure 23.2 illustrates these conditions. While

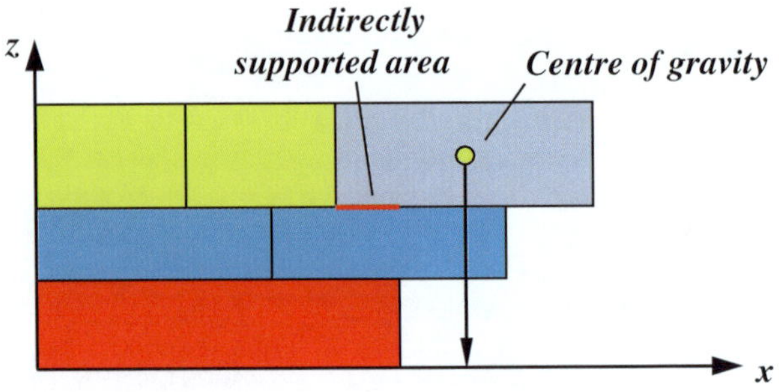

Fig. 23.2 Reference [19] centre of gravity condition example

the centre of gravity is directly above a supporting box, only a small part of its base is supported indirectly by the container floor and therefore it is considered not stable. Even though this condition avoids the generation of some unstable arrangements that could be generated if only the centre of gravity condition was enforced, it also prevents the possibility of generating some stable ones.

More recently, [25] incorporated a static stability constraint based on the static mechanical equilibrium conditions applied to rigid bodies, into a constructive heuristic placement procedure, used as part of a CLP algorithm. The proposed algorithm, a multi-population biased random-key genetic algorithm combines a genetic algorithm, responsible for the evolution of coded solutions (chromosomes), and a constructive heuristic algorithm responsible for decoding the chromosome, generating a solution and evaluating its fitness. The new static stability criterion is used to evaluate static stability during the filling of the container. The reported computational results show that, on average, the algorithm variant with the new stability criterion achieves a higher percentage of space utilisation than the variant with the classical full base support condition, while fully guaranteeing the static stability of the cargo.

Table 23.1 summarises the best results of existing CLP approaches with stability constraints enforcement. In the table Parallel_HYB.XL refers to the parallel hybrid local search algorithm of [19]. Presented in the table header are the spatial representation, the box arrangement strategy and the stability criterion used by each of the algorithms. The values from columns 2 to 7 of the table correspond to the average percentage of volume utilisation for the test instances of [4, 10] organised in 15 classes.

In the table, Parallel_HYB.XL refers to the parallel hybrid local search algorithm of [19], BRKGA refers to the multi-population biased random-key genetic algorithm of [15], HBMLS refers to the block-loading heuristic based on multi-layer search of [30] (two versions are presented, the AS version that uses simple blocks and the AC version that uses composite bocks), ID-GLTS refers to the iterative-doubling greedy-lookahead algorithm of [31], BSG-CLP refers to the Beam Search Approach of [2], CLA-SS refers to the multi-population biased random-key genetic algorithm with static stability of [25] (two versions are presented, the W version for weakly heterogeneous instances and the S version for strongly heterogeneous instances). Presented in the table header is the stability criterion used by each of the algorithms

(PBS - Partial Base Support; FBS - Full Base Support; SME - Static Mechanical Equilibrium). The values from columns 2 to 7 of the table correspond to the average percentage of volume utilisation for the test instances of [4, 10] organised in 15 classes, with a total of 100 instances per class.

An analysis of the results presented in Table 23.1 show that the CLA-SS(W) with the SME variant, had the best average performance for BR1 to BR8 classes of instances, while the CLA-SS(S) version with the SME variant outperformed all approaches for BR9 to BR15 classes of instances. Even thought the full base support constraint is the one that is mostly used to address static stability it is outperformed by the Static Mechanical Equilibrium approach.

23.2.1.1 Dynamic Stability

Dynamic stability has also been tackled in a simplified way in the CLP literature and only a handful of authors have addressed it. The majority of these approaches address dynamic stability as a soft constraint, by including a set of rules in the algorithms that contribute to improve the results of a set of metrics that intend to represent the degree of dynamic stability of a solution. These metrics however are over-simplified and do not provide a realistic representation of dynamic stability.

While in the static stability constraint there is no significant difference between the CLP and the pallet loading problem, that is not the case with the dynamic stability constraint. The difference lies in the physical limits of the large item. In the container the walls have a high resistance to deformation and can provide an effective lateral support (metal sheets, or side curtains with side boards) that contribute to the dynamic stability of the cargo. The pallet is limited by stretch film that has as main goal to convert a set of boxes into a single rigid body, but has a smaller contribution to the dynamic stability of the cargo.

Reference [24] proposed a classification to the approaches found in the CLP literature according to the type of container wall: flexible or rigid wall. A flexible container wall is regarded as not providing lateral support for the boxes, in opposition to rigid walls.

In problems with flexible container walls, authors consider that in order to achieve dynamic stability, the possibility of generating arrangements that allow guillotine cuts must be avoided. Guillotine cuts are considered to contribute to an unstable pallet. As such, authors focus on having boxes interlocking each other to avoid guillotine cuts. Two different approaches to interlocking can be found in the literature. The first, by [3, 8], treat interlocking as a soft constraint and use a set of criteria to evaluate the dynamic stability of generated layouts, while the second proposed by [1] enforces the interlocking of the column stacks but does not measure it.

Two stability criteria to measure dynamic stability were proposed by [8]. One criterion states that each box must have its base in contact with at least two boxes, ignoring contact surface areas with less than a predetermined percentage value of the base of the box, and aims to evaluate the degree of interlocking of the pallet. Another criterion considers the problems related with the pallet guillotine section cutting in

Table 23.1 Comparison of the best existing algorithms with static stability constraints

	Parallel_ HYB.XL (2004)	BRKGA (2012)	HBMLS (AS) (2012)	HBMLS (AC) (2012)	ID-GLTS (2012)	BSG (2014)	CLA-SS(W) (2016)	CLA-SS(S) (2016)	CLA-SS(W) (2016)
Stability criterion	55% PBS	FBS	FBS	FBS	FBS	FBS	FBS	SME	SME
BR1	93.70	94.34	94.30	93.95	94.40	94.50	93.86	93.16	**94.79**
BR2	94.30	94.88	94.74	94.39	94.85	95.03	94.55	94.73	**95.40**
BR3	94.54	95.05	94.89	94.67	95.10	95.17	94.75	95.21	**95.55**
BR4	94.27	94.75	94.69	94.54	94.81	94.97	94.63	95.29	**95.51**
BR5	93.83	94.58	94.53	94.41	94.52	94.80	94.38	95.15	**95.43**
BR6	93.34	94.39	94.32	94.25	94.33	94.65	94.24	95.09	**95.31**
BR7	92.50	93.74	93.78	93.69	93.59	94.09	93.82	94.93	**95.14**
BR8	–	92.65	92.88	93.13	92.65	93.15	93.16	94.69	**94.78**
BR9	–	91.90	92.07	92.54	92.11	92.53	92.62	**94.51**	94.45
BR10	–	91.28	91.28	92.02	91.60	92.04	92.09	**94.07**	93.95
BR11	–	90.39	90.48	91.45	90.54	91.40	91.56	**93.68**	93.38
BR12	–	89.81	89.65	90.91	90.35	90.92	91.28	**93.23**	92.61
BR13	–	89.27	88.75	90.43	89.69	90.51	90.93	**92.59**	91.64
BR14	–	88.57	87.81	89.80	89.07	89.93	90.38	**91.68**	90.71
BR15	–	87.96	86.94	89.24	88.36	89.33	90.08	**90.58**	90.10
Mean (BR 1–7)	93.78	94.53	94.46	94.27	94.51	94.74	94.32	94.79	**95.30**
Mean (BR 8–15)	–	90.23	89.98	91.19	90.55	91.22	91.51	**93.13**	92.70
Mean (BR 1–15)	–	92.24	92.07	92.63	92.40	92.87	92.82	93.91	**93.92**

The best values appear in bold

Table 23.2 Reference acceleration values (in g) for different transport modes [16]

Mode of transport		Forwards	Backwards	Sideways	Downwards
Road		0.8	0.5	0.5	1
Railway		0.5	0.5	0.5	1
Sea	Sea area A	0.3		0.5	1
	Sea area B	0.3		0.7	1
	Sea area C	0.4		0.8	1

the vertical direction. The criterion states that the guillotine cut must not exceed a predetermined percentage value of the maximum length or width of the stack. Reference [3] replaced the first criterion by another where each box positioned on the perimeter of the pattern has to be supported by at least two boxes in the layer below, ignoring contact surfaces areas with less than a predetermined percentage value of the base of the box.

In problems with rigid container walls, all the authors treat dynamic stability as a soft constraint. They consider that it is the interlocking and/or the limited lateral movement of boxes that contribute to dynamic stability. To evaluate interlocking, [5] presented two metrics. The first one is the average number of supporting boxes for each box that is not positioned on the container floor. The higher the value the better. The second is similar to the first but does not consider contact areas with less than 5% of the base area of a box. To evaluate the lateral movement [5] proposed the percentage of boxes that does not have at least three of its four lateral sides in contact with another box or with the container walls. The smaller the value the better. The first interlocking metric (M1) and the lateral movement metric (M2) are the metrics used in the CLP to evaluate dynamic stability.

Computational results for the M1 metric have been published by five authors and for the M2 metric by nine. All use in the computational tests the 700 problems proposed by [4]. Reference [24] analysed the results and observed that the algorithm with the best average percentage of occupied volume does not have the best average values for either M1 or M2. This observation is very significant since in CLP literature it is frequently assumed that there is a strong positive correlation between stability of the cargo and the volume occupation of the container [7].

If all the forces that cargo must withstand during transportation have to be taken into consideration, whether it is by road, railway or sea, the approach that has been followed is effectively over-simplified. Dynamic forces result from the handling and transport operations. During transportation, forces result from events like braking, turning or lifting in land transport or by rolling or vertical movements at sea. The measure of these dynamic forces is usually determined through the acceleration since these forces can be determined by the product of the mass of the cargo and its acceleration. The reference acceleration values for safety evaluation of cargo in different transport modes are presented in Table 23.2 and expressed as a product of the gravitational acceleration ($g = 9.81\,\text{m/s}^2$).

The impact of these forces on cargo can lead to different types of cargo movement. One of these movements is sliding. Sliding occurs when the force acting on a box overcomes the friction force generated between the box and its supporting surface. Another type of movement is tipping. Tipping occurs when a force acting on a box creates a rotational movement of the box. Sliding and tipping are the basic cargo movements. Other types of movement can occur in flexible loading arrangements. These are just a few examples that illustrate that stability is much more complex than the simplified approach generally used when considering the CLP.

Reference [24] developed new performance indicators for the evaluation of dynamic stability within the CLP, and new dynamic stability metrics for incorporation in container loading algorithms that translates real-world dynamic stability. The two performance indicators for dynamic stability evaluation where the number of fallen boxes and the number of boxes within the Damage Boundary Curve fragility test. The first one reflects the costly consequences of the fall of cargo during transportation, that is, the damaging of the cargo and the decrease of operations efficiency by the increase of the unloading time or by adding a quality inspection operation. The second performance indicator focuses on the evaluation of the damage that results from mechanical shock, whether as a result of impact between boxes, or between boxes and the container walls. The performance indicators of a cargo arrangement were measured using the physics simulation tool, StableCargo, developed to simulate the movement of a set of boxes inside a shipping container in road transport [27]. Due to the computational cost of using StableCargo within a container loading algorithm, two new dynamic stability metrics that model the performance indicators, were proposed. Derived through multiple linear regression analysis, these metrics were found more suitable to incorporate in a container loading algorithm than the existing metrics in the literature.

23.3 Discussion and Future Research

This review on the stability constraint in the CLP has highlighted the significative attention that has received from the research community. However, the majority of the proposed approaches still came short to be adopted in practice as cargo planning tools by transportation companies. The over simplified way stability constraints, which have a strong influence on cargo arrangements, have been modelled contributes to it. In an effort to overcome this issue, recent efforts have been develop to bring down the gap between research and practice using different tools and modelling approaches to stability constraints closer to real world stability conditions.

The static mechanical equilibrium approach is still in its infancy and more efficient algorithms can be devolved to enforce these conditions.

An important flaw of stability in the CLP is the lack of approaches to address dynamic stability as a hard constraint. With the growing number of cargo safety regulations the stability of the cargo during transportation is a pressing issue for the transportation industry.

The hybridisation of optimisation, that is, combining two or more techniques in an algorithm that contains the positive features of all of them, has also proven to be an effective way to tackle a complex problem, which grows on complexity every time a constraint is added or they are modelled closer to their real world characteristics.

Another relevant issue that needs to be addressed, and which is not exclusive of stability constraints, is the need for new formulations for the problem and its constraints. The majority of the approaches to the CLP uses heuristic methods that do not necessary require a formulation of the problem.

Other research opportunities lie on the integrating of stability with other different practical relevant constraints in the CLP. Constraints in the CLP are usually addressed in a reduced number and do not take into consideration the way they relate to each other.

Tackling these research opportunities would be a step forward in providing solutions that can be effectively used in practice.

Acknowledgements The research was partially supported by ERDF – European Regional Development Fund through the Operational Programme for Competitiveness and Internationalisation - COMPETE 2020 Programme within project "POCI-01-0145-FEDER-006961", and by National Funds through the Portuguese funding agency, FCT – Fundação para a Ciência e a Tecnologia as part of project "UID/EEA/50014/2013".

References

1. G. Abdou, M. Elmasry, 3D random stacking of weakly heterogeneous palletization problems. Int. J. Prod. Res. **37**(7), 1505–1524 (1999)
2. I. Araya, M.C. Riff, A beam search approach to the container loading problem. Comput. Oper. Res. **43**, 100–107 (2014)
3. E.E. Bischoff, Stability aspects of pallet loading. OR Spektrum **13**(4), 189–197 (1991)
4. E.E. Bischoff, M.S.W Ratcliff, Issues in the development of approaches to container loading. Omega **23**(4), 377–390 (1995)
5. E.E. Bischoff, F. Janetz, M.S.W. Ratcliff. Loading pallets with non-identical items. Eur. J. Oper. Res. **84**(3): 681–692 (1995)
6. A. Bortfeldt, H. Gehring, A hybrid genetic algorithm for the container loading problem. Eur. J. Oper. Res. **131**(1), 143–161 (2001)
7. A. Bortfeldt, G. Wäscher, Constraints in container loading-a state-of-the-art review. Eur. J. Oper. Res. **229**(1), 1–20 (2013)
8. H. Carpenter, W.B. Dowsland, Practical considerations of the pallet-loading problem. J. Oper. Res. Soc. **36**(6), 489–497 (1985)
9. S.G. Christensen, Container loading with multi-drop constraints. Int. Trans. Oper. Res. **16**(6), 727–743 (2009)
10. A.P. Davies, E.E. Bischoff, Weight distribution considerations in container loading. Eur. J. Oper. Res. **114**(3), 509–527 (1999)
11. J.L. de Castro Silva, N.Y. Soma, N. Maculan, A greedy search for the three-dimensional bin packing problem: the packing static stability case. Int. Trans. Oper. Res. **10**(2), 141–153 (2003)
12. G. Fuellerer, K.F. Doerner, R.F. Hartl, M. Iori, Metaheuristics for vehicle routing problems with three-dimensional loading constraints. Eur. J. Oper. Res. **201**(3), 751–759 (2010)
13. H. Gehring, A. Bortfeldt, A genetic algorithm for solving the container loading problem. Int. Trans. Oper. Res. **4**(5–6), 401–418 (1997)

14. M. Gendreau, M. Iori, G. Laporte, S. Martello, A Tabu search algorithm for a routing and container loading problem. Transp. Sci. **40**(3), 342–350 (2006)
15. J.F. Gonçalves, M.G. Resende, A parallel multi-population biased random-key genetic algorithm for a container loading problem. Comput. Oper. Res. **39**, 179–190 (2012)
16. IMO/ILO/UNECE, *IMO/ILO/UNECE Code of Practice for Packing of Cargo Transport Units*, vol. 4 (International Maritime Organization, 2014)
17. L. Junqueira, R. Morabito, On solving three-dimensional open-dimension rectangular packing problems. Eng. Optim. **49**(5), 733–745 (2017)
18. J.L. Lin, C.H. Chang, J.Y. Yang, A study of optimal system for multiple-constraint multiple-container packing problems, in *Advances in Applied Artificial Intelligence*, vol. 4031, Lecture Notes in Computer Science, ed. by M. Ali, R. Dapoigny (Springer, Berlin, 2006), pp. 1200–1210
19. D. Mack, A. Bortfeldt, H. Gehring, A parallel hybrid local search algorithm for the container loading problem. Int. Trans. Oper. Res. **11**, 511–533 (2004)
20. D. Männel, A. Bortfeldt, A hybrid algorithm for the vehicle routing problem with pickup and delivery and three-dimensional loading constraints. Eur. J. Oper. Res. **254**(3), 840–858 (2016)
21. A. Moura, J.F. Oliveira, A GRASP approach to the container-loading problem. IEEE Intell. Syst. **20**(4), 50–57 (2005)
22. B.K.A. Ngoi, M.L. Tay, E.S. Chua, Applying spatial representation techniques to the container packing problem. Int. J. Prod. Res. **32**(1), 111–123 (1994)
23. F. Parreño, R. Alvarez-Valdes, J.M. Tamarit, J.F. Oliveira, A maximal-space algorithm for the container loading problem. INFORMS J. Comput. **20**(3), 412–422 (2008)
24. A.G. Ramos, J.F. Oliveira, J.F. Gonçalves, M.P. Lopes, Dynamic stability metrics for the container loading problem. Transp. Res. Part C: Emerg. Technol. **60**, 480–497 (2015)
25. A.G. Ramos, J.F. Oliveira, J.F. Gonçalves, M.P. Lopes, A container loading algorithm with static mechanical equilibrium stability constraints. Transp. Res. Part B: Methodol. **91**, 565–581 (2016)
26. A.G. Ramos, J.F. Oliveira, M.P. Lopes, A physical packing sequence algorithm for the container loading problem with static mechanical equilibrium conditions. Int. Trans. Oper. Res. **23**(1–2), 215–238 (2016)
27. A.G. Ramos, J. Jacob, J.F. Justo, J.F. Oliveira, R. Rodrigues, A.M. Gomes, Cargo dynamic stability in the container loading problem - a physics simulation tool approach. Int. J. Simul. Process Model. **12**(1), 29–41 (2017)
28. C.D. Tarantilis, E.E. Zachariadis, C.T. Kiranoudis, A hybrid metaheuristic algorithm for the integrated vehicle routing and three-dimensional container-loading problem. IEEE Trans. Intell. Transp. Syst. **10**(2), 255–271 (2009)
29. G. Wäscher, H. Hauß ner, H. Schumann, An improved typology of cutting and packing problems. Eur. J. Oper. Res. **183**(3), 1109–1130 (2007)
30. D. Zhang, Y. Peng, S.C. Leung, A heuristic block-loading algorithm based on multi-layer search for the container loading problem. Comput. Oper. Res. **39**(10), 2267–2276 (2012)
31. W. Zhu, A. Lim, A new iterative-doubling Greedy–Lookahead algorithm for the single container loading problem. Eur. J. Oper. Res. **222**(3), 408–417 (2012)

Chapter 24
An Intercontinental Replenishment Problem: A Hybrid Approach

Elsa Silva, António G. Ramos, Manuel Lopes, Patrícia Magalhães and José Fernando Oliveira

Abstract This work addresses a case study in an intercontinental supply chain. The problem emerges in a company in Angola dedicated to the trade of consumable goods for construction building and industrial maintenance. The company in Angola sends the replenishment needs to a Portuguese company, which takes the decision of which products and in which quantities will be sent by shipping container to the company in Angola. The replenishment needs include the list of products that reached the corresponding reorder point. The decision of which products and in which quantity should take into consideration a set of practical constraints: the maximum weight of the cargo, the maximum volume the cargo and financial constraints related with the minimum value that guarantees the profitability of the business and a maximum value associated with shipping insurance. A 2-stage hybrid method is proposed. In the first stage, an integer linear programming model is used to select the products that maximise the sales potential. In the second stage, a Container Loading Algorithm is used to effectively pack the selected products in the shipping container ensuring the geometrical constraints, and safety constraints such as weight limit and stability. A new set of problem instances was generated with the 2DCPackGen problem generator, using as inputs the data collected in the company. Computational results for the

---

E. Silva (✉)
INESC TEC, Porto, Portugal
e-mail: emsilva@inesctec.pt

A. G. Ramos
INESC TEC and CIDEM, School of Engineering, Polytechnic of Porto, Porto, Portugal
e-mail: agr@isep.ipp.pt

M. Lopes
CIDEM, School of Engineering, Polytechnic of Porto, Porto, Portugal
e-mail: mpl@isep.ipp.pt

P. Magalhães
School of Engineering, Polytechnic of Porto, Porto, Portugal
e-mail: 111051@isep.ipp.pt

J. F. Oliveira
INESC TEC and Faculty of Engineering, University of Porto, Porto, Portugal
e-mail: jfo@fe.up.pt

A. I. F. Vaz et al. (eds.), *Operational Research*, Springer Proceedings
in Mathematics & Statistics 223, https://doi.org/10.1007/978-3-319-71583-4_24

algorithm are presented and discussed. Good results were obtained with the solution approach proposed, with an average occupation ratio of 92% of the container and an average gap of 4% for the solution of the integer linear programming model.

Keywords Inventory planning · Container loading · Load balance

24.1 Introduction

This work is based on a replenishment problem of an Angolan company based in Luanda, dedicated to the trade of products for the construction building and industrial maintenance sectors. Being one of the most expressive product categories in Portuguese exports to Angola, the commercial relations between organisations stimulates the creation of partnerships which share services and resources to boost trade.

In this context, and in logistic terms, the Angolan company created a partnership with a company in Portugal, to manage all purchases, preparation and shipping of merchandise to meet its supply needs. As the scope of business is wholesale and retail, 95% of purchases correspond to material imported from Europe and 5% represent purchases made locally in the Angolan market to meet special or urgent customer needs.

With increased operational difficulties in Angola, an improvement opportunity has been identified regarding inventory management, considering, not only the minimisation of replenishing cost, but also the impact of replenishing decisions on sales revenues. The case study presented in this work combines inventory management issues with intercontinental transport, where the main goal is to obtain an optimised replenishment solution that complies with all the specific constraints within the business context and transport mode.

In Sect. 24.2 the problem is formally defined, all the goals and constraints related to the business are presented. Section 24.3 is dedicated to the literature review related with management of inventory and resource level and container loading problems. The solution approach is presented in Sect. 24.4, the Integer Linear Programming model is defined and the Container Loading Algorithm considered for packing the products is also presented. In Sect. 24.5 the computational results obtained for the generated problem instances using the solution approach proposed are presented. Finally, in Sect. 24.6 the final conclusions are presented.

24.2 Problem Definition

The company focused in this study has two facilities in Angola, one retail store and one wholesale, that target consumers and businesses respectively.

In recent years the company has registered its greatest growth, however, with the increase of competition the market began to demand a higher service levels, a scenario propitious for the review of all internal processes and practises of the company, establishing new objectives and costumer service policies. The adoption of new supply chain management policies and models was the largest organisational restructuring achieved by the company, since previously the management of the supply chain depended on the feeling of the decision maker, on historical or on current demand for products.

The logistics process is one of the most complex processes that prevails among intercontinental trade. There are numerous and imponderable aspects related to bureaucracy, interests of shipping companies and the administrative and functional systems, which makes the logistics operation important in the replenishment process.

The inventory needs are issued by the Angolan company and sent to the purchasing centre that is based in Portugal, being a business partnership created for that purpose that manages the entire process of purchasing, and shipping of the needs. The warehouse in Angola stores the merchandise and replenishes the wholesale (businesses). It is also responsible for replenishing the store (customers).

The purchasing centre receives a list of products with replenishing needs, analyses the list and decides which products will be sent. The list of products is composed by the products that have reached the reorder point. The company establishes a biweekly cycle of shipment of merchandise (replenishing cycle). This management decision is based on the ability of the destination company to receive, verify, check and launch the merchandise for sale.

There is a transport capacity limitation and it is necessary to determine the volume and the weight of the products to evaluate if they meet the shipping constraints. The products are shipped in boxes with heterogeneous shapes. The maximum volume is a constraint directly related to the maximum capacity of the shipping container. The weight is also a constraint related to the type of shipping container, since the transport companies stipulate a maximum limit.

In this analysis another important element is the financing of the operation. A budget per merchandise shipping operation is defined by the company, considering the business activity and legal issues. On the one hand, a minimum value is defined to guarantee the profitability of the business. On the other hand, a maximum value is defined related to the insurance of the shipment, the maximum value that an insurance company is able to pay for the merchandise. These values, minimum and maximum, are defined by management and are guidelines that the purchasing process should follow. At this stage the problem arise, the analyst should take a decision considering the constant adjustment in the ideal quantities versus the need of replenishing.

Summing up, the replenishment process begins in Angola with the data collection from the computer system of the products that reached the reorder point. The company in Angola sends the list to the central purchasing office in Portugal that analyses and decides which products and in which quantities will be replenished. The central purchasing office analyses and manages the entire process, trying to ensure the availability of the merchandise in time for consolidation and shipping.

The merchandise is unloaded in Angola and received in the warehouse which, after the conference and launch in computer system, replenishes the store according to the needs.

The problem consists in determining the selection of products which will be replenished and in which quantities, considering financial constraints and the sales potential. Besides, it should be considered that only one shipping container will be sent and it is also required that the selected products geometrically fit within the container.

24.3 Related Work

This section is dedicated to review literature on inventory management related with the addressed replenishment problem and container loading problem.

Reference [3] present a review of the literature from 1989 to 2005 on Joint Replenishment Problem (JRP). Problems within the JRP use models to minimise total cost and satisfy demand. The total cost considers the production or preparation costs of products, transportation and shipping costs, inventory costs, taxes and insurance. In a multi-product problem, the decision is to determine which ideal quantities of a particular product should be bought, shipped or produced.

Reference [5] present a policy of joint replenishment of multiple products considering the sharing of resources (e.g. the same transport) with a view to improving and/or dividing fixed costs. It is a problem within the scope of the JRP aggregating the following factors: minimum quantities for each product, costs of preparation of production orders admitting minimum quantity and fixed costs of transportation. The case study considered depicts a company that sells animation products in Netherlands and Belgium and has a distribution centre in Netherlands. Production orders are issued to the manufacturer in China that in turn receives raw materials and components from other suppliers. This work presents the first approach considering joint replenishing with production orders and with minimum quantities per products in a given period. If the inventory is made by the replenishing quantities the system is controlled by the level of the product, using viable combinations for joint replenishment. The ordered quantities have to satisfy the minimum quantity for the product considering a fee for the level of service. If the replenishing quantity is determined by the transport, the transport capacity is potentiated against the previous scenario. The second approach considers joint replenishing without minimum order quantities, with intermediate stocks, shortening delivery times. For this, the company must have a supply contract with the manufacturer, reducing the production time to the minimum quantity allowed, and the system is controlled by a policy of maximum level of stock considering a certain level of service. The authors concluded that the use of intermediate stocks to relax the minimum order quantities facilitates the control of the supply chain and shortens the delivery time, achieving a 44% reduction of costs [5].

In [10] it is presented a system that supports the collection of merchandise in various geographically dispersed suppliers to the central warehouse where products are stored and finally, the distribution of the products to the retailers. The focus is on reducing inventory and transportation costs, since delivery (group of customers) or collection (group of suppliers) contains diverse merchandise. For a particular vehicle the suppliers are grouped, generating the collection orders per vehicle, and the products can not be divided by more vehicles. The focus is on the division of products into groups, vehicle collection routes and specification of replenishing quantities while minimising total shipping, vehicle routing, fixed orders and stock management costs. A branch-and-price algorithm is considered to solve small instances, however since the algorithm is not scalable, heuristics are used.

The container loading problem (CLP) can be interpreted as geometric assignment problem, in which three-dimensional boxes have to be packed into three-dimensional, rectangular containers, such that a given objective function is optimized and two basic geometric feasibility conditions hold, i.e. all boxes lie entirely within the container and the boxes do not overlap. The CLP has been extensively studied in the literature with 163 papers published until 2011 [1].

A state-of-the-art review work on constraints in CLP was proposed in [1], from a exhaustive analysis of the literature, the authors concluded that there is a lack of realistic representation of practical relevant constraints. An important contribution to the representation of practical constraints was given in [6–8], which consider stability constraints closer to reality, by separating static from dynamic stability, that is, stability of the cargo during loading operations from stability during transportation.

24.4 Intercontinental Replenishment Algorithm

24.4.1 Overview

The description of the Intercontinental Replenishment algorithm is presented. The proposed algorithm selects the products that will be replenished and proposes a feasible cargo arrangement of the products in the container, maximising the products sales potential and ensuring maximum weight limit of the cargo and cargo stability.

The products selected for replenishment are packed in boxes that will be afterwards packed in the container. Since the CLP is known as NP-hard, it is difficult to solve, and considering that besides packing boxes additional decisions have to be made, we propose a solution approach considering two steps. Firstly, an Integer Linear Programming (ILP) model is defined with the propose of selecting the products that will be replenished and in which quantities, considering as constraints the minimum and maximum financing of the company, the maximum volume that can be packed and the maximum weight with the objective of maximising the products sales potential. Secondly, a Container Loading Algorithm (CLA) is used to try to pack the selected products in the container considering that the boxes can not overlap and ensuring the stability of the cargo arrangement.

The ILP model considers only the volume of the boxes and not effectively the three physical dimensions of the boxes, which means that the CLA algorithm in some cases will not be able to pack all the selected products.

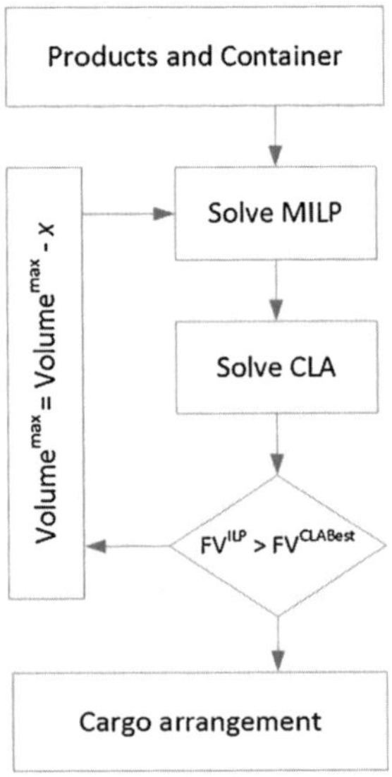

Fig. 24.1 Architecture of the intercontinental replenishment algorithm

In Fig. 24.1 the architecture of the algorithm is presented. The algorithm starts by gathering information about the products, parameters and container characteristics, then the ILP model is solved and a set of products is selected to be shipped. The CLA tries to find a feasible packing for the selected products. At each iteration the function value (FV) of the CLA is saved, the best value is update (FV^{best}) and compared with FV^{ILP} the current value of the solution obtained with ILP. If $FV^{ILP} > FV^{best}$, the volume constraint in the ILP model is updated by reducing the maximum volume a given x, in the computational experiments was considered 1%. The algorithm ends when the value of the sales potential of the solution obtained by the CLA is greater than the value of the sales potential obtained by ILP model with the maximum volume reduced.

24.4.2 *Integer Linear Programming Model*

The ILP model (24.2)–(24.8) is defined to maximise the sales potential of the products by selecting the products to replenish and the corresponding quantities ensuring the problem constraints. The objective function of the optimisation model considers a function called Function-Value (FV). This function is used to value products and differentiate them.

As the supply of replenishing needs is not immediate, it is important to know in advance the potential that an product represents in term of sales, in order to sustain the decision in the supply process.

Therefore, when a product reaches the reorder point, it is guaranteed that, on average, there will be sufficient stock to cover the demand until the next replenishment. In the case a product is not selected for shipping in that period, the FV informs the potential value of lost sales.

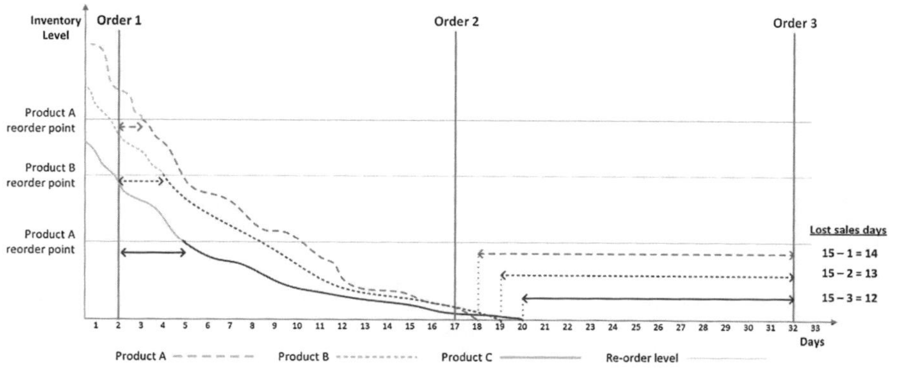

Fig. 24.2 Replenishment example

Since the period set to send products is an input parameter, the value resulting from the FV is always analysed until the next replenishing cycle.

A - Order delivery period (days)

RP_i - Number of days to reach the reorder point (days)

$\bar{d}_i$ - Average daily sales (€)

$$FV_i = (A - RP_i) \times \bar{d}_i \tag{24.1}$$

In Fig. 24.2 a replenishment example of the function FV is illustrated based on the stock levels of three different products. Considering the reorder date at day two, the reorder point of products A, B and C are at one, two and three days distance. Since the replenishment period is aligned with the delivery lead time, i.e., 15 days, the number of lost sales days $(A - RP_i)$ of products A, B and C are 14, 13 and 12, if the products are not replenished at order 1.

The ILP parameters and decision variables definition are presented below.

Parameters:

n - Total number of product types

FV_i - Value of the product type i

wgt_i - Weight of the item type i

QR_i - Maximum replenishment quantity of product type i

l_i - Length of product type i

w_i - Width of product type i

h_i - Height of product type i

v_i - Volume of product type i

F_i - Financing of product type i

P_{max} - Maximum weight admissible in the container

V_{max} - Maximum volume of the container

F_{min} - Minimum funding

F_{max} - Maximum funding

Decision variables:

x_i - Quantity to replenish of product $i, \forall i = 1, ..., n$

The ILP model (24.2)–(24.8) can be viewed as a knapsack formulation with additional constraints. The objective function (24.2) intends to maximise the sales potential. Constraints (24.3) ensures that the maximum weight of the container is not exceeded and, the maximum volume constraint is represented in (24.4), the range for the minimum and maximum value for the operation are ensured by constraint (24.5) and (24.6) and the maximum replenishment of each product is ensured by constraint (24.7).

$$\text{Maximize Z} = \sum_{i=1}^{n} FV_i \cdot x_i \tag{24.2}$$

$$\text{subject to:} \quad \sum_{i=1}^{n} wgt_i \cdot x_i \leq P_{max} \tag{24.3}$$

$$\sum_{i=1}^{n} v_i \cdot x_i \leq V_{max} \tag{24.4}$$

$$\sum_{i=1}^{n} F_i \cdot x_i \leq F_{max} \tag{24.5}$$

$$\sum_{i=1}^{n} F_i \cdot x_i \geq F_{min} \tag{24.6}$$

$$x_i \leq QR_i \qquad \forall i = 1, ..., n \tag{24.7}$$

$$x_i \geq 0 \text{ and integer } \forall i = 1, ..., n \tag{24.8}$$

24.4.3 Container Loading Algorithm

In accordance with the typology proposed by [11] for C&P problems, the CLP addressed in this paper can be characterised as the 3-Dimensional Single Knapsack Problem (3D-SKP), with static stability constraint.

The container loading algorithm (CLA) used within the Intercontinental Replenishment Algorithm is the multi-population biased random-key genetic algorithm with static stability proposed by [7]. This algorithm was chosen since it has reported excellent benchmarking results for the CLP test instances of [2], one of the main data sets used for the 3D-SKP problem.

The CLA is a multi-population biased random-key genetic algorithm that combines a genetic algorithm, responsible for the evolution of coded solutions (chromosomes), and a constructive heuristic algorithm responsible for decoding the chromosome, generating a solution and evaluating its fitness. This constructive heuristic uses a *maximal-space* representation for the management of empty spaces, and a layer building strategy to fill those *maximal-spaces*, i.e., identical boxes are grouped along two axes in order to fill the empty space. The maximal-spaces concept, was introduced by [4], in which empty space is described as a set of non-disjoint spaces, corresponding to the largest parallelepiped shapes that can be defined in that empty space. The CLA also determines the actual physical sequence by which each box can be actually loaded inside the container by incorporating the loading algorithm proposed by [8].

A detail description of the used CLA is presented in [7].

24.5 Computational Results

This section presents the results of the computational experiments run to evaluate the performance of the proposed algorithm.

The algorithm was coded in Visual C++ and run on an computer with 2 Intel Xeon CPU E5-2687W at 3.1 Ghz with 128 Gigabytes of RAM running the Windows 7 Pro 64 bit operating system and the ILP model was solved with IBM ILOG Cplex 16.3 solver.

The genetic algorithm parameters used in the algorithm are similar to the ones used in [7] (Table 24.1).

24.5.1 Test Instances

The problem tests used to evaluate the effectiveness of the algorithm were generated using an adaptation of 2DCPackGen problem generator proposed by [9]. In 2DCPackGen the test problem properties are controlled by a beta probability distribution, reflecting real-world characteristics. 2DCPackGen generates problem instances for the 2D and 3D Cutting and Packing problems, therefore it was adapted to include the other characteristics that are relevant for the problem considered in this case study.

The parallelepiped boxes dimensions were generated by, firstly, defining the size and shape of one of its facings, called base, and then the third dimension being

Table 24.1 Genetic algorithm parameters used in the CLA

Parameters	Values
Top	15%
Bottom	15%
Crossover probability	0.7
Population size	2000
Number of populations	2
Exchange between pop.	Every 15 generations
Fitness function	Maximize the % of packed container volume
Stopping criteria	After 1000 generations

Table 24.2 Ranges for the generation of problem instances

Parameter	Minimum value	Maximum value
n (SKU)	1000	1000
QR_i	1	30
V_i (cm^3)	1000	216 000
wgt_i (Kg)	1	20
F_i (€)	1	100
FV_i (€)	−100	500
P_{max} (Kg)	–	21 770
F_{min} (€)	40 000	–
F_{max} (€)	–	75 000
V_{max} (cm^3)	–	30 089 620

materialised was the height of the box. Sixteen different types of size and shape were considered for the dimensions of the items and the other parameters were generated considering the uniform distribution.

The ranges used for the generation of the different parameters that describe the studied problem are based on the experience and knowledge of the business activity. In Table 24.2 the ranges used in the generation of each problem parameter is defined. A total of sixteen test problems with one thousand boxes were generated.

24.5.1.1 Performance of the Algorithm

In Table 24.3 the computational results are presented and mainly the table is divided in three sections. The first six columns state the optimal solution obtained with the ILP model, the total number of different box types selected for replenishment (n), the total quantity of boxes sent to replenish ($\sum_i^n x_i$), the total weight (wgt_{total}), the total finance for the operation (F_{total}) and the value of the objective function (Z^{*FV}). The second section of the table (Best solution CLA), presents the results for the container loading problem, column (V_{max}) represents the value of the maximum volume considered in the ILP model (100% represents the total volume of the container), the percentage of volume of the container occupied by the packing arrangement obtained in the best solution of the CLA (V_{best}), the objective function value of the best solution of the CLA (Z_{best}^{FV}) and the total number of boxes packed in the container solution (n_{total}). The last section of the table is dedicated to the characterisation of the stopping iteration, the column (V_{ILP}) represents the maximum volume V_{max} considered in the constraint of the maximum volume in the ILP model that obtained a solution value smaller than the best solution for the container loading with the CLA and column (Z_{ILP}^{FV}) represents the value of the optimal solution of the corresponding ILP. The last two columns represent the percentage of volume occupied (V_{CLA}) and the value of the solution (Z_{CLA}^{FV}) obtained by the CLA in the stopping iteration.

Table 24.3 Computational results

Instance	ILP (V_{max} = 100%)					Best solution CLA				Stopping iteration			
	n	$\sum_i^n x_i$	wgt_{total}	F_{total}	Z^{*FV}	$V_{max}(\%)$	$V_{best}(\%)$	Z_{best}^{FV}	n_{total}	$V_{ILP}(\%)$	Z_{ILP}^{FV}	$V_{CLA}(\%)$	Z_{CLA}^{FV}
SRLO_1	165	2504	21768	75000	959581	94	93.1	932915	2421	91	932413	90.5	927619
SRLO_2	121	1839	20223	75000	732437	93	92.6	700181	1739	92	698604	89.9	678977
SRLO_3	146	2170	21769	74997	859883	92	91.1	820401	2067	90	820079	88.7	807196
SRLO_4	156	2280	21762	74991	896838	89	88.0	860560	2195	88	860560	88.0	860560
SRLO_5	130	1911	19827	74998	732272	98	94.0	698223	1767	91	696665	89.3	684462
SRLO_6	71	1195	12605	54703	465117	96	93.0	439194	1090	92	438466	90.9	433994
SRLO_7	101	1587	17888	74988	631147	95	92.6	599119	1476	91	598600	89.8	591553
SRLO_8	130	1824	20449	74999	679251	93	92.5	652572	1754	92	651260	91.0	644203
SRLO_9	141	2142	21764	75000	841740	92	91.3	802143	2041	89	795335	87.2	786130
SRLO_10	107	1668	18556	75000	667380	94	92.8	636809	1571	92	636799	88.0	597557
SRLO_11	123	1923	19987	74988	746460	93	89.5	702096	1780	89	701956	86.4	688521
SRLO_12	137	1989	19402	74986	755623	97	94.6	729230	1891	92	726553	89.7	708523
SRLO_13	155	2430	21759	74996	919404	92	90.6	883079	2336	89	879538	86.8	867789
SRLO_14	121	1788	20919	74999	698907	95	93.9	670338	1692	92	668483	89.1	646692
SRLO_15	134	2052	21747	74998	760006	91	90.9	730438	2008	90	727961	89.0	715851
SRLO_16	145	2189	21770	74999	867029	92	91.9	844840	2139	91	842507	91.0	842507

Fig. 24.3 SRLO_1 CLA solution representation

In the solutions obtained by the ILP model the constraint related with the volume has no slack and the constraint related with maximum value for funding is almost in all instances near the maximum allowed (75 000). The exception is problem instance number 6, and the reason is related with the geometric characteristics of the boxes of this instance, which were classified as big and square and for this reason has a smaller number of items to replenish in comparison with the other problem instances.

Regarding the best solutions obtained by the CLA, the average volume occupied in the container is 92% and the total number of items packed in the container is very high, the solution with smaller number has a total of 1090 boxes. In Fig. 24.3 it is represented the best CLA solution for problem instance SRLO_1 with an occupation of 93.1% of the volume of the container.

In what concerns the quality of the value of the best solution obtained for the container loading problem with the CLA, the average difference to the optimal solution value obtained by the ILP model is 4.3% with a minimum of 2.6% in problem instance 16 and a maximum of 5.9% in problem instance 11.

24.6 Conclusions

In this paper we proposed a new hybrid algorithm to deal with an intercontinental replenishment problem that takes into consideration transport constraints. The ILP model selects the products that will be replenished and the respective quantities with the aim of potentiating sales. The CLA algorithms deals with the problem of packing the products in the container to be shipped. The good performance of the proposed algorithm was demonstrated by the results obtained using a set of problem instances generated considering the data encountered in practice. The results obtained also highlight the benefits of hybrid approaches to address problems with business and geometric constraints.

Acknowledgements The first author was supported by FCT – Fundação para a Ciência e a Tecnologia within the grant SFRH/BPD/98981/2013. The research was partially supported by ERDF European Regional Development Fund through the Operational Programme for Competitiveness and Internationalisation - COMPETE 2020 Programme within project "POCI-01-0145-FEDER-006961", and by National Funds through the Portuguese funding agency, FCT – Fundação para a Ciência e a Tecnologia as part of project "UID/EEA/50014/2013".

References

1. A. Bortfeldt, G. Wäscher, Constraints in container loading a state-of-the-art review. Eur. J. Oper. Res. **229**(1), 1–20 (2013)
2. A. Paul Davies, E.E. Bischoff, Weight distribution considerations in container loading. Eur. J. Oper. Res. **114**(3), 509–527 (1999)
3. M. Khouja, S. Goyal, A review of the joint replenishment problem literature: 1989–2005. Eur. J. Oper. Res. **186**(1), 1–16 (2008)
4. K.K. Lai, J.W.M. Chan, Developing a simulated annealing algorithm for the cutting stock problem. Comput. Ind. Eng. **32**(1), 115–127 (1997)
5. E.P. Musalem, R. Dekker, Controlling inventories in a supply chain: a case study. Int. J. Product. Econ. **93**, 179–188 (2005)
6. A. Galrão Ramos, J.F. Oliveira, J.F. Gonçalves, M.P. Lopes, Dynamic stability metrics for the container loading problem. Transp. Res. Part C Emerg. Technol. **60**, 480–497 (2015)
7. A. Galrão Ramos, J.F. Oliveira, J.F. Gonçalves, M.P. Lopes, A container loading algorithm with static mechanical equilibrium stability constraints. Transp. Res. Part B Methodol. **91**, 565–581 (2016)
8. A. Galrão Ramos, J.F. Oliveira, M.P. Lopes, A physical packing sequence algorithm for the container loading problem with static mechanical equilibrium conditions. Int. Trans. Oper. Res. **23**(1–2), 215–238 (2016)
9. E. Silva, J.F. Oliveira, G. Wäscher, 2DCPackGen: a problem generator for two-dimensional rectangular cutting and packing problems. Eur. J. Oper. Res. **237**(3), 846–856 (2014)
10. S. Sindhuchao, H.E. Romeijn, E. Akçali, R. Boondiskulchok, An integrated inventory-routing system for multi-item joint replenishment with limited vehicle capacity. J. Glob. Optim. **32**(1), 93–118 (2005)
11. G. Wäscher, H. Haußner, H. Schumann, An improved typology of cutting and packing problems. Eur. J. Oper. Res. **183**(3), 1109–1130 (2007)

Chapter 25
Use of Analytic Hierarchy Process (AHP) to Support the Decision-Making About Destination of a Batch of Defective Products with Alternatives of Rework and Discard

João Cláudio Ferreira Soares, Anabela Pereira Tereso and Sérgio Dinis Teixeira Sousa

Abstract This study discusses the application of Analytic Hierarchy Process (AHP) to support the decision-making regarding the destination of a batch of defective products. The alternatives of destination are rework or discard. Six criteria of analysis and comparison were used. The mathematical development of the model was performed in Excel, which allowed several interactions and simulations, giving greater reliability to its application. The study was developed in a Brazilian plant of a Japanese auto parts industry which supplies a world-renowned Japanese motorcycle manufacturer. The defective product is the steering column of one of the models that presented the weld bead displaced from the correct position. From a flow of analysis of quality problems, the AHP method was adapted and applied in this case study, using evaluation questions to establish the criteria for comparison. The evidence generated by the problem analysis promotes answers and determination of criteria weights according to the influences of the answers on the cost and the quality of the product in case of rework or disposal. The AHP method assisted the systematization of the decision process, allowing the developed system to be used in other quality problems involving the destination of defective products. The contribution of this work is the adaptation of the AHP method to the application of problems of this type, using questions and answers (information already existent in the analysis of quality problems).

J. C. F. Soares (✉) · A. P. Tereso · S. D. T. Sousa
Centre ALGORITMI, University of Minho, Campus de Azurém,
4804-533 Guimarães, Portugal
e-mail: id6293@alunos.uminho.pt

A. P. Tereso
e-mail: anabelat@dps.uminho.pt

S. D. T. Sousa
e-mail: sds@dps.uminho.pt

A. I. F. Vaz et al. (eds.), *Operational Research*, Springer Proceedings
in Mathematics & Statistics 223, https://doi.org/10.1007/978-3-319-71583-4_25

In continuation of this specific application, the format can be adapted to the reality of other companies with inclusion or exclusion of criteria and weightings as necessary, justified, either by the characteristic of the problem or by internal policies. The applied method assisted in the decision to discard the parts of the study.

Keywords AHP · Cost · Decision-making · Quality

25.1 Introduction

Inspection points in manufacturing are designed to identify defective products and components to separate them from the planned quality flow, and to reach the next process or customer by promoting quality and lowest cost. Companies should find methodologies to systematically evaluate the defective items, and decide to rework or discard them, so that this decision promotes the lowest cost while fulfilling quality requirements. In this context, this work proposes to use the AHP method to assist in the decision of destination of defective products. The case study was developed from a Japanese auto parts industry with plant in the Polo Industrial of Manaus (PIM) – Brazil, which supplies a world-renowned Japanese motorcycle manufacturer. The defective product is the steering column of one of the models that presented the weld bead displaced from the correct position in 669 pieces in March 2016. The problem was visually detected by the quality control inspector. Gas Metal Arc Welding (GMAW) is the process to join the steering board and the steering column of the motorcycle. This welding process involves operators and equipment. The defect of weld bead displacement occurs due to failure of the welder that needs high expertise for handling the welding torch. The information was collected through the engineering manager and the criteria weighted in consensus with specialists in the areas of manufacturing, quality and engineering.

The work begins with a literature review on decision-making, AHP method and concepts involving quality and costs. There follows a chapter devoted to the development of the case study and finally the presentation of the results and the final considerations.

25.2 Review of Literature

This chapter presents a review of the literature of the following topics: Decision-making process; Analytic Hierarchy Process (AHP); and Quality and Costs. These are the central theoretical topics involved in this research, because it focuses on the use of these theories to present a practical and applicable solution in decision-making in relation to the destination of defective products.

25.2.1 Decision-Making Process

The decision-making process involves several aspects. Among them, the most important one is the understanding of the reality in which it is necessary to select a certain action, coming from the choice between the various alternatives constructed along the formulation of the problem, and the objectives of the decision. The choice for a certain alternative is not only to achieve the objectives, but to do it in the best way, in the context of each reality, considering the limitations and means available. This implies maximizing the resources available and the time taken to implement the actions [8, 23].

In decision-making, the available information is the subject of analysis for future definitions, formulations, considerations and conclusions by decision-makers. This information, in addition to being reliable in its acquisition methods, should also be free from partiality by means of construction and reading. Some information is constructed from data that alone does not specify anything, but when organized and compared generates information for decision processes, as is the case of control charts. Analyzing this aspect it is understood that the quality and level of the information contributes significantly to the decision process [18].

Qualitative information data is directly linked to subjective criteria, both in establishing the metrics and in its collection. They are usually used because of the difficulty in quantifying all the aspects. In many cases, the creation of weights for subjective criteria ends up making the assignment of a quantitative value even more complex [18].

Information and data are used in the decision-making process as evidence of the facts, since the ideal is that decision-making is based on facts and not hypotheses, assumptions and conjectures. This would make the process totally subjective, which, by scientific standards, is unacceptable. Considering this factual aspect, the reliability of an information or data is in its level of consistency with the facts, since they must express these as closely as possible. Thus, information must guarantee an acceptable level of reliability, to be considered as an active part of the decision-making process, otherwise, the decision-maker or team should choose to declassify the information or accept it under a restrictive condition. The analysis of the reliability of information and data is an important part of the decision making process, because if not observed it can significantly compromise the best choice and the achievement of the objectives. The factual analysis can be done by means of an evaluation to the mechanisms, methods, sources and agents of collection that must complement the files and reports. In case of inconsistencies, an analysis and on-site monitoring by specialists is indicated [18].

There is an imminent need in organizations to minimize subjectivity, providing professionals with capacity and experience that interfere subjectively in the decision-making processes, so that this capacity and experience contribute to the choice of the best alternatives [36].

In order to systematize the decision-making processes through the organization of information, maximization of resources and time spent in the analysis and decision stages, methods and models to support decision-making were developed. The advent of information systems, improved storage, processing and analysis of data, gave organizations greater ability to relate data in decision-making, streamlining considerations and conclusions. In spite of the existence of such advanced computer models, capable of defining the best choices by the decision maker, using mathematical formulations, as in the case of operational research, it should be emphasized that the premises in the formulation of the mathematical models are not exempt from subjectivity, since there is a previous decision for the prioritization and weight of the multiple criteria to be related. There is also a constant need to verify and check the operation of the models, as well as the analysis and comparison of the results provided [36].

25.2.2 Analytic Hierarchy Process (AHP)

The AHP (Analytic Hierarchy Process), developed by Thomas L. Saaty in the 1970, is one of the methods most used by managers, and its main objective is to assist in the structuring of a decision making process [22]. AHP starts by decomposing the decision problem into a hierarchy, normally composed by the goal, at the top, the criteria at the second level and the alternative at the final level. This allows to easily analyze sub-problems independently and reaching a final solution by putting everything back together. AHP is a method that is based on pairwise comparisons. The comparison of criteria and alternatives occurs with the attribution of an evaluation of each element in comparison to the others at the same level (two at a time), considering their impact on the element above them in the hierarchy, using judgment of teams and specialists. This comparison between pairs defines priorities, aiding in the systematization of complex decisions. With this, the AHP allows to consider a great amount of criteria and alternatives in the decision process. The criteria or attributes are the factors or variables selected by decision makers to assist in choosing the best alternative. Alternatives are the selected decision options [31].

The definition of the criteria and alternatives is something specific to the decision context and is influenced by the people involved in the process. Also, the comparison between pairs has a degree of subjectivity relative to the decision makers' perception of alternatives and criteria. Saaty Scale is normally used for pairwise comparisons, although other approaches are possible and currently used. The range of values proposed in this scale go from 1, meaning equal importance, to 9, meaning extreme importance, when comparing two items in relation to achieving a criteria or goal. In several AHP application studies, problems are structured (hierarchy of alternatives and criteria) and comparisons are performed by consensus teams, or by several experts, who combine their observations in the synthesis and decision evaluation phase [13, 16, 30]. The consistency ratio is a measure determined by AHP to evaluate the result of the comparisons. To improve consistency if necessary, decision makers should review some comparison, which can normally be achieved by using intermediate values from the Saaty Scale (1–9) as 2, 4, 6, and 8 [32].

Currently, several techniques and scales are used to assist AHP in the definition of weights, preferences, evaluation of alternatives and criteria such as Fuzzy Logic, TOPSIS, GRA, COPRAS, ELECTRE and VIKOR [1, 5, 11, 12, 19, 20, 26, 28–30, 32, 39]. The Fuzzy Logic is one of the most used and treats the subjective imprecision of the AHP extending the concept of false or true to deal with bias. Then, the fuzzy inference on the scale of 1–9 (Saaty) expresses the degree of imprecision of the judgment establishing lower and upper limits, it becomes extremely useful in the quantification of criteria expressly qualified by human sensitivity, that is, with very high degrees of subjectivity [4, 7, 21].

Decision histories that collect and retain factual information and information should be maintained. They should preferably be maintained in electronic media or computerized systems for later consultation, both to safeguard the decision of possible consequences and to assist other similar decisions [3, 6, 13, 24, 30].

The steps of structuring the decision-making process using the AHP will be described later in the presentation section of the case study. The criticisms to AHP are usually linked to the relation between evaluation and scale (random subjectivity). Each organization should seek to perform this connection based on tested or evident relationships with the risk of generating inconsistencies, or worse, that this relationship is so subjective that it invalidates the application of the decision-making process.

Inconsistencies in the comparisons can happen. For example, if A is considered 9 times more important than B, and B is 5 times more important than C, then A should be 27 times more important than C, but this is not possible because the scale finishes at 9 (nine). Therefore, managers should have a broad comparative view, since each comparison will also influence other comparisons [25, 41].

This method may require a large number of comparison matrices, in proportion to the number of alternatives and criteria, which makes the method difficult to apply. In these cases, it is interesting that computational resources are used.

There are several studies that relate the use of AHP to the quality area, namely: in the selection of improvement projects [43]; in the prioritization and classification of TQM implementation critical factors [40]; in the process of evaluation of priorities in service quality strategies [17]; in the evaluation and comparison of competitive differences in the quality of airlines [37]; in the identification of criteria for the suppliers selection process through product quality [14, 45]; and in the evaluation of the reliability of the collection and analysis of qualitative sensorial data of products' quality attributes [15].

25.2.3 *Quality and Costs*

Quality can be defined as the product's compliance with technical requirements and specifications, provided that these specifications are known to be desirable from the customer's perspective and actually functional to the product objectives [18].

Other perspectives of the term quality can be used considering different points of view, since the meaning of the word is not definitive being applied in several contexts. In relation to products and services, customer satisfaction is related to adequacy of use, fulfillment of requirements, cost-benefit relation, value added and differentials in relation to the competition. Customers evaluate based on a number of specific features such as dimension, color, durability, appearance, aesthetics, functions and safety, relating them to the price they are paying. Some clients have difficulty defining their own quality criteria which contributes to the complexity of the term definitions. However, they know how much a product satisfies them or meets their expectations, so companies could elicit customer feedback through specific satisfaction surveys [44].

In manufacturing, the focus is to plan, control and improve processes in order to reduce or eliminate errors that originate defective products, which is a deviation from a characteristic that causes customer dissatisfaction with financial consequences. In designing the quality strategy, the organization must consider the nature of its business, its size and its specific characteristics, identifying the differences between customer requirements and the current capabilities and limitations of its production system to meet the requirements. Mobilizing efforts and resources to promote customer satisfaction is directly associated with costs that can translate into profits when properly managed, and customers want to pay for it [18, 38, 44].

The concept that involves quality costs management has the objective of facilitating the classification of activities related to these costs, and through quality improvement programs contribute to reduction of such costs [9].

The most well-known and used model of quality costs is the PAF, usually employed in the categorization of Prevention, Appraisal, Internal and External Failure activities. Seeks to reduce the costs of internal and external failures with prevention and appraisal. The PAF supports that an optimal level of quality costs is achieved before reaching 100% compliance (Classic Model). This concept is contested by many authors that assume an optimal level of quality costs in the condition of zero defect (Modern Model) [2, 27, 33, 35].

Part of the quality costs includes expenditures for defective products, including rework, scrap, missed opportunity for not producing, shipping, non-selling, replacement under warranty and loss of sales and customers [9, 10, 34]. The rework consists in process a product again that has not reached the required quality level at the first time. So, the rework can be a problem or a solution of technical, financial and management dimensions. Rework or discard, the decision should promote the reduction of the cost of failure [9].

The theory of cost of quality does not highlight the reduction of the cost of the failure through the decision of destination of defective products. There are no studies with a theoretical and practical analysis of economic aspects and criteria involving various categories of costs, including opportunities lost due failures. Including the times of the activities and deadlines involved, respective financial consequences and the customer satisfaction, besides political and philosophical issues of the company that wants to include in this decision-making process.

25.3 Case Study

This case study proposes the application of the AHP to support decision-making regarding the disposal or rework of defective products aiming at reducing quality costs with internal failures (and fulfilling technical requirements) [31, 42]. Six decision criteria were used in the form of objective questions with "Yes" or "No" answers. The answers to the questions of the criteria are obtained from the evidence collected and verified in the technical analysis of the problem. Each criterion undergoes a change of importance (weight) according to the answer (yes or no) of the respective question. The product is the steering column of a motorcycle. The information was collected through the engineering manager and the criteria weighted in consensus with specialists in the areas of manufacturing, quality and engineering.

25.3.1 Problem Definition

The steering column is the tube at the top of the motorcycle chassis, where the front fork assembly of the suspension is attached. Its main function is to allow the driver of the motorcycle to turn the handlebar left or right, allowing directional control and facilitating its balance when the motorcycle is in motion. Figure 25.1 shows an exploded view of the complete steering column assembly and its attachment to the motorcycle chassis.

The defect found in the parts refers to the welding process, specifically the displacement of the contour of the weld bead, at the point of attachment of the tube column and the steering board as shown in Fig. 25.2 (left side).

The displacement of the weld bead occurs due to the displacement of the fork holes made in the boring machining process because the fixing of the steering column in the welding equipment occurs through the holes of the fork, which is the reference of the position of the part. The tolerance of ± 0.2 mm from the holes of the fork combined with the gap in the welding torch attachment contributes to the problem

Fig. 25.1 Complete steering column assembly

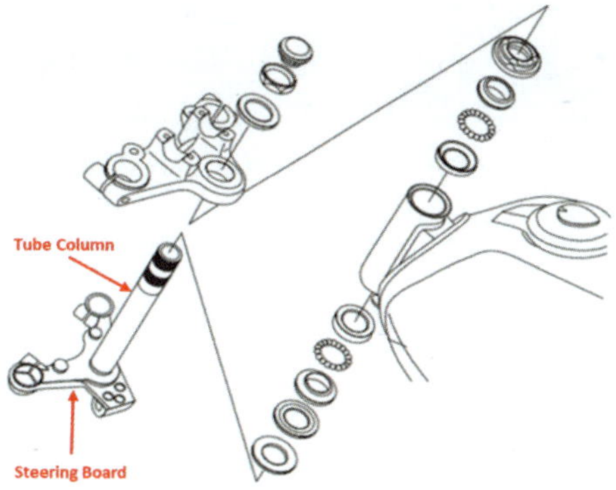

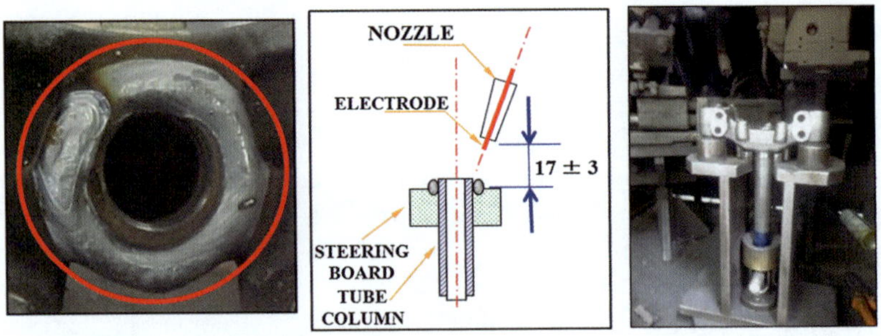

Fig. 25.2 Weld bead of the displaced steering column (left), diagram of the welding process with alignment by the fork holes (center and right)

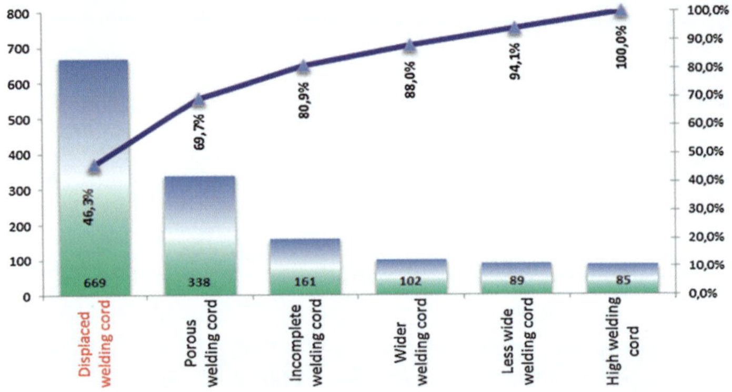

Fig. 25.3 Type of defects in the welding process of the steering column in March 2016

of the displaced weld bead. In March 2016, 3% of the welded parts presented the displaced cord problem, corresponding to 669 pieces according to the defect chart of Fig. 25.3. The unitary cost, the times and the reworking process indicated are shown in Table 25.4.

25.3.2 Definition of Decision Criteria

Figure 25.4 shows the proposed flow for the development of the analysis, from the detection of the problem to the evaluation of the costs with rework and disposal. The analysis is composed of six phases: detection, confirmation, clarification of the causes, impact on the client, proposal of solution of the problem and proposal of rework or discard. In the six phases, the analysis focuses on the quality and cost dimensions to define the decisions and evaluate the necessary resources and employees.

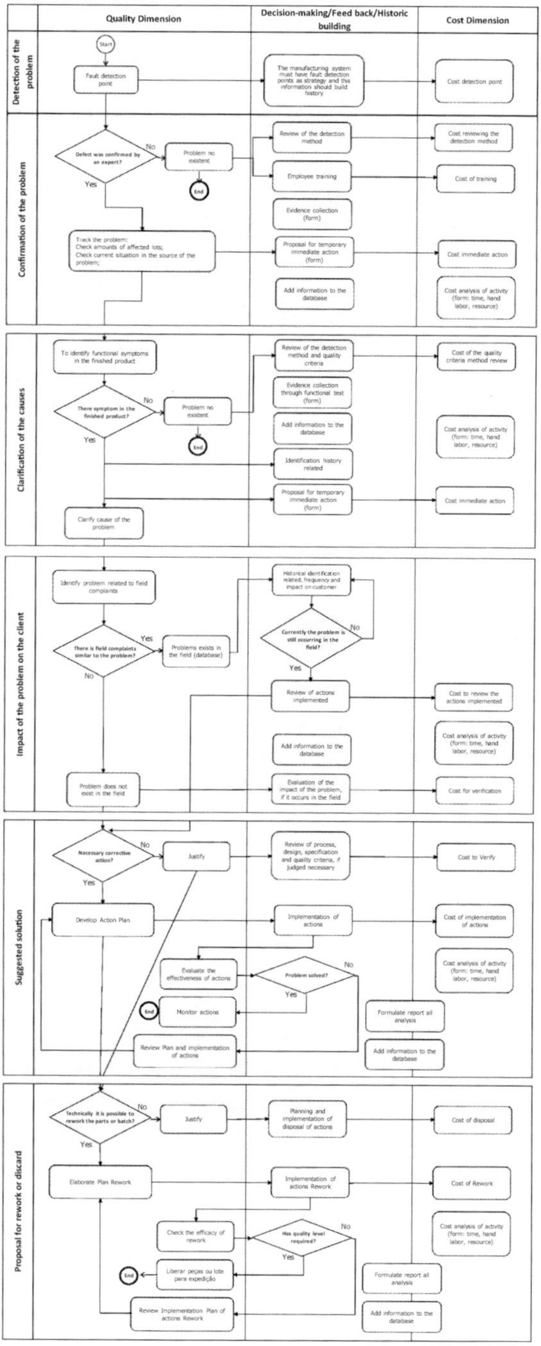

Fig. 25.4 Flow of defective parts

Table 25.1 Criteria with two possible conditions and respective tendencies

Attributes/Criteria	Responses of flow analysis	
	Yes	No
Problem solved?	Tendency to rework	Tendency to discard
History of occurrence in the final customer?	Tendency to discard	Tendency to rework
Is currently occurring in the final customer?	Tendency to discard	Tendency to rework
Rework plan approved?	Tendency to rework	Tendency to discard
Company has all the capabilities to rework?	Tendency to rework	Tendency to discard
Rework economically viable?	Tendency to rework	Tendency to discard

The decision criteria were defined based on the Analysis Flow of Fig. 25.4 and presented in Table 25.1. The analysis of the problem is composed of several steps and these suggest the questions in Table 25.1 that were answered through the facts observed and the evidence collected during the analysis. The analysis flow responses suggest trends in the choice of alternatives as presented in Table 25.1. (Tendency to Rework and Tendency to Disposal according to the "Yes" or "No" criterion.) The application of the AHP method was adapted to this problem, because criteria are questions that may have two answers: "yes" or "no". Each answer will score differently for the decision-making process as it will be shown.

The Criteria are in the form of objective questions that depend on the problem analysis to define the answer (yes or no), and these interfere differently (different weight) in the decision result. Follows a brief discussion of each Criterion and its relationship to the Analysis Flow of Fig. 25.4.

The Criterion "Problem Solved?" is related with the first five phases of the Analysis Flow and has its definition in the "Suggested solution" phase, when an Action Plan based on the root cause of the problem is verified.

The Criterion "History of occurrence in the final customer?" is related with "Impact of the problem on the client" phase of the Analysis Flow and has its definition in the question "There is field complaints similar to the problem?" within the Quality Dimension.

The Criterion "Is currently occurring in the final customer?" is related with "Impact of the problem on the client" phase of the Analysis Flow and has its definition in the question "Currently the problem is still occurring in the field?" within the Decision making/Feedback/Historic building Dimension.

The Criterion "Rework plan approved?" is related with the "Proposal for rework or discard" phase of the Analysis Flow and has its definition within the "Elaborate Plan Rework" of Quality Dimension.

The Criterion "Company has all the capabilities to rework?" is related in the "Proposal for rework or discard" phase of the Analysis Flow and has its definition

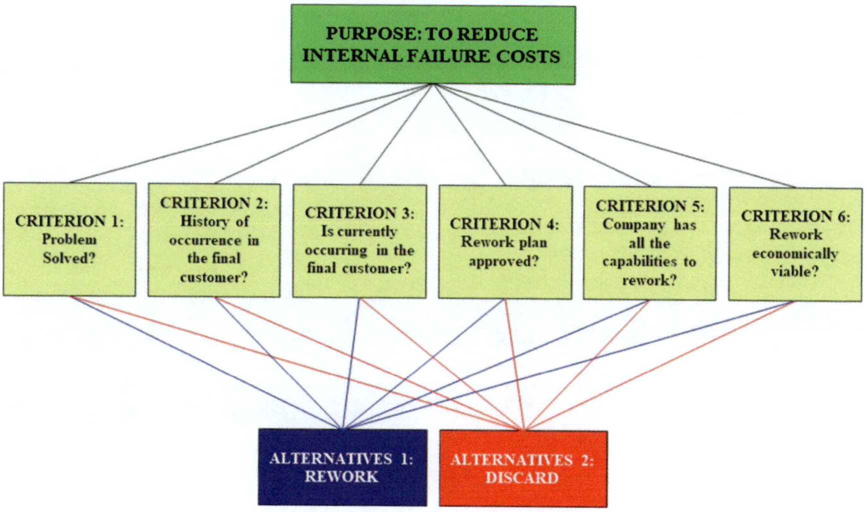

Fig. 25.5 Hierarchical problem structuring

within the "Elaborate Plan Rework" part, from the structure and materials adequate for the rework.

The Criterion "Rework economically viable?" is related in the "Proposal for rework or discard" phase of the Analysis Flow and has its definition from the comparison of the cost of rework and disposal shown in Cost Dimension.

These criteria have been defined considering that they are key points in the analysis of the defective products. However, other criteria can be identified and included in the evaluation.

Figure 25.5 presents the Hierarchical Structuring of the Problem in the form of the AHP Method with the objective in the first level, six decision criteria in the second level and two alternatives in the third level.

The Criteria are classified within the quality and cost dimensions. Criteria 1, 2, 3 and 4 are form the quality dimension. Criteria 5 and 6 are form the cost dimension.

25.3.3 Weight and Relationship of Criteria with Alternatives

The weight of each criterion was defined from Table 25.2 [31], in agreement with the specialists of the quality area of the company, according to each response of the Analysis Flow and its respective tendency for the alternatives, as presented in the Table 25.3. To meet the objectives of reducing quality costs with internal failures, six decision criteria were used in the form of objective questions with "Yes" or "No" answers. The answers to the criteria questions were obtained from the evidence collected and verified in the technical analysis of the problem (described in the

Table 25.2 Saaty fundamental scale – AHP

Intensity scale of importance - AHP		
Intensity scale of importance	Definition	Explanation
1	Equal importance	Two elements contribute equally to the objective
3	Weak importance of one over another	Experience and judgment moderately favor one element over another
5	Strong importance	Experience and judgment strongly favor one element over another
7	Very strong importance	One element is favored very strongly over another; its dominance is demonstrated in practice
9	Absolute importance	Evidence favors one activity over another, with the highest degree of certainty
2, 4, 6, and 8	Median of both neighboring judgments	When compromise is needed

Table 25.3 Weight of the criteria in the possibilities of answers "Yes" and "No"

Attributes/Criteria	Responses of flow analysis		Weight AHP	
	Yes	No	Yes	No
Problem solved?	Tendency to rework	Tendency to discard	2	9
History of occurrence in the final customer?	Tendency to discard	Tendency to rework	9	2
Is currently occurring in the final customer?	Tendency to discard	Tendency to rework	9	3
Rework plan approved?	Tendency to rework	Tendency to discard	4	9
Company has all the capabilities to rework?	Tendency to rework	Tendency to discard	5	9
Rework economically viable?	Tendency to rework	Tendency to discard	9	9

Analysis Flow of Fig. 25.2). The comparison of the alternatives with respect to each criterion undergo change of importance (weight) according to the answer (yes or no) of the respective criterion, as shown in the last two columns of Table 25.2. For the criterion "Solved Problem?", the answer "Yes" assigns weight of importance 2 (intermediate value - between equal importance and weak importance, by the Saaty Scale of Table 25.2) of the Alternative Rework in comparison to Alternative Discard. The answer "No" to this same criterion assigns weight of importance 9 (absolute importance by the Saaty Scale of Table 25.2) of the Alternative Discard in comparison to Alternative Rework. Only in the Criterion "Rework economically viable?" the specialists chose to allow the weights to remain the same in any of the answers (yes or no) and with absolute importance (9), considering the high influence of this criterion in reducing the costs of the defect and consequently in the decision that seeks to minimize them.

The objective of differentiating the weights of each response was to cover the specificities of these in the decision. This analysis was carried out jointly by a group of experts from the various areas involved in the problem.

There is an assumption in the application of this methodology using the AHP that must be defined previously: the rework must provide the required level of quality, otherwise, there is no reason to apply the method, because the decision should be discarded. If the rework provides the required level of quality, then the method must be applied to support the decision.

25.3.4 Hierarchy, Criteria Analysis and Weight Assignment for Alternatives

From the analysis of the problem according to the flow presented in Fig. 25.4, we reached the results presented in Table 25.4, that aided in the evaluation of the assumed condition of the quality of the rework (Rework must provide the required level of quality) and Criterion 6 ("Rework economically viable?"). As seen in Table 25.4, there are two possible types of rework for the problem. The detailed analysis of the development of the AHP Method is presented in the paper from the comparison of the Discard Alternative with Rework Alternative (1) (Rework to remove the cord for new welding), because the rework (2) "Rework fill with welding" does not provide the required level of quality (presupposed) according to Table 25.4.

Table 25.4 Result of the evaluation of required quality level and cost of rework options

Activity	Welding production piece with cord displacement	(1) Rework to remove the cord for new welding	(2) Rework fill with welding
Time	28 s	83 s	16 s
Condition of cost	22.82 BRL	43.31 BRL	3.92 BRL
Visual inspection	Not satisfy quality	Satisfy quality	Satisfy quality
Rupture test	Satisfy maximum load	Satisfy maximum load	Satisfy maximum load
Test macrography	Not satisfy penetration	Satisfy penetration	Not satisfy penetration
Appraisal report	Necessary to rework or dispose of the part	High cost, bigger than to produce a new piece	Lack of penetration possible premature fatigue
		Not satisfy cost	Not satisfy quality

The difference between the two Rework alternatives is presented in the last line of Table 25.4. In the first Rework option, the cost is higher than to manufacture a new part and in the second option the welding macrography reveals that there is not adequate penetration of the welding which compromises the quality standards, which can lead to premature fatigue during the use of the motorcycle by the final customer. The question seems to be defined in this case study, because the rework alternatives do not promote lower cost and quality. The first by the economic aspect and the second by the quality aspect. However, comparing AHP analysis results with these real perceptions of a quality problem allows to propose improvements to the method where there are limitations and to develop new formulations.

Table 25.5 presents the whole hierarchy of the problem from the defect, objectives, dimensions and criteria. From the analysis of the evidences and observed facts, the answers were assigned and included in Table 25.5. This analysis was performed by a group of experts from the various areas involved in the problem. From an Excel spread sheet all results were automatically computed from the criteria responses (yes or no), and according to the systematization of the AHP method. Table 25.5 is an adaptation to the AHP method, because it compiles the comparison of the alternatives according to the criteria and adds two conditions for the same criterion (yes or no), and still assigns different weight to each of the conditions. To facilitate calculation in Excel the "yes" and "no" answers to the criteria (questions) are indicated by the numeral one (1: evidenced) or zero (0: not evidenced), according to the column "Answers in the Analysis Flow" of table. To the answer "Yes" or "No" of the Criterion, assign the weight of the respective answer (column "Weight AHP" in Table 25.5) to the alternative indicated in the tendency in Table 25.2. The alternative that did not obtain the tendency is assigned weight 1 (one) as shown in the "Alternatives Column" of Table 25.5. As the Criterion "Problem Solved?" shows, the answer "Yes" assigned the weight 2 (two) to Alternative Rework in comparison to Alternative Discard indicating its weak intermediate importance. This analysis was meticulously scaled to promote the result of lower cost and higher quality.

In this case, tendencies to Discard are found in the viability of rework (Criterion 6), rework plan not approved considering complexity of rework (Criterion 4) and historical finding of the problem in the final customer (Criterion 2), with registration of non-fatal accident. The other criteria collaborate with rework.

25.3.5 Construction of the Preference Matrices of the Alternatives for Each Criterion

The matrices presented in Table 25.6 are the result of the previous analysis in which the alternatives were compared for each criterion in an individualized way. These were developed in Excel spreadsheet automatically from the answers of the criteria of Table 25.6, considering the respective weights of the relation. This stage was developed considering the systematization of the AHP method including the responses

Table 25.5 Problem hierarchy and assignment of analysis flow responses

Problem: weld bead of the steering column moved

Goal	Dimension	Attributes/Criteria	Responses of flow analysis		Weight AHP		Alternatives	
			Yes	No	Yes	No	Rework (1)	Discard
Reduce the cost of quality, mainly with internal and external flaws (depending on the external impact of rework in the field)	Quality	Problem solved?	1	0	2	9	2	1
		History of occurrence in the final customer?	1	0	9	2	1	9
		Is currently occurring in the final customer?	0	1	9	3	3	1
		Rework plan approved?	0	1	4	9	1	9
	Cost	Company has all the capabilities to rework?	1	0	5	9	5	1
		Rework economically viable?	0	1	9	9	1	9

Table 25.6 Matrices of preference of the alternatives for each criterion

Preference for Criterion 1	Result of the analysis		Preference for Criterion 2	Result of the analysis	
Problem solved?	Yes	No	History of occurrence in the final customer?	Yes	No
	1	0		1	0
C1	Rework	Discard	C2	Rework	Discard
Rework	1	2	Rework	1	1/9
Discard	1/2	1	Discard	9	1
Preference for Criterion 3	Result of the analysis		Preference for Criterion 4	Result of the analysis	
Is currently occurring in the final customer?	Yes	No	Rework plan approved?	Yes	No
	0	1		0	1
C3	Rework	Discard	C4	Rework	Discard
Rework	1	3	Rework	1	1/9
Discard	1/3	1	Discard	9	1
Preference for Criterion 5	Result of the analysis		Preference for Criterion 6	Result of the analysis	
Company has all the capabilities to rework?	Yes	No	Rework economically viable?	Yes	No
	1	0		0	1
C5	Rework	Discard	C6	Rework	Discard
Rework	1	5	Rework	1	1/9
Discard	1/5	1	Discard	9	1

of the results of the analysis (yes or no for each criterion). Numeral 1 (um) at the top of Table 25.6 indicates which response (Yes or No) the criterion obtained in the Analysis Flow. In the lower part, the respective comparisons of the alternatives with respect to the criterion are presented.

25.3.5.1 Normalization of Each Criterion

Table 25.7 shows the normalization of the weights assigned in the previous step. It is a kind of verification of correspondence between the grades given to alternatives within each criterion. Table 25.7 was obtained in Excel spreadsheet from the data of the previous tables. In this way, each score is divided by the sum of the scores in the respective column, the result of this sum must be equal to 1 as shown in Table 25.7.

25.3.5.2 Average of the Alternatives for Each Criterion

From the previous step that is in the fraction format, the matrices of the averages of the alternatives of each criterion are constructed in decimal format. The resulting averages for each alternative within each criterion is an index that represents the preference percentage of the alternative as presented in the criteria matrices of Table 25.8. Table 25.8 was obtained automatically in Excel spreadsheet from the data of the previous tables.

25.3.5.3 Definition of Preferences for Each Criterion

From the results found in the previous step, Table 25.9 presents a matrix with all averages of the criteria for each alternative. Then, the preferences of the decision makers for each alternative within each criterion is identified, expressed through the index relative to the percentage of preference. The sum of the indices within each criterion will be equal to 1, indicating 100%. Table 25.9 was obtained automatically in Excel spreadsheet from the data of the previous tables.

25.3.6 Comparison Between Criteria

The criteria were compared one by one from the Saaty Scale presented in Table 25.3. A group of experts from various areas involved in the problem assigned the comparative values as set forth in the matrix in Table 25.10. Each value expresses how much the row criterion is more important than the column criterion. When the same criterion is compared (row and column of the matrix) the assigned value must equal 1. Then, when the importance of the row criterion is smaller than that the column one, this inversion of the comparison is represented by the inverse fraction. Inconsistencies generated in this step are common, because there must be a corresponding relationship in all comparisons.

25.3.7 Normalization and Average of the Criteria

From the previous step, the value of the pairwise comparisons of the criteria is divided by the corresponding column sum and included in Table 25.11 in decimal format. The sum of the columns must be equal to 1 by the AHP method which indicates that the process was executed correctly. The objective of this step is to obtain the

Table 25.7 Normalization of criteria

Normalize the Criterion 1			Normalize the Criterion 2		
Problem solved?			History of occurrence in the final customer?		
C1 - Criterion 1	Rework	Discard	C2 - Criterion 2	Rework	Discard
Rework	1	2	Rework	1	1/9
	+	+		+	+
Discard	1/2	1	Discard	9	1
	=	=		=	=
	1 1/2	3		10	1 1/9
Normalization			Normalization		
Rework	2/3	2/3	Rework	1/10	1/10
	+	+		+	+
Discard	1/3	1/3	Discard	9/10	9/10
	=	=		=	=
	1	1		1	1
Normalize the Criterion 3			Normalize the Criterion 4		
Is currently occurring in the final customer?			Rework plan approved?		
C3 - Criterion 3	Rework	Discard	C4 - Criterion 4	Rework	Discard
Rework	1	3	Rework	1	1/9
	+	+		+	+
Discard	1/3	1	Discard	9	1
	=	=		=	=
	1 1/3	4		10	1 1/9
Normalization			Normalization		
Rework	3/4	3/4	Rework	1/10	1/10
	+	+		+	+
Discard	1/4	1/4	Discard	9/10	9/10
	=	=		=	=
	1	1		1	1
Normalize the Criterion 5			Normalize the Criterion 6		
Company has all the capabilities to rework?			Rework economically viable?		
C5 - Criterion 5	Rework	Discard	C6 - Criterion 6	Rework	Discard
Rework	1	5	Rework	1	1/9
	+	+		+	+
Discard	1/5	1	Discard	9	1
	=	=		=	=
	1 1/5	6		10	1 1/9
Normalization			Normalization		
Rework	5/6	5/6	Rework	1/10	1/10
	+	+		+	+
Discard	1/6	1/6	Discard	9/10	9/10
	=	=		=	=
	1	1		1	1

Table 25.8 Matrices of the averages of the alternatives for each criterion

Calculation of the average of the Criterion 1				Calculation of the average of the Criterion 2			
Problem solved?				History of occurrence in the final customer?			
C1- Criterion 1	Rework	Discard	Average	C2- Criterion 2	Rework	Discard	Average
Rework	0.667	0.667	0.667	Rework	0.100	0.100	0.100
Discard	0.333	0.333	0.333	Discard	0.900	0.900	0.900
Calculation of the average of the Criterion 3				Calculation of the average of the Criterion 4			
Is currently occurring in the final customer?				Rework plan approved?			
C3- Criterion 3	Rework	Discard	Average	C4- Criterion 4	Rework	Discard	Average
Rework	0.750	0.750	0.750	Rework	0.100	0.100	0.100
Discard	0.250	0.250	0.250	Discard	0.900	0.900	0.900
Calculation of the average of the Criterion 5				Calculation of the average of the Criterion 6			
Company has all the capabilities to rework?				Rework economically viable?			
C5- Criterion 5	Rework	Discard	Average	C6- Criterion 6	Rework	Discard	Average
Rework	0.833	0.833	0.833	Rework	0.100	0.100	0.100
Discard	0.167	0.167	0.167	Discard	0.900	0.900	0.900

Table 25.9 Averages of the alternatives for each criterion which is the array of preferences

Alternatives	Criteria					
	C1	C2	C3	C4	C5	C6
Rework	0.667	0.100	0.750	0.100	0.833	0.100
Discard	0.333	0.900	0.250	0.900	0.167	0.900

"Average of the Criteria" shown in the last column of Table 25.11. These values will be used to calculate the final result. Table 25.11 was obtained automatically in Excel spreadsheet from the data of the previous tables.

25.3.8 *Calculation to Obtain the Preference Index for the Alternatives*

This is the decision step for choosing the best alternative. Comparisons between criteria and between alternatives for each criterion were executed through the listed criteria. Using the matrix of preference in Table 25.9, a column with the comparative averages of the criteria, obtained in Table 25.11, is included on the right. Finally, the calculation of the final result of each alternative is done by the sum of the products of the averages, of the comparisons of the alternatives for each criterion, with the average of the comparisons between the criteria. This result is a decimal number that indicates the percentage of choice of the alternative, so the sum of the indexes of the alternatives must be equal to 1. Table 25.12 was obtained automatically in Excel spreadsheet worksheet considering the previous data of the tables.

Table 25.10 Matrix of comparison between the criteria

Criteria		C1	C2	C3	C4	C5	C6
		Problem Solved?	History of occurrence in the final customer?	Is currently occurring in the final customer?	Rework plan approved?	Company has all the capabilities to rework?	Rework economically viable?
C1	Problem solved?	1	4	2	2	2	1/5
C2	History of occurrence in the final customer?	1/4	1	1/4	1/4	1/4	1/5
C3	Is currently occurring in the final customer?	1/2	4	1	1	1	1/5
C4	Rework plan approved?	1/2	4	1	1	1	1/5
C5	Company has all the capabilities to rework?	1/2	4	1	1	1	1/5
C6	Rework economically viable?	5	5	5	5	5	1
SUM		7.75	22.00	10.25	10.25	10.25	2.00

Table 25.11 Normalization and average of the comparison between the criteria

	C1		C2		C3		C4		C5		C6		Average of the criteria
C1	0.129	+	0.182	+	0.195	+	0.195	+	0.195	+	0.100	=	0.166
C2	0.032	+	0.045	+	0.024	+	0.024	+	0.024	+	0.100	=	0.042
C3	0.065	+	0.182	+	0.098	+	0.098	+	0.098	+	0.100	=	0.107
C4	0.065	+	0.182	+	0.098	+	0.098	+	0.098	+	0.100	=	0.107
C5	0.065	+	0.182	+	0.098	+	0.098	+	0.098	+	0.100	=	0.107
C6	0.645	+	0.227	+	0.488	+	0.488	+	0.488	+	0.500	=	0.473
	=		=		=		=		=		=		=
Totals	1.000		1.000		1.000		1.000		1.000		1.000		1.000

Table 25.12 Indexes of preference of the alternatives from the averages of the alternatives by criterion and average of the comparison between the criteria

Alternatives	Criteria							Average of the criteria	Result
	C1	C2	C3	C4	C5	C6			
Rework	0.667	0.100	0.750	0.100	0.833	0.100	X	0.166	**0.34**
Discard	0.333	0.900	0.250	0.900	0.167	0.900		0.042	**0.66**
								0.107	1.00
								0.107	
								0.107	
								0.473	

The preference index of the alternatives, identified in the Result column of Table 25.12, was obtained through the calculations of the equations:

$$\begin{aligned}\text{Rework} &= (0.667 \times 0.166) + (0.100 \times 0.042) + (0.750 \times 0.107)\\ &\quad + (0.100 \times 0.107) + (0.833 \times 0.107) + (0.100 \times 0.473) = 0.34;\end{aligned}$$

and

$$\begin{aligned}\text{Discard} &= (0.333 \times 0.166) + (0.900 \times 0.042) + (0.250 \times 0.107)\\ &\quad + (0.900 \times 0.107) + (0.167 \times 0.107) + (0.900 \times 0.473) = 0.66.\end{aligned}$$

25.3.9 *Consistency Check*

In this step the consistency of the results was evaluated in comparison to the size of the matrix. This step is justified for matrices equal to or greater than 3 × 3, which have 3 criteria or more, as in this case study (6 criteria).

25.3.9.1 Totalizing the Entries of the Criteria

The totals of the criteria entries were obtained from the calculations shown in equation:

$$\begin{aligned}C1 &= (1 \times 0.166) + (4 \times 0.042) + (2 \times 0.107)\\ &\quad + (2 \times 0.107) + (2 \times 0.107) + (0.200 \times 0.473) = 1.0668,\end{aligned}$$

Table 25.13 Total of the entries from the comparison between the criteria and the average of the comparison between the criteria

Criteria	C1	C2	C3	C4	C5	C6	X	Average of the criteria	=	Totals
C1	1.000	4.000	2.000	2.000	2.000	0.200	X	0.166	=	1.0668
C2	0.250	1.000	0.250	0.250	0.250	0.200	X	0.042	=	0.2577
C3	0.500	4.000	1.000	1.000	1.000	0.200	X	0.107	=	0.6643
C4	0.500	4.000	1.000	1.000	1.000	0.200	X	0.107	=	0.6643
C5	0.500	4.000	1.000	1.000	1.000	0.200	X	0.107	=	0.6643
C6	5.000	5.000	5.000	5.000	5.000	1.000	X	0.473	=	3.1094

Table 25.14 Maximum eigenvalue from the totals of the entries and average of the comparison between the criteria

Calculation of the maximum eigenvalue (λmax)				
Totals		Average of the criteria		Result
1.0668	/	0.1660	=	6.4253
0.2577	/	0.0418	=	6.1637
0.6643	/	0.1065	=	6.2375
0.6643	/	0.1065	=	6.2375
0.6643	/	0.1065	=	6.2375
3.1094	/	0.4726	=	6.5788
		Sum	=	37.8803
		Average (λmax)		6.3134

which exemplifies the calculation of the total of C1 (Criterion 1) as shown in Table 25.13.

25.3.9.2 Calculation of the Maximum Eigenvalue (λmax)

The Maximum eigenvalue is obtained according to Table 25.14, where the totals of the criteria obtained in the previous step are divided by the average of the criteria. The average of the sum of the results from Table 25.14 is the Maximum eigenvalue (λmax). Table 25.14 was obtained automatically in Excel spreadsheet from the data of the previous tables.

Table 25.15 Random index according to the number of criteria

Random index (RI)

Dimension of the array	1	2	3	4	5	**6**	7	8	9	10	11	12	13	14	15
Random consistency	0.00	0.00	0.58	0.90	1.12	**1.24**	1.32	1.41	1.45	1.49	1.51	1.48	1.56	1.57	1.59

Table 25.16 Consistency result

Consistency index	CI	0.0627
Consistency ratio	CR	0.0445
Consistency		

25.3.9.3 Consistency Result

The Consistency Index (CI) is obtained by the equation: $CI = (\lambda max - n)/(n - 1)$, where n is the number of criteria. The Consistency Ratio (CR) was obtained from the division of the Consistency Index (CI) by the Random Index (RI) of Table 25.15 (Saaty 2008) based on the 6 criteria of the case study. The Consistency Result: $C = 0.0627/1.24 = 0.0445$ less than 0.10 is acceptable in the definitions of the AHP method. Small adjustments were required in the comparisons between criteria, weights 4 and 3 were changed to 3 and 2 respectively. Table 25.16 was obtained automatically in Excel spreadsheet from the study values.

25.4 Results and Discussion

Table 25.17 was obtained automatically from the data of the previous tables. It represents the main aspects of problem analysis: objective, criteria, criteria answers, criteria weights and respective contributions to the decision. As shown in the last line of Table 25.17, the AHP defined Discarding as the best option. The main aspects that influenced the result was that the Rework Alternative is not economically viable and the Rework Plan was not approved due to the complexity of the actions. These issues rightly influenced the results in the AHP. Opportunity costs and other criteria costs are not included in the comparative cost analysis of this case study, which can be done in future research.

In this case study, the decision obtained from the adapted AHP did not change, in other words, the application of the decision method maintained the same result of discard of the defective parts, according practiced before. However, the great contribution of the use of AHP in this case study is verified, since decisions are now based on expert-weighted criteria (decision together with shared knowledge) and influenced by quantitative and qualitative evidence of cost and analysis of quality.

Table 25.17 Final result of the application of the AHP method in the case of study considering the comparison of the alternative discard with the alternative (1) rework to remove the cord for new welding

Problem: weld bead of the steering column moved

Goal	Dimension	Attributes/criteria	Responses of flow analysis		Weight AHP		Alternatives	
			Yes	No	Yes	No	Rework (1)	Discard
Reduce the cost of quality, mainly with internal and external flaws (depending on the external impact of rework in the field)	Quality	Problem solved?	1	0	2	9	2	1
		History of occurrence in the final customer?	1	0	9	2	1	9
		Is currently occurring in the final customer?	0	1	9	3	3	1
		Rework plan approved?	0	1	4	9	1	9
	Cost	Company has all the capabilities to rework?	1	0	5	9	5	1
		Rework economically viable?	0	1	9	9	1	9
		Index/result					**0.34**	0.66
							Discard	

25.5 Conclusions and Considerations

From a flow of quality problem analysis, the AHP Method developed and applied in this case study, used six criteria in the form of objective questions, with answers of "Yes" or "No", with attribution of different weights according to the each answer. The evidence generated by the analysis of the problem promotes the answers of criteria and defines weights according to the influences of the answers on the cost and the quality of the product in case of rework or disposal. The AHP aided the decision to destination a batch of 669 defective parts. The mathematic development of the matrices and data were performed in Excel spreadsheet which provided several interactions and simulations of results giving greater reliability to the application. The welding process was automated so as not to depend on the skill of the welder avoiding the defect of displacement of the weld bead. The contribution of this work is the developing of questions which answers already exist in the analysis of quality problems, and uses AHP to decide the destination of defective products. The AHP Method assisted the systematization of the decision process. This model can be adapted to the reality of other companies with inclusion or exclusion of criteria and weightings as necessary. Future research can and should include other criteria, weights and comparisons, according to the need and specificity of the decision problem and the policies of each company. The company was practicing the decision to discard of the pieces based on philosophical criteria of quality that repudiate the practice of rework in a systematic way. Therefore, there was no change in the outcome of the discard decision that was practiced before for this problem. However, now the decision is based on qualitative and quantitative criteria involving the cost and quality dimensions duly weighed by experts and based on existing and collected evidence in the company. A rational decision based on an adapted methodology (AHP) that will allow other evaluations like this. This application presents consistent results that can and should be improved in other quality problems in the company and in other cases of study involving other companies that intend to test it.

References

1. M.P. Amiri, Project selection for oil-fields development by using the AHP and fuzzy TOPSIS methods. Expert Syst. Appl. **9**(37), 6218–6224 (2010). https://doi.org/10.1016/j.eswa.2010.02.103
2. T. Arabian, M. Jourabchi, Z. Leman, M. Ismail, A research on the impact of cost of quality models and reporting system on managing cost of quality. Int. Proc. Econ. Dev. Res. **36**(55), 178–183 (2013). https://doi.org/10.7763/IPEDR
3. A.V. Bentes, J. Carneiro, J.F. da Silva, H. Kimura, Multidimensional assessment of organizational performance: integrating BSC and AHP. J. Bus. Res. **12**(65), 1790–1799 (2012). https://doi.org/10.1016/j.jbusres.2011.10.039
4. E. Bulut, O. Duru, T. Keçeci, S. Yoshida, Use of consistency index, expert prioritization and direct numerical inputs for generic fuzzy-AHP modeling: a process model for shipping asset management. Expert Syst. Appl. **2**(32), 1911–1923 (2012). https://doi.org/10.1016/j.eswa.2011.08.056

5. G. Büyüközkan, G. Çifçi, A combined fuzzy AHP and fuzzy TOPSIS based strategic analysis of electronic service quality in healthcare industry. Expert Syst. Appl. **3**(39), 2341–2354 (2012). https://doi.org/10.1016/j.eswa.2011.08.061
6. U. Cebeci, Fuzzy AHP-based decision support system for selecting ERP systems in textile industry by using balanced scorecard. Expert Syst. Appl. **5**(36), 8900–8909 (2009). https://doi.org/10.1016/j.eswa.2008.11.046
7. D.-Y. Chang, Applications of the extent analysis method on fuzzy AHP. Eur. J. Oper. Res. **3**(95), 649–655 (1996). https://doi.org/10.1016/0377-2217(95)00300-2
8. N.N. Cheek, B. Schwartz, On the meaning and measurement of maximization. Judgm. Decis. Mak. **2**(11), 126–146 (2016). https://doi.org/10.1007/BF02722112
9. A. Chopra, D. Garg, A research on the impact of cost of quality models and reporting system on managing cost of quality. TQM J. **5**(23), 510–515 (2011). https://doi.org/10.1108/17542731111157617
10. P.B. Crosby, *Quality is Free: The Art of Making Quality Certain* (New American Library Edition, New York, 1979). https://doi.org/10.2172/1025774
11. M. Dağdeviren, S. Yavuz, N. Kilinç, Weapon selection using the AHP and TOPSIS methods under fuzzy environment. Expert Syst. Appl. **4**(36), 8143–8151 (2009). https://doi.org/10.1016/j.eswa.2008.10.016
12. M.C. Das, B. Sarkar, S. Ray, A framework to measure relative performance of Indian technical institutions using integrated fuzzy AHP and COPRAS methodology. Socio-Econ. Plan. Sci. **3**(46), 230–241 (2012). https://doi.org/10.1016/j.seps.2011.12.001
13. M.A.P. Davies, A multicriteria decision model application for managing group decisions. J. Oper. Res. Soc. **45**(1), 47–58 (1994), http://www.jstor.org/stable/2583950
14. F. Dweiri, S. Kumar, S.A. Khan, V. Jain, Designing an integrated AHP based decision support system for supplier selection in automotive industry. Expert Syst. Appl. **62**, 273–283 (2016). https://doi.org/10.1016/j.eswa.2016.06.030
15. F.S. Fogliatto, S.L. Albin, An AHP-based procedure for sensory data collection and analysis in quality and reliability applications. Food Qual. Prefer. **5–6**(14), 375–385 (2003). https://doi.org/10.1016/S0950-3293(03)00006-5
16. E. Forman, K. Peniwati, Aggregating individual judgments and priorities with the analytic hierarchy process. Eur. J. Oper. Res. **1**(108), 165–169 (1998). https://doi.org/10.1016/S0377-2217(97)00244-0
17. R. Gopalan, Sreekumar, B. Satpathy, Evaluation of retail service quality – a fuzzy AHP approach. Benchmarking Int. J. **6**(22), 1058–1080 (2015). https://doi.org/10.1108/BIJ-05-2013-0052
18. J.M. Juran, A.B. Godfrey, Juran's quality handbook. Deutsche medizinische Wochenschrift **1** (1946) (1999). https://doi.org/10.1055/s-0031-1280544
19. T. Kaya, C. Kahraman, An integrated fuzzy AHP-ELECTRE methodology for environmental impact assessment. Expert Syst. Appl. **7**(38), 8553–8562 (2011). https://doi.org/10.1016/j.eswa.2011.01.057
20. T. Kaya, C. Kahraman, Fuzzy multiple criteria forestry decision making based on an integrated VIKOR and AHP approach. Expert Syst. Appl. **6**(38), 7326–7333 (2011). https://doi.org/10.1016/j.eswa.2010.12.003
21. O. Kilincci, S.A. Onal, Fuzzy AHP approach for supplier selection in a washing machine company. Expert Syst. Appl. **8**(38), 9656–9664 (2011). https://doi.org/10.1016/j.eswa.2011.01.159
22. S. Kumar, O.S. Vaidya, Analytic hierarchy process: an overview of applications. Eur. J. Oper. Res. **1**(169), 1–29 (2006). https://doi.org/10.1016/j.ejor.2004.04.028
23. H. Mintzberg, D. Raisinghani, A. Thérêt, The structure of "unstructured" decision processes. Adm. Sci. Q. **2**(21), 246–275 (1976). https://doi.org/10.2307/2392045
24. E.W.T. Ngai, E.W.C. Chan, Evaluation of knowledge management tools using AHP. Expert Syst. Appl. **4**(29), 889–899 (2005). https://doi.org/10.1016/j.eswa.2005.06.025
25. I. Ognjanović, D. Gašević, E. Bagheri, A stratified framework for handling conditional preferences: an extension of the analytic hierarchy process. Expert Syst. Appl. **4**(40), 1094–1115 (2013). https://doi.org/10.1016/j.eswa.2012.08.026

26. S. Opricovic, G.H. Tzeng, Compromise solution by MCDM methods: a comparative analysis of VIKOR and TOPSIS. Eur. J. Oper. Res. **2**(156), 445–455 (2004). https://doi.org/10.1016/S0377-2217(03)00020-1
27. M. Plewa, G. Kaiser, E. Hartmann, Is quality still free?: empirical evidence on quality cost in modern manufacturing. Int. J. Qual. Reliab. Manag. **9**(33), 1270–1285 (2016). https://doi.org/10.1108/IJQRM-11-2014-0189
28. G.R. Pophali, A.B. Chelani, R.S. Dhodapkar, Optimal selection of full scale tannery effluent treatment alternative using integrated AHP and GRA approach. Expert Syst. Appl. **9**(38), 10889–10895 (2011). https://doi.org/10.1016/j.eswa.2011.02.129
29. R. Rostamzadeh, S. Sofian, Prioritizing effective 7ms to improve production systems performance using fuzzy AHP and fuzzy TOPSIS (case study). Expert Syst. Appl. **5**(38), 5166–5177 (2011). https://doi.org/10.1016/j.eswa.2010.045
30. R.D.F.S.M. Russo, R. Camanho, Criteria in AHP: a systematic review of literature. Procedia Comput. Sci. **55**, 1123–1132 (2015). https://doi.org/10.1016/j.procs.2015.07.081
31. T.L. Saaty, Decision making with the analytic hierarchy process. Int. J. Serv. Sci. **1**(1) (2008), http://www.colorado.edu/geography/leyk/geog_5113/readings/saaty_2008.pdf
32. T.L. Saaty, L.T. Tran, On the invalidity of fuzzifying numerical judgments in the analytic hierarchy process. Math. Comput. Model. **7–8**(46), 962–975 (2007). https://doi.org/10.1016/j.mcm.2007.03.022
33. A. Sailaja, P. Basak, K. Viswanadhan, Hidden costs of quality: measurement and analysis. Int. J. Manag. Value Supply Chains **2**(6), 13–25 (2015). https://doi.org/10.5121/ijmvsc.2015.6202
34. D. Sandoval-Chávez, M. Beruvides, Using opportunity costs to determine the cost of quality: a case study in a continuous-process industry. Eng. Econ. **2**(43), 107–124 (1998). https://doi.org/10.1080/00137919808903192
35. A. Schiffauerova, V. Thomson, A review of research on cost of quality models and best practices. Int. J. Qual. Reliab. Manag. **6**(23), 647–669 (2006). https://doi.org/10.1108/02656710610672470
36. H.A. Simon, *The New Science of Management Decision* (Prentice Hall PTR, Upper Saddle River, 1977)
37. A.K. Singh, Competitive service quality benchmarking in airline industry using AHP. Benchmarking: Int. J. **4**(23), 768–791 (2016). https://doi.org/10.1108/BIJ-05-2013-0061
38. J.C. Soares, S.D. Sousa, E. Nunes, Application of the three realities approach to customer complaints analysis in the motorcycles industry. Int. Conf. Ind. Eng. Oper. Manag. 1–10 (2012), http://hdl.handle.net/1822/21749
39. C.C. Sun, A performance evaluation model by integrating fuzzy AHP and fuzzy TOPSIS methods. Expert Syst. Appl. **12**(37), 7745–7754 (2010). https://doi.org/10.1016/j.eswa.2010.04.066
40. F. Talib, Z. Rahman, Identification and prioritization of barriers to total quality management implementation in service industry. TQM J. **5**(27), 591–615 (2015). https://doi.org/10.1108/TQM-11-2013-0122
41. R.R. Tan, K.B. Aviso, A.P. Huelgas, M.A.B. Promentilla, Fuzzy AHP approach to selection problems in process engineering involving quantitative and qualitative aspects. Process Saf. Environ. Prot. **5**(92), 467–475 (2014). https://doi.org/10.1016/j.psep.2013.11.005
42. J. Wallenius, J.S. Dyer, P.C. Fishburn, R.E. Steuer, S. Zionts, K. Deb, R.E. Steuer, Multiple criteria decision making, multiattribute utility theory: recent accomplishments and what lies ahead. Manag. Sci. **54**(7), 1–32 (2008). https://doi.org/10.1287/mnsc.1070.0838
43. H. Water, J. Vries, Choosing a quality improvement project using the analytic hierarchy process. Int. J. Qual. Reliab. Manag. **4**(23), 409–425 (2006). https://doi.org/10.1108/02656710610657602
44. A. Weckenmann, G. Akkasoglu, T. Werner, Quality management – history and trends. TQM J. **3**(23), 281–293 (2015), http://dx.doi.org/10.1108/TQM-11-2013-0125
45. X. Xi, Q. Qin, Product quality evaluation system based on AHP fuzzy comprehensive evaluation. J. Ind. Eng. Manag. **1**(6), 356–366 (2013). https://doi.org/10.3926/jiem.685

Chapter 26
Determinants of Nursing Homes Performance: The Case of Portuguese Santas Casas da Misericórdia

André S. Veloso, Clara Bento Vaz and Jorge Alves

Abstract This study aims to evaluate the economic efficiency of Nursing Homes owned by 96 Santas Casas da Misericórdia (SCM) and the determinants that influenced their efficiency in 2012 and 2013. The SCM are the oldest non-profit entities, which belong to Third Sector in Portugal, provide this social response and receive significant financial contributions annually from the state. The study is developed in two stages. In the first stage, the efficiency scores were calculated through the non-parametric DEA technique. In the second stage, Tobit regression is used to verify the effect of certain organizational variables on efficiency, namely the number of users and existence of Nursing Home chains. The results of the DEA model show that the efficiency average is 81.9%, and only 10 out of 96 Nursing Homes are efficient. Tobit regression shows that the number of users has a positive effect on the efficiency of Nursing Homes, whereas the existence of Nursing Home chains affects their efficiency negatively.

Keywords Data envelopment analysis · Efficiency · Nursing homes · Third sector

26.1 Introduction

According to [12] in 2013, the average old age dependency ratio of the 28 Member States of the European Union (the ratio between the number of persons aged 65 and over, the age when they are generally economically inactive, and the number of

A. S. Veloso (✉)
Polytechnic Institute of Bragança, Bragança, Portugal
e-mail: asveloso@ipb.pt

C. B. Vaz
Centre for Management and Industrial Engineering (CEGI/INESC TEC),
Polytechnic Institute of Bragança, Bragança, Portugal
e-mail: clvaz@ipb.pt

J. Alves
OBEGEF/UNIAG; Polytechnic Institute of Bragança, Bragança, Portugal
e-mail: jorge@ipb.pt

A. I. F. Vaz et al. (eds.), *Operational Research*, Springer Proceedings
in Mathematics & Statistics 223, https://doi.org/10.1007/978-3-319-71583-4_26

persons aged between 15 and 64) was 27.5, which means that there were 28 people over 65 for every 100 people of working age, 17% more than in 2001. In 2060, the prediction point to a ratio of 51.6 (88% more than in 2013) [13]. In 2013, Portugal was the fifth country of the 28 Member States with the highest ratio (29.4) being surpassed by Sweden, Greece, Germany and Italy [12]. Between 2001 and 2013 there was an increase of 21% and in 2001 Portugal was already the 9th Member State with the highest ratio [12]. Predictions for 2060 show that the ratio will continue to rise. In fact, Portugal will be the third Member State with the highest ratio (64.9), being surpassed only by Latvia (65.7) and Greece (67.9) [13].

The observed growth of the elderly population in Portugal was followed by an increase in social responses to this target population, namely Nursing Homes, day care centers, home cares services. In 2013, there were more than 7,000 social responses to elderly people, 47% more than in 2000 [16]. Firstly, the home care services were the social response that presented the greatest number of requests and evolution. In 2013, there were more than 2,500 home care services with a relative increase of 66% over the year 2000 [16]. Secondly, the Nursing Homes registered close to 2,500 social responses and an increase of 55% in relation to the year 2000 [16]. Given the predictions of the elderly population by 2060, it enables us to conclude that the number of Nursing Homes will tend to increase. Thus, it is important for regulator, in this case Instituto da Segurança Social, IP (ISS, IP), to see if the efficient Institutions manage the assigned financial contributions. This information is also relevant to the central management of the SCM and to the administration of the Institutions, since it allows comparing Nursing Homes practices in the management of allocated resources, identifying efficient and inefficient units and becoming a benchmarking tool. Finally, the performance of Nursing Homes is important for the users (and relatives) of the Institutions, as they want to know if their payments are properly applied, which ultimately affects the image of the services provided by the Nursing Homes. This study aims to evaluate the economic efficiency of Nursing Homes owned by SCM in 2012 and 2013. In this stage, the efficiency assessment is performed through the non-parametric technique DEA. Besides the study of Nursing Homes efficiency, it is important to know the determinants, namely organizational factors that influence their efficiency. This information is relevant for the regulator and Institutions since solutions should be identified to provide financial sustainability to the Institutions. At this stage, the Tobit regression is used to verify the influence of certain determinants on economic efficiency, by investigating the influence of organizational variables such as the number of users of the entity, the integration into a chain of Nursing Homes and the possibility of the entity in providing other social responses.

This study is organized according to the following. Section 26.2 presents a literature review on the evaluation of efficiency in Nursing Homes. Section 26.3 describes the Nursing Homes sector, introducing the case study. Section 26.4 presents the methodology based on the DEA method and the Tobit regression. Section 26.5 presents and discusses the obtained results. Finally, Sect. 26.6 presents the conclusions and future research.

26.2 Literature Review on the Evaluation of Efficiency in Nursing Homes

The Nursing Homes efficiency has been studied worldwide [29]. It is widely accepted that non-profit Nursing Homes have less efficiency than profit Nursing Homes e.g. [2, 10, 22]. However, concerning profit Nursing Homes, the financial and economic results appear as priorities. In what non-profit Nursing Homes are concerned, those aspects are contemplated in a secondary way, being the quality of the services provided to users the main focus [7]. Reference [31] interviewed a group of individuals involved in the provision of care, such as administrators, nurses, regulators, and other collaborators about the quality of care. The authors identified two models of Nursing Homes, one with good quality of care and the other with poor quality of care. In Nursing Homes with good quality of care, the main focus was the satisfaction of users. In Nursing Homes with poor quality of care, the main focus is not defined, which may be the financial survival of Nursing Home and financial results, regardless of the user needs. As [7] found in non-profit Nursing Homes, the objective is to provide the highest quality of services to users, regardless of the associated costs. This does not occur in profit Nursing Homes because financial results prevail, which invest less in the quality of care [24].

According to [18], who conducted a survey about the techniques used in the investigation of the efficiency of medical care units (among which the equivalents to Nursing Homes), the author verified that the non-parametric DEA technique was the most used. According to the author, the use of efficiency scores as a dependent variable in Tobit regression has been increasingly used. The objective is to verify the influence that determinants have on efficiency. Two characteristics commonly used in regression, besides ownership status, are the number of users and the integration into a chain of Nursing Home e.g. [1, 28].

Concerning the Nursing Home chains, the relationship with efficiency has led to mixed results [1]. References [3, 14, 22] found that Nursing Home chains affect efficiency positively. The fact that Nursing Homes are integrated into a chain can increase efficiency through the effect of economies of scale, in particular by sharing resources (e.g. human resources such as nurses and doctors) which leads to lower general and administrative costs [2]. In addition, Nursing Home chains move faster in the learning curve through the sharing and adoption of new information among Nursing Homes [2]. However, the increase in operational activity in Nursing Home chains is not always positive [14]. In fact, increased operational activity can increase maintenance costs and slow down decision making process, which ultimately can reduce efficiency [21]. In the study by [1], Nursing Home chains were less efficient than independent Nursing Homes. The explanation may lie in the size of the Nursing Homes, since they tend to waste resources on bureaucracies instead of benefiting from economies of scale and experience curve.

Regarding the relationship between Nursing Homes efficiency and the number of users, [2, 28, 30, 32] concluded that the number of users influences positively their efficiency. The economies of scale explain, once again, the relationship found, through average expenditures with users being smaller, [4, 6], particularly with human resources expenditure [30].

This is a general literature review on the evaluation of efficiency in Nursing Homes, but this study focuses on economic efficiency following [2].

In terms of performance assessment of Portuguese Nursing Homes, the regulator (ISS, IP) has information derived from reports of inspections on the quality of services for profit and non-profit Nursing Homes (Article 19 from Regulatory Ordinance (RO) n.° 67/2012 of 21st March) and the Income Statement of non-profit Nursing Homes [19]. In this case, the Income Statement consolidates the financial information for each non-profit independent Nursing Home or Nursing Home chain. Regarding non-profit Nursing Homes, until 2014, this information is not available for public consultation in Portugal. From 2015, non-profit Nursing Homes have to publish yearly this information on their website (n.° 2 from Article 14-A from Decree-Law (DL) n.° 172-A/2014 of 14th November) but some of them do not fulfill this obligation.

As far as we know, no study was published in applying the DEA methodology to Portuguese Nursing Homes. In the next section we contextualize the Nursing Homes area to introduce the case study.

26.3 Nursing Homes Case Study

26.3.1 The Nursing Homes of Portugal

Nursing Homes are establishments for collective housing, "for temporary or permanent use, in which social support activities and nursing care are provided" (n.° 2 from Article 1 from RO n.° 67/2012 of 21st March). The objectives of the Nursing Homes are "Provide permanent and adequate services to the biopsychosocial problems of the elderly; Contribute to the stimulation of an active aging process; Create conditions that allow preserving and encouraging the intra-family relationship; Promote social integration" (Article 3 from RO n.° 67/2012 of 21st March). The maximum capacity of Nursing Homes is 120 residents while the users of Nursing Homes, as a rule, are persons over 65 years of age, and in certain cases, persons under 65 may be admitted (Articles 5 and 6 from RO n.° 67/2012, of 21st March).

According to the [17], currently,[1] in Portugal, there are 2,381 Nursing Homes managed by profit and non-profit entities, as summarized in Table 26.1. The installed capacity is 93,373 beds and the number of users is 85,569, which corresponds to an occupancy rate of 92%. About 70% of the total Nursing Homes are managed by non-profit entities and more than 80% of the users are installed in these entities. The occupancy rate in non-profit organizations is 95%, while in profitable entities it is 79%.

[1]There is no information available for previous years.

Table 26.1 Nursing Homes of Portugal (2016)

Sector		No. entities	No. Nursing Homes	Nursing Homes (%)	No. beds	% beds	No. users	% users
Profit			695	29	19,537	21	15,519	18
Non-profit			1,686	71	73,836	79	70,050	82
	Total: SCM and Nursing Homes	300	467	20	26,645	29	25,375	30
	SCM: Independent Nursing Home	198	198					
	SCM: Nursing Home Chain	102	269					
Total			2,381	100	93,373	100	85,569	100

In case of non-profit sector, there are 300 SCM (out of a total of 358[2]) that manage 467 Nursing Homes, representing 20% of total Nursing Homes. The number of beds and number of users correspond to 26,645 and 25,275, respectively, which show an occupancy rate of 95%. There are 198 SCM that only have one Nursing Home (Independent Nursing Home) while the remaining SCM (102) hold more than one (Nursing Home chain). It should be noted that SCM have more beds and more users than that in the entire profit sector which indicates the high importance that SCM has in Nursing Homes area. The next section explores in detail the Particular Case of Nursing Homes owned by SCM.

26.3.2 The Particular Case of Nursing Homes Owned by SCM

The SCM entities are "associations recognized in the canonical legal system, with the purpose of satisfying social needs and practicing acts of Catholic worship, in harmony with their traditional spirit, guided by the principles of Christian doctrine and morality" (n.° 1 from Article 68 from DL n.° 172-A/2014 of 14th November). The SCM are also non-profit entities, which have the status of Private Institutions

[2]Reference [33].

of Social Solidarity (IPSS). The IPSS are non-profit collective persons with the "purpose of giving organized expression to the moral duty of justice and solidarity, contributing to the realization of the social rights of citizens, provided they are not administered by the state or another public entity" (n.° 1 from Article 1 from DL n.° 172-A/2014 of 14th November). The objectives referred in the previous article are mainly for the provision of services in several areas, namely in the support of the elderly, disabled people, children and youth, among others. It should be noted that holding the IPSS status confers a set of tax advantages, namely being exempt from taxation (e.g. Value-Added Tax, Income Tax) [25]. In addition, the Institutions receive from the ISS, IP a monthly payment per user that is covered by a cooperation agreement settled between the ISS, IP and the entity.

The activities developed by SCM are commonly referred to as social responses. A social response shall be deemed to be the support services provided to persons and families whose purpose is: "(1) To prevent and remedy situations of social and economic deprivation and inequality, social dependence and social dysfunction, exclusion or vulnerability; (2) The community integration and promotion of people and the development of their abilities and skills; (3) Special protection for the most vulnerable groups, including children, young people, the disabled and the elderly" (Article 3 from DL n.° 64/2007 of 14th March). Thus, the activities developed by SCM are branched out by several intervention areas: (a) Children and youth; (b) Children, youth and adults with disabilities; (c) Elderly people; d) Family and community; (e) People with drug addiction; (f) People infected by HIV/AIDS; (g) People with mental disease; (h) People in a dependency situation. In each of these areas there are several services provided for this specific population. For example, in the area of the elderly people there are host families, day care center, night care center, home care services and Nursing Homes. In the area of children and young people there are kindergarten, pre-school, after school activities center, children and youth households, among others [16]. In the case of SCM, for the year 2016, the most developed intervention areas are at children/young and elderly people levels. In fact, of the total 2,041 social responses involving the various intervention areas, 1,216 belong to the elderly people area and 684 belong to the children and youth area [17].

In the specific case of Nursing Homes, adequate food services are provided to the needs of the users, personal hygiene care, clothing treatment, space hygiene, socio-cultural activities, support in the users daily life activities, nursing care, as well as access to health care and administration of medicines. The infrastructure of the Nursing Home is organized in 9 areas: (a) Reception area; (b) Management area, technical and administrative services; (c) Facilities area for personnel; (d) Area of activities; (e) Dining area; (f) Accommodation area; (g) Kitchen and laundry area; (i) Nursing services area. The provision of services is ensured by staff 24 h a day. In addition, each Nursing Home is obliged to comply with the staff ratios per user summarized in Table 26.2 (Article 12 from RO n.° 67/2012 of 21st March). If there are users with higher dependency level, each entity has to change the ratios of nurses, auxiliary nurse and assistant employee to 1/20, 1/5 and 1/15, respectively.

The operational activity of each SCM is regulated by ISS, IP. The cooperation agreements are established between IPSS and ISS, IP (n.° 4 from Legal Standard (LS)

Table 26.2 Staff ratios per user

Staff	Schedule	No. users
Sociocultural animator/Educator/Geriatrics technician	Part-time	40
Nurses	Full-time	40
Auxiliary nurse	Full-time	8
Auxiliary nurse-night schedule	Full-time	20
Housekeeping clerk	Full-time	≥40
Chief cook	Full-time	By Nursing Home
Assistant cook	Full-time	20
Assistant employee	Full-time	20

I from Legislative Order (LO) 75/92 of 20th May). The agreements have the objective of "the pursuit of actions by the Institutions aimed at supporting children, young people, the disabled, the elderly and the family, as well as preventing and remedying situations of lack, dysfunction and social marginalization and the development of communities and social integration and promotion." (LS III from LO 75/92 of 20th May). The conclusion of cooperation agreements implies the allocation of financial contributions by ISS, IP. The allocation of the contributions is made by the user and varies according to the social response. Usually, the conclusion of cooperation agreements occurs when a social response, for example a Nursing Home is opened. Suppose that a Nursing Home has 60 beds. Only a proportion of these 60 beds is covered by the cooperation agreement.

The purpose of the contributions is to "subsidize the running expenses of the equipment or services" (n.° 2 from LS XXII from LO 75/92 of 20th May). The amounts of contributions granted from the ISS, IP are annually defined by the Cooperation Protocol established between the ISS, IP and each representative of non-profit organizations (União das Misericórdias Portuguesas, Confederação Nacional das Instituições de Solidariedade and União das Mutualidades Portuguesas) (n.° 4 from LS XXII from LO 75/92 of 20th May). In 2012 and 2013, the monthly value covered by the Nursing Home cooperation agreement was €351.83 and €355.00 per user, respectively. This financial contribution per month may be subject to variations, particularly if the users are in a dependent situation or the vacancies are filled out by the competent services (namely the ISS, IP) [26, 27].

In addition to the financial contribution of the ISS, IP, the IPSS can receive payments from users and their families. The contribution of the users corresponds to the application of a rate (maximum limit 85%). This rate is defined internally by the Institution and should be proportional to the per capita income of the household [11], taking into account the economic and financial capacity of the users and their descendants.

To sum up, the main source of IPSS income corresponds to the sum of the following amounts: payments from users and family descendants plus the contributions granted by the ISS, IP. It should be noted that for users who are covered by cooperation agreement there is a limit of income that the Institution may receive. The limit corresponds to the product of the reference value per user, by the number of users in a cooperation agreement, plus 15%. For example, a Nursing Home with 60 beds covered by a cooperation agreement, the annual sum for 60 users could not exceed €770,089.68 [(938.43 × 60 × 12) × 115%] in 2012 and €777,020.04 in 2013. The reference value per user is defined in each year between the ISS, IP and the representative organizations of the Institutions. The reference value was €930.06 and €938.43 in 2012 and 2013, respectively.

26.4 Methodology

This study aims to evaluate the economic efficiency of Nursing Homes owned by SCM and the determinants that influence their efficiency in 2012 and 2013. The study is developed in two stages. In the first stage, the efficiency scores were calculated through the non-parametric DEA technique. Only economic variables were used based on the Income Statement variables. In the second stage, a Tobit regression was performed to verify the effect of determinants on efficiency. The number of users, the number of social responses and the existence of Nursing Home chains were the variables used. The information used in this study was collected from the ISS, IP and involves 96 SCM in 2012 and 2013 whose Income Statement was validated by ISS, IP until the third quarter of 2015.

This financial information is only available for the entities themselves and the ISS, IP. In order to obtain the information about the entities, diligences were established with the ISS, IP and only the information from the Nursing Homes developed by SCM was obtained. Regarding the number of users, this information is available on the site developed by [16] which is yearly updated. Although, it is not possible to establish, for a given year, a match between the organizational characteristics of the profit Nursing Home, namely the number of users and its Income Statement, since their time reference has a lag of 1 year and information from previous years is not available in the site.

It is verified that the quality of the services provided by the Nursing Home is a major aspect in the management of the units. In the literature, there are many authors who use different aspects of quality (for example users who received the flu vaccination, users with depression symptoms or weight loss) [10]. These aspects are mostly based on inspections carried out by accredited entities (for example, state entities) and by ISS, IP, in our case study. Globally, these quality assessment results are displayed to the public (e.g. Nursing Home Compare; Online Survey, Certification, and Reporting Systems), although this does not happen in Portugal. Specifically, the reports from inspection of the Nursing Home, carried out by ISS, IP, are not available for public consultation. It should be noted that this information has already been requested to ISS, IP. However, it was not yet possible to obtain such information, hindering the study of the quality of the services provided by

Table 26.3 Final sample of case study

	2012	2013
Number of SCM	96	96
Number of Nursing Homes	149	150

the Nursing homes. In this study we assume that Nursing Homes meet the quality standards established by ISS, IP and it is *ceteris paribus*.

This study involves a sample of 96 (represents 32% of total Nursing Homes in 2016) Nursing Homes owned by SCM during the year 2012 and 2013 and their data is achieved from Income Statement after validated by ISS, IP. Note that the Income Statement for each SCM can include the data concerning one Nursing Home or consolidates the data concerning all Nursing Homes chain affiliated. Thus, the SCM corresponds to the decision making unit (DMU). Thus, the 96 SCM have 149 Nursing Homes in 2012 and 150 Nursing Homes in 2013 (one SCM opened a new Nursing Home in 2013), as summarized in Table 26.3.

26.4.1 Data Envelopment Analysis (1st Stage)

The DEA technique, initially introduced by [8] enables to assess the relative efficiency of homogenous organizational units, Decision Making Units (DMU), which use multiple inputs to produce multiple outputs. Considering the input orientation, the efficiency value of each DMU is calculated through the ratio between inputs of the efficient unit producing similar outputs to the evaluated unit over the inputs of this unit [5]. In the case of the DMU using multiple inputs to produce multiple outputs, the DEA technique assigns weights in order to maximize the ratio of the sum of its weighted outputs to the sum of its weighted inputs, subject to the restriction that all DMUs, for the same set of weights, hold a maximum ratio of 100% [5]. Thus, the DEA technique identifies the frontier of the set of production possibilities (PPS) defined by the efficient DMUs and the segments that connect them. The remaining DMUs are inefficient, being evaluated by reference to the obtained frontier.

Consider a set of n Nursing Homes named by j ($j = 1, \ldots, n$) that use m inputs x_{ij} $(x_{1j}, \ldots, x_{mj}) \in R^m_+$ to obtain s outputs y_{rj} $(y_{1j}, \ldots, y_{sj}) \in R^s_+$. In the DEA model, input orientation was used, since Nursing Homes have more control over inputs than outputs [29]. The relative efficiency of the Nursing Home$_o$ can be evaluated considering the frontier with variable returns to scale and the orientation of the inputs, using the linear programming model (26.1), obtaining an optimal solution, θ^*_o.

$$\min\left\{\theta_o \middle| \theta_o\, x_{io} \geq \sum_{j=1}^{n} \lambda_j\, x_{ij},\; i = 1, \ldots, m;\; y_{ro} \leq \sum_{j=1}^{n} \lambda_j\, y_{rj},\; r = 1, \ldots, s; \sum_{j=1}^{n} \lambda_j = 1;\; \lambda_j \geq 0;\; \forall_{j,i,r}\right\} \quad (26.1)$$

The pure technical efficiency (θ_o^*) of Nursing Home$_o$ corresponds to the minimum factor with which all its inputs can be reduced taking into account the obtained results. The efficiency measure will reach 100% when the evaluated Nursing Home$_o$ is considered efficient, while lower values will indicate the existence of inefficiencies. The variable returns to scale frontier is considered since there are marked differences in SCM size [9].

In the evaluation of efficiency, only economic variables were used based on the Income Statement. Since the Income Statement is composed of income and expenses, we used indicators of expenditure on inputs and income on outputs. It was decided to aggregate the variables of expenditures and income with the objective of reflecting that efficient entities had positive net results. Bearing in mind this perspective, the DEA model is constructed using two inputs (Operating Expenses and Other Costs) and a single output (Total Revenue). Operating Expenses include Costs of Goods Sold and Consumed, External Supplies and Services and Personnel Expenses. Other Costs include, among others, other expenses related to expenses with depreciation and losses in inventories. It was considered a single output (Total Revenue) resulting to the income of each unit, i.e. Sales (user and family payments), Subsidies received from the ISS, IP, and other subsidies (e.g. European Community funds and funds of national programs for non-profit Institutions). Note that the choice of the previous variables was restricted by the available information obtained from ISS, IP.

Table 26.4 shows the averages, coefficients of variation, maximum and minimum of the inputs and output used. There is a large discrepancy in the data (a coefficient of variation of more than 50% in Operating Costs and Revenues and over 100% in Other Expenses). This means that the size of Nursing Homes under analysis is quite different, as is also shown by maximum and minimum values.

Table 26.4 Descriptive statistics of inputs and output

	2012				2013			
	Mean	C.V. (%)	Maximum	Minimum	Mean	C.V. (%)	Maximum	Minimum
Operating expenses (€)	838,692.3	57	2,401,267.3	137,748.9	864,356.8	57	2,469,842.5	134,876.5
Other costs (€)	82,122.2	131	911,463.0	9,208.1	79,410.1	127	897,019.3	7,520.4
Total revenue (€)	897,167.8	57	2,229,108.8	159,450.4	924,890.8	56	2,282,380.7	138,586.1

26.4.2 Tobit Methodology (2nd Stage)

In the second stage, a Tobit regression is performed with three independent variables: (X_1) number of users; (X_2) existence of Nursing Home chains (binary variable with value 0 if it has a single Nursing Home and value 1 otherwise); (X_3) number of social responses. The variables number of users and existence of associated Nursing Homes in chains are often used in efficiency studies e.g. [2, 28]. Regarding the inclusion of the variable number of social responses, the SCM can develop other activities besides the Nursing Homes, such as day care center and home care service for elderly, kindergarten, pre-school, among others. The Tobit regression is appropriate for this study since the dependent variable, the efficiency score ranges from 0 and 1 e.g. [14, 15, 20, 23, 34]. The regression model is presented in (26.2).

$$\theta_j = \alpha + \beta_1 X_{1j} + \beta_2 X_{2j} + \beta_3 X_{3j} + \varepsilon_j \tag{26.2}$$

where α is the intercept, β_1, β_2 and β_3 are estimated coefficients of regression, ε_j is the error term $\varepsilon_j \sim N(0, 1)$ and X_{1j}, X_{2j} and X_{3j} are the dependent variables, number of users, number of social responses and the existence of Nursing Home chains. This is a binary variable with value 0 if it is an independent Nursing Home and 1 otherwise. There are 61 independent Nursing Homes and 35 Nursing Home Chains. Table 26.5 presents the descriptive statistics of independents variables X_{1j}, X_{2j} and X_{3j}. For each indicator, there is a large discrepancy in the data. This is symptomatic of the variability of the SCM dimension, as observed for example in the maximum and minimum of social responses, ranging from 1 to 53.

Table 26.5 Descriptive statistics of variables used in the Tobit regression

	2012				2013			
	Mean	C.V. (%)	Maximum	Minimum	Mean	C.V. (%)	Maximum	Minimum
Number of users	83	54	232	12	85	56	244	12
Number of Nursing Homes	2	58	5	1	2	59	5	1
Number of social responses	10	80	53	1	10	75	53	1

As seen in the literature, Nursing Home chains can increase efficiency through economies of scale [2]. However, the reverse may occur due to management problems, such as resource wastage. The increasing firm size may increase monitoring costs and slow decision making process e.g. [1, 21]. In addition, non-profit Nursing Homes focus on the quality of care provided, and financial results are considered secondary [7]. Given the Portuguese context, Nursing Homes must meet staff ratios for each Nursing Home (Article 12 from RO n.° 67/2012 of 21st March). That is, the opening of new equipment always involves the hiring of at least 8 people, and thus there is no possibility of rationalization of human resources. The same applies in other activities developed by the SCM, namely in the areas of the elderly people (e.g. home care service) and children and youth (e.g. kindergarten). It is necessary to comply with the staff ratios established in RO n.° 67/2012 of 21st March. Moreover, Nursing Homes in the analysis are non-profit, so the focus on quality is the main objective, which implies a greater consumption of resources. Thus, it is expected that the variables existence of Nursing Home chains and number of social responses have a negative effect on efficiency. Conversely, the number of users is expected to have a positive effect on efficiency due to the possibility to have scale economies through the rationalization of resources, namely human resources, which reduces the average costs [30]. Therefore, there are three research hypotheses:

H_1: The number of users has a positive effect on efficiency;

H_2: The existence of Nursing Home chains has a negative effect on efficiency;

H_3: The number of social responses has a negative effect on efficiency.

26.5 Results and Discussion

26.5.1 Efficiency Assessment of Nursing Homes (1st Stage)

In the first stage, the economic efficiency of a Nursing Home in a given year is estimated through the model (26.1), considering the best practices observed during 2012 and 2013. The results show that the average efficiency is 81.9% which means that each Nursing Home can reduce its expenses (inputs), on average, 18.1%, given the level of revenues obtained. 10 out of 96 Nursing Homes are considered efficient: 3 units are efficient in 2012, 6 are efficient in 2013 and only one is efficient in both years. These Nursing Homes are considered as benchmarks for inefficient Nursing Homes. The efficient Nursing Homes have different characteristics, namely in the organizational dimension. In fact, 5 are independent Nursing Homes and 5 are Nursing Home chains (2 chains are composed by 2 Nursing Homes, 1 chain is composed by 3 Nursing Homes, 1 chain is composed by 4 Nursing Homes and 1 chain is composed by 5 Nursing Homes). The same occurs for the number of users, ranging from a minimum of 12 to a maximum of 244. Furthermore, the number of social responses from the corresponding SCM has also a large variance. There are 5

Table 26.6 Average of efficiency of Nursing Home Chains and independent Nursing Homes

	Independent Nursing Homes			Nursing Home Chains		
	Efficiency (%)	Standard deviation (%)	Users range	Efficiency (%)	Standard deviation (%)	Users range
All DMUs	81.9	9	12–150	81.8	12	51–244
First quartile	83.0	13	12–43	77.7	10	51–87
Second quartile	79.4	7	44–60	78.9	14	88–106
Third quartile	78.0	7	61–76	81.9	11	107–153
Fourth quartile	87.0	7	77–150	89.3	8	154–244

SCM with less than 10 social responses, 4 SCM with social responses ranging from 11 to 16 and one SCM with 43 social responses.

Globally, the average efficiency of independent Nursing Homes is 81.9% (standard deviation is 9%),while in Nursing Home chains the average efficiency is 81.8% (standard deviation is 12%) as shown by Table 26.6. In addition there are 75 Nursing Homes that have slacks in the input Other Expenses and one with slack in the input Operating Expenses. These SCM should reduce their expenses without disturbing the provision of social services. Table 26.6 details also the average efficiency considering the different quartiles of the number of users by Nursing Homes group.

For independent Nursing Homes, it is not observed any trend between efficiency and number of users. In fact, the average efficiency decreases in second and third quartiles of users. In Nursing Home chains, there seems to be a positive association between efficiency and number of users. Indeed, the efficiency has increased in all quartiles. The Tobit regression presented in next section will us to reach more conclusive results in terms of the determinants of Nursing Homes performance.

26.5.2 Tobit Regression (2nd Stage)

In the second stage, the Tobit regression is used to verify the influence of organizational variables on efficiency. Specifically, we investigate the influence of organizational variables such as the number of users of the entity, the integration into a chain of Nursing Homes and the possibility of the entity to provide other social responses. The results are presented in Table 26.7. Concerning the model's goodness of fit, the Variance Inflation Factor (VIF) for each independent variable does not present problems of collinearity and residuals, according to the $p = 0.658$ value of the chi-square test, follow a normal distribution (the null hypothesis of the error dis-

Table 26.7 Results of Tobit regression

	Expected sign	Coefficient	Standard error	Z	$P-Value$	VIF
Intercept		0.774	0.016	48.080	<0.001	
Number of social responses	–	−0.001	0.001	−0.697	0.486	1.324
Integration into a chain of Nursing Homes	–	−0.040	0.019	−2.043	0.041	1.761
Number of users	+	0.001	0.000	4.107	<0.001	1.568

tribution is not rejected). The Kolmogorov–Smirnov test was run for each variable and the results show that the independent variables do not follow a Normal distribution. However, it is not an obstacle due to the Central Limit Theorem, which justifies the asymptotic normality for large samples ($n \geq 30$). Thus, the model's goodness of fit is considered acceptable [35].

The Tobit regression shows that number of users and the existence of Nursing Home chains are statistically significant. The chain affiliation of Nursing Homes negatively affects efficiency. The results confirm the conclusions of [1]. That is, Nursing Home chains have lower efficiency than independent Nursing Homes. Such as [1] concluded the greater size of the entities may make it difficult to manage them. The non-profit status and compliance with staff ratios explain the differences found. Given that the Nursing Homes in question are non-profit, the focus is on the quality of care for the users, which leads to a greater expenditure of resources. Since Nursing Homes have to meet staff ratios for each Home, with at least 8 employees, entities that have more than one Nursing Home, in turn, have to hire more staff. That is, there is no possibility of rationing human resources, which means more expenses, which can cause difficulties in the management of Nursing Homes. On the other hand, the number of users positively affects efficiency. The results are in agreement with the literature e.g. [28, 30, 32], that is, the greater the number of users, the more efficient the Nursing Homes are due to economies of scale. The Nursing Homes can benefit from lower average costs, namely human resources, feeding and energy. It is interesting to note that the fact that a SCM has more than one Nursing Home does not mean that it is more efficient than another SCM that has only one unit. That is, SCM are more efficient with only one Nursing Home and with the largest number of users. Therefore hypotheses H_1 and H_2 are validated.

Regarding the number of social responses of SCM, no statistical significance was obtained. However, the negative coefficient should be highlighted, which indicates that the number of social responses negatively influences efficiency, which can support the regulator in allocating the social responses to each SCM. Furthermore, as SCM increases its social activities, the associated costs, such as human resources, also increase, which may hinder the management of activities. However, this hypothesis (H_3) was not validated.

26.6 Conclusions

The objective of this study is to assess the economic efficiency of non-profit Nursing Homes owned by SCM and the organizational variables that can influence their efficiency. In the first stage, a DEA model (26.1) is used to assess the efficiency of the Nursing Homes, considering the best practices observed during 2012 and 2013. In the second stage, a Tobit regression is used, using as the dependent variable the efficiency scores obtained from the DEA and as independent variables the number of users, the existence of Nurse Homes chains and the number of social responses.

The DEA model showed that the efficiency average of the Nursing Homes is 81.9% and 10 out of 96 Nursing Homes are efficient. Furthermore, the efficiency average between independent Nursing Homes and Nursing Home chains is practically the same. It should be noted that there were 76 Nursing Homes with slacks in the inputs, so it is inferred that these entities should reduce their inputs without disturbing the provision of social services.

The Tobit regression reveals that number of users and the existence of Nursing Home chains are statistically significant, influencing their efficiency. The number of users has a positive effect while the existence of a chain of Nursing Homes affects negatively their efficiency. This means that although Nursing Homes can benefit from economies of scale by increasing the number of users e.g. [4, 6], but when this increase implies the opening of new Nursing Home may lead to difficulties in management, reducing their efficiency [1].

In terms of future research, it would be interesting to compare the efficiency among non-profit to profit Nursing Homes and verify the impact that certain aspects concerning the quality of social services (e.g. safety conditions, hygiene and comfort conditions, food safety, among others).

Acknowledgments This work is financed by the ERDF – European Regional Development Fund through the Operational Programme for Competitiveness and Internationalisation - COMPETE 2020 Programme within project "POCI-01-0145-FEDER-006961", and by National Funds through the Portuguese funding agency, FCT - Fundação para a Ciência e a Tecnologia as part of project "UID/EEA/50014/2013".

References

1. R. Anderson, D. Lewis, J. Webb, The Efficiency of nursing home chains and the implications of non-profit status. J. Real Estate Portf. Manag. **5**(3), 235–245 (1999)
2. R. Anderson, H. Weeks, B. Hobbs, J. Webb, Nursing home quality, chain affiliation, profit status and performance. J. Real Estate Res. **25**(1), 43–60 (2003)
3. G. Arling, R. Nordquist, J. Capitman, Nursing home cost and ownership type: evidence of interaction effects. Health Serv. Res. **22**, 255–269 (1987)
4. J. Banaszak-Holl, W. Berta, D. Bowman, C. Joel, B. Mitchell, The rise of human service chains: antecedents to acquisitions and their effects on the quality of care in US nursing Homes. Manag. Decis. Econ. **23**(4), 261–282 (2002)

5. R. Banker, A. Charnes, W. Cooper, Some models for estimating technical and scale inefficiencies in data envelopment analysis. Manag. Sci. **30**(9), 1078–1092 (1984)
6. J. Bazzoli, B. Chan, The financial performance of hospitals belonging to health networks and systems. Inquiry **37**(3), 234–252 (2000)
7. A. Ben-Ner, T. Ren, Does Organization Ownership Matter? Structure and Performance in For-Profit, Non-profit and Local Government Nursing Homes, in *Annual Conference Industry Studies* (Boston MA, USA, 2008)
8. A. Charnes, W. Cooper, E. Rhodes, Measuring the efficiency of decision making units. Eur. J. Op. Res. **2**(6), 429–444 (1978)
9. T. Coelli, D. Rao, C. O'Donnell, G. Battese, *An Introduction to Efficiency and Productivity Analysis* (Springer Science Business Media, New York, 2005)
10. N. DeLellis. *Determinants of Nursing Home Performance: Examining the Relationship Between Quality and Efficiency*. Ph.D. thesis, Virginia Commonwealth University, Richmond, Virginia, USA (2009)
11. DGSS, Circular de Orientação Normativa. Technical Report 3, Direção Geral da Segurança Social (1997)
12. Eurostat, Old Age Dependency Ratio. Technical report (2017), http://ec.europa.eu/eurostat/web/products-datasets/-/tsdde510
13. Eurostat, Projected Old Age Dependency Ratio. Technical report (2017b), http://ec.europa.eu/eurostat/web/products-datasets/-/tsdde511
14. J. Fizel, T. Nunnikhoven, The efficiency of nursing home chains. Appl. Econ. **25**, 49–55 (1993)
15. G. Garavaglia, E. Lettieri, T. Agasisti, S. Lopez, Efficiency and quality of care in nursing homes: an italian case study. Health Care Manag. Sci. **14**, 22–35 (2011)
16. GEP-MSESS, Carta Social - Rede de Serviços e Equipamentos. Technical report, Gabinete de Estratégia e Planeamento do Ministério de Solidariedade (2013)
17. GEP-MSESS, Rede de Serviços e Equipamentos. Technical report, Gabinete de Estratégia e Planeamento do Ministério de Solidariedade (2016), http://www.cartasocial.pt/index2.php
18. B. Hollingsworth, Non-parametric and parametric applications measuring efficiency in health care. Health Care Manag. Sci. **6**, 203–218 (2003)
19. Instituto da Segurança Social, IP. Guia Prático - Orçamento e Contas de Instituições Particulares de Solidariedade Social. Technical report (2015)
20. J. Iparraguirre, R. Ma, Efficiency in the provision of social care for older people. A three-stage data envelopment analysis using self-reported quality of life. Socio-Econ. Plan. Sci. **49**, 33–46 (2015)
21. M. Jensen, W. Meckling, Theory of the firm: managerial behavior, agency cost and ownership structure. J. Financ. Econ. **3**(4), 305–360 (1976)
22. K. Knox, E. Blankmyer, J. Stutzman, Organizational efficiency and quality in texas nursing facilities. Health Care Manag. Sci. **6**, 175–188 (2003)
23. P. Kooreman, Nursing home care in the Netherlands: a nonparametric efficiency analysis. J. Health Econ. **13**(3), 301–316 (1994)
24. Y. Li, X. Cai, Z. Ye, L. Glance, C. Harrington, D. Mukamel, Satisfaction with Massachusetts nursing home care was generally high during 2005–2009, with some variability across facilities. Health Aff (Millwood) **32**(8), 1416–1425 (2013)
25. M. Marques, IPSS: Uma Abordagem Fiscal. TOC **120**, 45–47 (2010)
26. MSSS and UMP, Protocolo de Cooperação 2011–2012. Technical report, Ministério da Solidariedade e da Segurança Social and União das Mutualidades Portuguesas (2012)
27. MSSS, UMP, CNIS, UMP, Protocolo de Cooperação 2013–2014. Technical report, Ministério da Solidariedade e da Segurança Social, União das Misericórdias Portuguesas, Confederação Nacional das Instituições de Solidariedade and União das Mutualidades Portuguesas (2012)
28. J. Nyman, D. Bricker, Profit incentives and technical efficiency in the production of nursing home care. Rev. Econ. Stat. **71**(4), 586–594 (1989)
29. Y. Ozcan, *Health Care Benchmarking and Performance Evaluation - An Assessment using Data Envelopment Analysis (DEA)* (Fred Hillier, Stanford, 2008)

30. Y. Ozcan, S. Wogen, L. Mau, Efficiency evaluation of skilled nursing facilities. J. Med. Syste. **22**(4), 221–224 (1998)
31. M. Rantz, D. Mehr, L. Popejoy, M. Stauffacher, L. Hicks, V. Grando, V. Conn, R. Porter, J. Scott, M. Maas, Nursing home care quality: a multidimensional theoretical model integrating the views of consumers and providers. J. Nurs. Care Quality **12**(3), 47–54 (1998)
32. D. Rosko, A. Chilingerian, S. Zinn, E. Aaronson, The effect of ownership, operating environment and strategic choices on nursing homes efficiency. Med. Care **33**(10), 1001–1021 (1995)
33. UMP, Misericórdias Nacionais. Technical report, União das Misericórdias Portuguesas (2017), http://www.ump.pt/misericordias
34. Y. Wang, L. Chou, The Efficiency of Nursing Homes in Taiwan: An Empirical Study Using Data Envelopment Analysis. Technical report (2003)
35. J. Wooldridge, *Introductory Econometrics - A Modern Approach* (South Western College Publishing, Ohio, 2006)

Chapter 27
Design and Planning of Sustainable Supply Chains: The Case Study of a Tissue Paper Business

Bruno van Zeller, Bruna Mota and Ana Barbosa Póvoa

Abstract While planning to expand its tissue paper business, a Portuguese company aims to explore options regarding the design and planning of its supply chain accounting not only for economic objectives but also environmental and social concerns (the three pillars of sustainability). A multi-objective mixed integer linear programming model (MOMILP) is developed and applied to the company supply chain focusing on the supply of tissue paper to the United Kingdom market. Decisions to be taken include the network structure, entity location and capacity definition, transportation network definition, production and storage levels, and material flow planning. Additionally, it is important to decide whether or not to postpone the conversion to the final product to a location closer to the clients. The augmented epsilon-constraint method is applied to study the trade-off between the three pillars of sustainability and important managerial insights are derived from this study.

Keywords Supply chain · Design and planning · Sustainability · Supply chain modelling · Optimization · Postponement

27.1 Introduction

The production cycle of an industry is completed when its products are sold to the end customers in order to be consumed in the correct quantity, location and time. For this to happen, it is necessary that companies ensure a distribution network that takes its products from the place of production to the place of consumption in an efficient way. This trip can often be complex, suffering several stops at the hub terminals – places where the goods are transferred from one mean of transportation

---

B. van Zeller · B. Mota (✉) · A. Barbosa Póvoa
CEG-IST, University of Lisbon, Av. Rovisco Pais, 1049-001 Lisboa, Portugal
e-mail: bruna.mota@tecnico.ulisboa.pt

B. van Zeller
e-mail: bmvanzeller@gmail.com

A. Barbosa Póvoa
e-mail: apovoa@tecnico.ulisboa.pt

A. I. F. Vaz et al. (eds.), *Operational Research*, Springer Proceedings in Mathematics & Statistics 223, https://doi.org/10.1007/978-3-319-71583-4_27

to another –, going through several entities and complying with rules and regulations that often vary from country to country. These complex operations often lead to high distribution costs. Adding to this complexity, consumers, governments and NGOs are increasingly pressuring industries to become more sustainable. Unfortunately, environmental and social objectives are more conflicting with the economic ones. One of the solutions to deal with such problem is finding the most sustainable solution among the alternatives [6]. This balance between economic development, environmental management and social equality is called the triple bottom line [1]. Optimization models are being developed by the academic community to assist on these decisions. However, very few include the three pillars of sustainability [5]. One of the exceptions is the work of Mota et al. [3] that developed a Multi-Objective Mixed Integer Linear Programming (MOMILP), which accounts for the three pillars of sustainability. The environmental assessment is performed based on Life Cycle Assessment (LCA) which has been described as the most scientifically reliable method to assess the environmental impacts of a particular product, process or activity [4]. The European Commission itself declared LCA as the best tool for environmental impact assessment, and included it in its Sustainable Development Strategy document, in order to standardize methodologies for LCA and to facilitate its implementation by companies. This methodology allows the quantification of the amount of relevant emissions and resources consumed, as well as the health impacts associated with a certain product or service, and takes into account the entire product life cycle, from extraction of resources until recycling, through production and utilisation [2]. Mota et al. [3], implemented a LCA methodology called ReCiPe 2008, which allows the impact analysis of the products and/or services throughout their life cycle, exploring all impact categories, instead of considering only CO2 emissions, as it is typically done in the literature. Impact categories characterize the various aspects of the environmental effects that should be taken into consideration where performing a life cycle assessment. The ReCiPe 2008 method uses two types of impact categories: midpoint (18 in total) – greater detail, clearer units and more practical sense – and endpoint (3) – broader and with higher level of uncertainty. On the social dimension of sustainability, Mota et al. [3] propose the application of a Social Benefit Indicator, which seeks to benefit the less developed regions, using inputs such as unemployment rate or income distribution by population. This indicator aims for the stimulation of job creation in areas most needed over areas that already have a favourable social performance and is adopted in this work.

The model developed by Mota et al. [3] is here applied to the case-study of a pulp and paper Portuguese company in its expansion to the tissue paper market in the United Kingdom, aiming to define the most sustainable supply chain network to be defined and the associated usage of resources at an aggregated level- planning.

27.2 Case Study

The company in study took, in 2015, the strategic decision of entering the tissue paper market. The acquisition of a paper-mill located in the north of Portugal was the first step towards this decision, reinforced with the intention of creating a new production line for the tissue paper in this plant. The production from this plant will be mostly exported, being the UK area one of the main destinations. The supply chain in study has only one factory producing the product in analysis, being located in the north of Portugal, near the coast. It has a production capability of 70 thousand tons of tissue paper per year. The company expects to have 38 retailer customers or delivery points in the UK, as can be seen in Fig. 27.1.

The product to be delivered to the clients is generically referred to as final product (FP). There are several final product forms (toilet paper, kitchen roll, paper napkin, etc.) but this work does not distinguish them as we are working at a strategic-tactical level. Before the formation of the final products, the production process includes an intermediate step in which jumbo rolls (JR) are formed. These tend to be more dense, and therefore easier to transport and store.

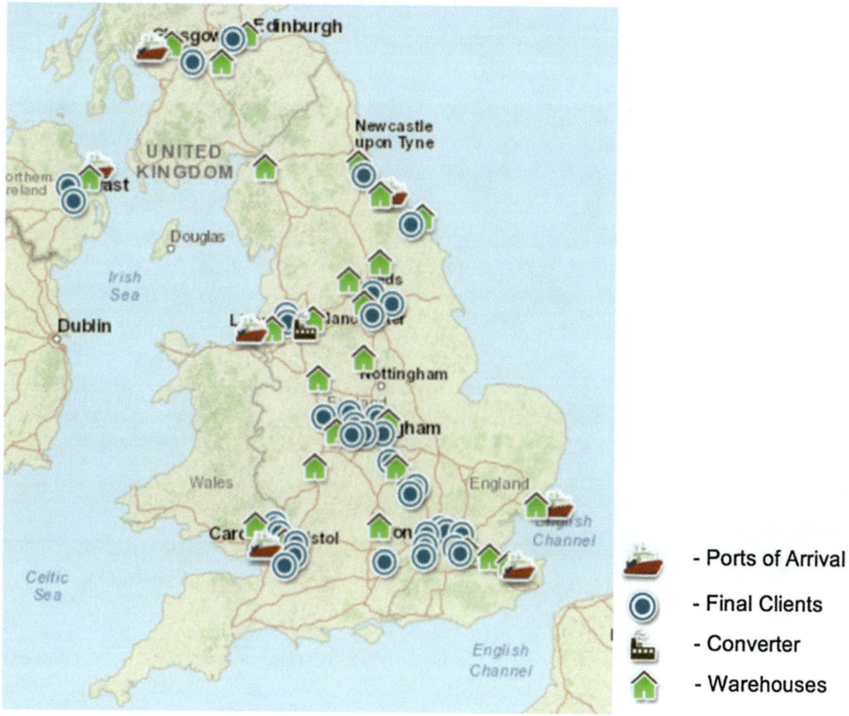

Fig. 27.1 Superstructure in the UK

One of the main strategic decisions to be taken from this work is which of the product forms should be sent from the plant, or in other words, if conversion should or should not be postponed. If the model decides to send the JR, these must necessarily pass through the converter in the UK in order to be transformed into FP. If not, conversion will be performed at the plant in Portugal. The converter will only be installed if the product to ship from Portugal is the JR. In this case, the converter, by company option, will be located in the city of Manchester, and will have a converting capability of 400 weekly tons.

Both the FP and the JR products can be transported by ship or by truck to the UK. In the case of the FP, when arriving in the UK they can go directly to the clients or be stored in warehouses. In the case of the JR, they are transported to the converter, where they are converted into FP, to be sent to the final customers or to the warehouses. From the warehouses, the FP can only be shipped to the end clients. Figure 27.2 describes all these possible flows. The green arrows represent the flows exclusively of FP, the red arrows symbolize the JR flows, and blue arrows may represent either the FP or the JR flows.

Transportation wise, all the distribution activities are outsourced. It is assumed that the transportation can be performed by road or sea. Depending on the product sent from the plant, the type of vehicle may differ. Leaving the plant, regardless of the products that are being shipped, they can be taken to the ports of departure, or directly to the UK by road. If is the FP that is being shipped, it goes on a container ship. If it is the JR that is being shipped, it goes on a regular cargo ship.

The company intends to consider only two departing ports in Portugal capable of handling containers: Figueira da Foz and Leixoes. Regarding the transportation of JR, it is assumed that only the port of Aveiro deals with regular cargo ships. For the arrival of the product to the UK it is only worth taking into account the ports with regular shipping lines from Portugal, at least weekly. From the port of Leixoes, the container ships can travel to the following UK ports: Belfast, Felixstowe, Greenock, Liverpool,

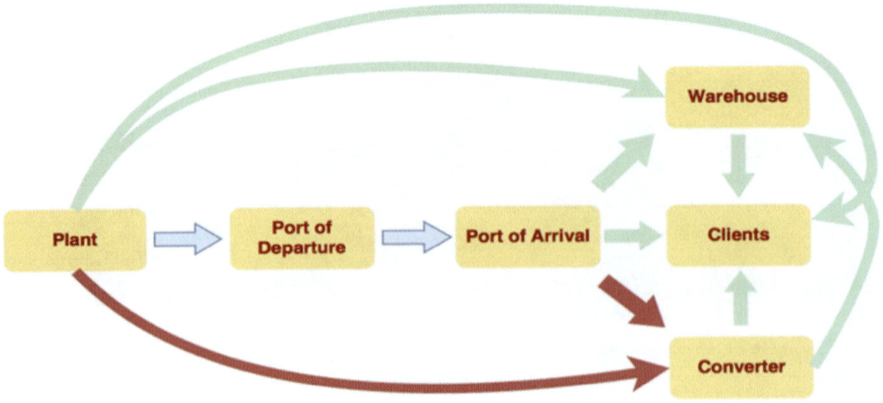

Fig. 27.2 Possible product-entity connections

Bristol, Teesport and Tilbury. Leaving the port of Figueira da Foz, the destinations to consider are only four: Belfast, Felixstowe, Bristol and Liverpool. Regarding the hypothesis of sending JR to the converter, the port of Aveiro reported that there are no regular shipping lines, and the transport is usually carried on chartered ships by customers to their port of interest.

The location of the warehouses has no limitations imposed by the company, so they can be located in any UK location that would be beneficial. In order to create a set of realistic locations, some parameters were taken into consideration such as (1) the proximity to the ports of entry, (2) the proximity to the end customers, (3) strategic locations that reduce the distances between entities aiming not only to reduce costs but also to reduce environmental impacts, and finally (4) a warehouse close to the converter.

Northern Ireland is an exception in the case study. The fact that it is separated from the rest of the United Kingdom by the Irish Sea means that there is no road transportation between the entities in this country and the other countries. The product to be delivered to its clients must necessarily be the finished product, and be sent by maritime transportation to the nearest port, which is Belfast.

27.3 Problem Definition

As previously mentioned, the problem is solved by adapting the Multi-Objective Mixed Integer Linear Programming (MOMILP) proposed by Mota et al. [3]. This is a generic model which includes several interconnected supply chain design and planning decisions along with three objective functions, representing the three pillars of sustainability. Being a generic model it was designed to accommodate more complex supply chains, such as closed-loop supply chains. For this case-study the model is simplified to a forward supply chain since the reverse logistic is not strategically relevant to the decision making process of the company. The problem can be summarized as follows.

Given: a superstructure with the potential locations of the entities (may differ depending on the product to be sent); production capacities and costs; distances between entities; considered products; estimated demand for each client, in each time interval; possible transportation modes between each pair of entities and for each mode its capacities, outsourcing costs, and handling costs in the hub terminals; stock keeping costs; required storage area for the flow of products; storage cost per unit of product; environmental impacts of entities and transportation modes; social impacts of entities and transportation modes. **The goal is to determine**: the network structure; capacity to be installed in each entity; transportation modes used; type of product to be transported between each level of the network; amount of product to be carried in each time interval; production and storage levels. **So as to**: Minimize the total cost of the supply chain; Minimize environmental impacts; Maximize social benefits.

Overall, the economic objective function is given by the total costs which include costs of storage, transformation and transportation. The environmental objective function accounts for the environmental impact of transportation and entity installation. The social objective function is simplified from the one presented by Mota et al. [3] to account for the number of jobs created.

27.4 Results

Given the different objectives, and based on the company requisites, three main macro scenarios were defined – A, B and C. Scenario A is obtained minimizing costs; scenario B is obtained minimizing the environmental impact; and scenario C is obtained maximizing job creation. Within each of these macro scenarios three different options were analysed restricting the form of the product to be shipped. Option 1 refers to the option of shipping either products in their final form (FP) or as jumbo rolls (JR). Option 2 refers to the option of shipping only FP. Option 3 refers to the option of shipping only JR which are then converted into FP in a converter located in the UK. All the 9 scenarios are summarized in Table 27.1.

27.4.1 Single Objective Optimization Results

The summary of these results is presented in Table 27.2. The unitary cost can be described as the distribution cost, per ton, of the supply chain. The environmental impact is the normalized sum of all the impacts of each activity considered, and the jobs created regard the number of new jobs that each option allows to create. As expected, the macro scenario A (represented in the first row of the table) is the one with the best results in the economic dimension, especially in options 1 and 3. The macro scenario B shows slightly higher costs, but the environmental impact experienced improvements in both the options 1 and 2. The macro scenario C presents

Table 27.1 Single-objective scenarios studied

		Macro scenarios		
		A - Objective function: minimize costs	B - Objective function: minimize environmental impact	C - Objective function: maximize job creation
Options	1: shipping both FP or JR	A1	B1	C1
	2: shipping only FP	A2	B2	C2
	3: shipping only JR	A3	B3	C3

Table 27.2 Results for the single objective optimization

			Shipping options		
			1	2	3
Macro scenarios	A	Unitary cost (Euro/ton)	221	350	221
		Environmental impact (Pts)	5329	4038	5239
		Jobs created	85	84	85
	B	Unitary cost (Euro/ton)	297	360	221
		Environmental impact (Pts)	3738	3949	5239
		Jobs created	86	86	85
	C	Unitary cost (Euro/ton)	652	776	652
		Environmental impact (Pts)	23 910	38 506	23 910
		Jobs created	376	376	376

extremely high costs and environmental impacts, which will become unviable at the operational level. Scenarios A1, A3 and B3 present the best economic result – the cost is about 26% lower than the second best result (B1). The three scenarios result in the same solution, presenting the exact same structure. Taking into account the indication that the company is planning to ship only one type of product (option 1 becomes invalid), we can conclude that it is clearly more efficient to ship jumbo rolls - solutions A1/A3/B3 (37% less expensive compared to the best scenario of shipping only finished product – A2). Furthermore, the environmental impact of shipping the JR (A1/A3/B3) is 24.5% higher compared to sending FP (A2). This is caused by the installation of the converter in scenario A2.

27.4.2 Multi Objective Optimization

This section aims to explore the possibility of building a compromise solution between the three dimensions of sustainability. As in the work by Mota et al. [3] the augmented epsilon-constraint method is used to determine solutions of compromise in the Pareto Frontier. Simply put, having decided which objective is more important, the model is designed so as to optimize that objective. In this case, the economic objective is considered to be the most important since without it the business becomes unviable. This way, the objective function of cost is firstly minimized. The remaining objectives are included as restrictions in incremental intervals of variation. These intervals are defined within specific boundaries. These boundaries are obtained from the lexicographic optimization of each objective. For example, to obtain the boundaries for the environmental objective function, the model is solved for environmental impact minimization. The value obtained for this variable is then included in a second run as a restriction and the model is solved for minimum cost. Having the values of minimum environmental impact and minimum cost as restrictions the model is then run a third time towards the maximization of the social benefit. This is performed for each of the objective functions. In this case it results in nine runs so as to obtain the so called payoff table which sets the mentioned boundaries. The augmented epsilon-constraint method was implemented to generate Pareto Solutions for the three objective functions (economic, environmental and social), for the three different options: allowing shipment of both products (option 1), shipping only finished product (option 2) and shipping only jumbo rolls (option 3). Analysing the results of the multi-objective analysis (results not shown), one can conclude that there are no solutions that improve the social performance without worsening both economic and environmental dimensions. In each of the three hypothesis analysed, the most interesting results are in the trade-off between economic and environmental dimensions. Thus, the solutions to be considered end up being those which relate to the single-objective scenarios previously exposed in Table 27.2. A comparison between the supply chains with the best performance in the combination of the three goals is now performed. Figure 27.3 shows the physical flows of goods in the logistics networks A1/A3/B3 and B1, as those with the best arrangement between the three pillars of sustainability, and which best suit the interests of the company. The blue arrows relate to the flow of finished products between the port of Leixoes and the arrival ports in the United Kingdom. The green arrows relate to the shipping of JR from Aveiro to Liverpool. The direct shipping of finished products between the ports of arrival and the final customers are shown with the orange arrows, while the supply of the warehouses is visible in the grey arrows. The black arrow indicates the flow warehouse – customers, and the purple arrow indicates the flow converter – customers. JR traveling from the port of Liverpool to the converter are shown with the yellow arrow.

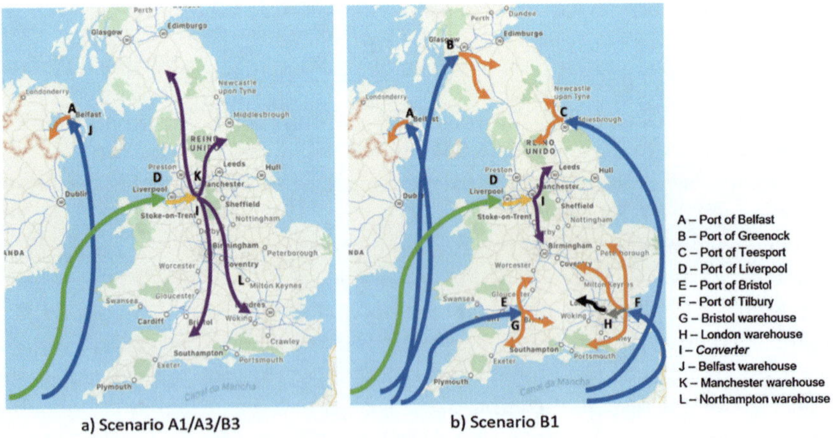

Fig. 27.3 Comparison between the two best network options: **a** Scenario A1/A3/B3 and **b** Scenario B1

The logistics chain represented on the left – scenario A1/A3/B3 – ships exclusively JR to the port of Liverpool, proceeding these towards the converter, where they are converted into finished products. Finally, the finished products are shipped directly to the end customers. On the other hand, the distribution network B1, on the right, reveals a mixed shipping of jumbo rolls and finished products. As it can be seen, finished product is shipped to 5 ports in the UK, and from there to the final customers and to one warehouse. There is also a shipment of jumbo rolls to the port of Liverpool, to be then converted and sent to the clients. Combining the difference in unitary costs between the two scenarios (cost of B1 is 35% higher than the A1), with the greater complexity of the distribution chain of B1 (which increases the likelihood of unexpected events along the chain), it can be concluded that not even the greater environmental impact (40%) of solution A1, will hinder it from being the best possible solution to respond to the current case study. Reinforcing this idea is the desire of the company to ship only one type of product towards the UK. Another issue faced during the development of this work was that the company required the creation of safety stock in the UK. However, due to the excessive costs of warehouse rental and order reception, as well as the large area that the finished product of tissue occupies, in scenario A1, the model chooses not to use any warehouse for the regular distribution of the product. However, the company's need for safety stock in the UK forced the implementation of a strategy to define the best location for the warehouses. The warehouses must be located in a site that minimizes the inflow and outflow costs when this stock is requested. The model then opens a warehouse in Belfast (to provide both customers of Northern Ireland), one in Northampton (supplies 8 clients) and another in Manchester (supplies 28 clients). In scenario B1 it is not necessary to use this strategy, since the main concern of minimizing the environmental impact is the reduction of transportation distances. Thus, the model naturally opens a warehouse in London, where all the safety stock is to be kept.

27.4.3 *Sensitivity Analysis*

A sensitivity analysis of critical parameters took place, in order to ensure the robustness of the solutions and understand what kind of variations may alter the design and planning decisions. The sensitivity analysis (results not shown) revealed that the JR solution is clearly more efficient than the FP shipment, being solid enough to withstand major cost increases. Considering the hypothesis that the company wishes to send only one type of product to the UK, the JR scenario becomes even more prominent, since all parameter variations continue to suggest the major shipment of this product, remaining the costs of option 2 (shipment of finished product) much higher.

All in all, the selected solution was the one obtained in scenario A1/A3/B3, depicted in Fig. 27.4.

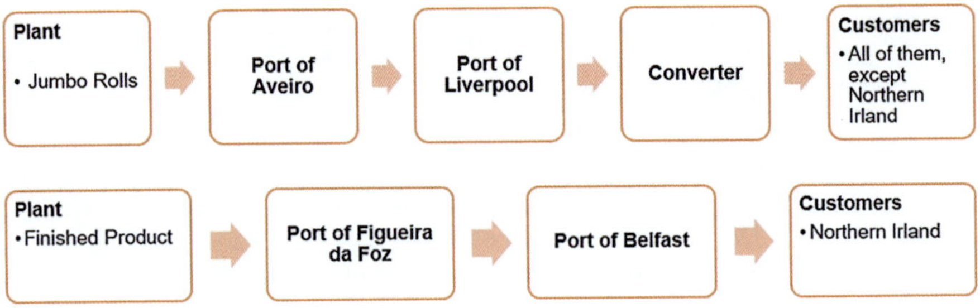

Fig. 27.4 Flows obtained with scenarios A1/A3/B3

27.5 Conclusions

The main goal of this work was to study the best structure for a Portuguese company's logistics network for one of its main products - the tissue paper to be exported to the UK, and to analyse different alternatives, in order to minimize the total costs, minimize the environmental impacts and maximize the social impacts, while guaranteeing high service levels.

To achieve these objectives a multi-objective mathematical optimization model, presented in Mota et al. [3], was used allowing the test of several scenarios, first considering a single objective and then performing a multi-objective analysis through the implementation of the augmented epsilon-constraint method. Within the options considered the possibility of shipping two types of products (finished products and jumbo rolls) was included along with the possible installation of a converter in the UK. From the results it was concluded that solutions that improve the social performance without worsening too much both economic and environmental dimensions could not be obtained. The most interesting results are in the trade-off between economic and environmental dimensions.

As a main result, it was concluded that the strategy that best suits the economic, environmental and social interests of the company is the expedition of jumbo rolls. Accordingly, a converter was installed in Manchester. As for the warehouses to be installed in the UK, the locations of Belfast (35 tons), Northampton (130 tons) and Manchester (435 tons) are suggested. A sensitivity analysis was performed to the main critical parameters and it was concluded that a robust solution was provided by the model.

Acknowledgements The authors acknowledge the Portuguese National Science Foundation (FCT) for Project PTDC/AGR-FOR/2178/2014.

References

1. J. Elkington, *Cannibals with Forks. The Triple Bottom Line of 21st Century*, (1997), p. 73
2. I.L.C.D. Handbook, *Recommendations for Life Cycle Impact Assessment in the European Context* (Publications Office of the European Union, Luxemburg, 2011)
3. B. Mota, M.I. Gomes, A. Carvalho, A.P. Barbosa-Povoa, Towards supply chain sustainability: economic, environmental and social design and planning. J. Clean. Prod. **105**, 14–27 (2015)
4. B. Ness, E. Urbel-Piirsalu, S. Anderberg, L. Olsson, Categorising tools for sustainability assessment. Ecol. Econ. **60**(3), 498–508 (2007)
5. S. Seuring, A review of modeling approaches for sustainable supply chain management. Decis. Support Syst. **54**(4), 1513–1520 (2013)
6. Z. Wu, M. Pagell, Balancing priorities: decision-making in sustainable supply chain management. J. Oper. Manag. **29**(6), 577–590 (2011)

Chapter 28
Reference Points in TOPSIS Methods for Group Decision Makers & Interval Data: Study and Comparison

Chergui Zhor and Abbas Moncef

Abstract In this paper, two new extensions of TOPSIS method for Group decision makers and Interval data are presented. In particular, the behavior of some past contributions when using Nadir point at the place of anti ideal point is studied. Otherwise, through simulation studies and simulation, which are mainly based upon smart random instances, a comparison between four algorithms is carried out, its purpose is to show the most effective one.

Keywords Group decision makers · Interval data · Pareto optimal area · Reference points · TOPSIS methods · Normalization forms

28.1 Introduction

The multicriteria decision making (MCDM) approach [1–3, 7, 15–17] appears as alternative to the classical optimization. It consists to resolve models based on the consideration of several conflicting criteria of different natures.

This approach allows us to generate an answer (one solution or more) to the decision problem, without necessarily turning the whole set of criteria into a single function. In this case, instead of searching an optimum, the analyst tries to define a compromise solution which can be reached using various forms: choice, sorting or ranking [2, 16, 17]. This approach allows us to distinguish between several MCDM methods [3, 5–7, 9, 10, 13], to each one its theoretical framework and methodological aspect.

The introduction of reference points concept [2, 15] in the ranking procedures of MCDM Methods dates back to the early seventies. A new reflection which was quickly implemented by some researchers of this epoch. We cite in particular, Paul

C. Zhor (✉)
Dpt. Mechanics, ENST, Algeria
e-mail: chergui_zhor@hotmail.fr

A. Moncef
Dpt. Mathematics, USTHB, Algeria
e-mail: moncef_abbas@yahoo.com

A. I. F. Vaz et al. (eds.), *Operational Research*, Springer Proceedings
in Mathematics & Statistics 223, https://doi.org/10.1007/978-3-319-71583-4_28

Yoon and Ching Lai Hwang by the proposition of the Technique for Order Preference by Similarity to Ideal Solution (TOPSIS) method [7] in the early eighties. It is essentially based on some concepts of the multiobjective programming approach [15].

Some years later an improved version of this method, called the revised TOPSIS, was proposed by Deng et al. [4]. Indeed, in the classical TOPSIS, the normalized decision matrix [7] is weighted by multiplying each column of the matrix by its associated weight. The overall performance of an alternative is then determined by the Euclidean distance from the Ideal point and the Negative ideal point (anti ideal solution) established from this new matrix. However, Deng and al. in [4] show that these distances are interrelated with the attribute weights, and should be incorporated in the distance measurement. On this basis they presented the weighted Euclidean distances, rather than creating a weighted decision matrix, in this process, the Positive ideal solution and the Negative ideal solution are not dependent on the weighted decision matrix.

The existence of more than one decision maker (DM) forms an extracted research area called group decision makers (GDM) in which the MCDM methods for one decision maker is not reliable [12–14]. Indeed, we need using tools that can satisfy all the group at the same time. Several adaptations of MCDM methods was presented in literature, we cite in particular those proposed by Jahanshahloo et al. [8, 11] and Zavadskas et al. [18] in which they adapt the original and revised TOPSIS in order to resolve some GDM problems.

In this paper, Two new extensions to TOPSIS method, in group decision makers for interval data, are presented. In particular, the behavior of the past contributions when changing the Negative ideal point by the Nadir point is studied. On this basis, two new procedures are proposed. Otherwise, through simulation study, based mainly on smart random instances, a comparison between the new procedures is carried out, its purpose is to show the most effective one.

28.2 TOPSIS Methods for Interval Data

In this section, the evolution through time of TOPSIS method for interval data is reviewed. To do that, some mathematical notations that will be useful in all parts of this document are introduced.

Let us consider a decision problem (A, F) composed from a set $A = \{A_1, A_2, ..., A_m\}$ of m actions and a family of n criteria F, $|F| = n$. For each criterion j a numerical function g in the set of real numbers R is defined, such that: $g_j(A_i) = a_{ij}, i \in \{1, \ldots, m\}, j \in \{1, \ldots, n\}$. In addition, the components a_{ij} form an interval, where $a_{ij} = [\underline{a_{ij}}, \overline{a_{ij}}]$, $(\underline{a_{ij}}, \overline{a_{ij}}) \in R, i \in \{1, \ldots, m\}, j \in \{1, \ldots, n\}$.

To each criterion, a weight w_j is evaluated which increases with the relative importance of the criterion. For this family of methods the sum of weights must be equal to one.

28.2.1 *TOPSIS Methods for Interval Data: One Decision Maker*

The crisp data TOPSIS method [7, 10] (the original TOPSIS for one DM) was extended by Jahanshahloo et al. [8, 11] to be applied in interval data cases. The proposed procedure is described by the following steps.

From the decision matrix associated to one decision maker (A, F), one makes normalization to each bound of the interval a_{ij}, such that:

$$\underline{r_{ij}} = \frac{\underline{a_{ij}}}{\sqrt{\sum_{i=1}^{m} \underline{a_{ij}}^2 + \sum_{i=1}^{m} \overline{a_{ij}}^2}}$$

and

$$\overline{r_{ij}} = \frac{\overline{a_{ij}}}{\sqrt{\sum_{i=1}^{m} \underline{a_{ij}}^2 + \sum_{i=1}^{m} \overline{a_{ij}}^2}} \tag{28.1}$$

where $i \in \{1, \ldots, m\}$, $j \in \{1, \ldots, n\}$ Consequently, the interval $[\underline{r_{ij}}, \overline{r_{ij}}]$ is the normalized interval of $[\underline{a_{ij}}, \overline{a_{ij}}]$.

Multiply each interval of the decision matrix by the corresponding weight w_j, $j \in \{1, \ldots, n\}$. So we create a new matrix V, such that:

$$\underline{v_{ij}} = w_j \underline{r_{ij}}$$

and

$$\overline{v_{ij}} = w_j \overline{r_{ij}} \tag{28.2}$$

Determine the Ideal solution V^+ and the anti ideal solution V^- relative to the normalized matrix V, often, it should be noted that these two actions are fictive for the decision problem:

$$V^+ = (< max_{i \in \{1,\ldots,m\}}(\overline{v_{ij}})/j \in J >, < min_{i \in \{1,\ldots,m\}}(\underline{v_{ij}})/j \in J' >)$$

and

$$V^- = (< min_{i \in \{1,\ldots,m\}}(\underline{v_{ij}})/j \in J >, < max_{i \in \{1,\ldots,m\}}(\overline{v_{ij}})/j \in J' >) \tag{28.3}$$

where J (resp. J') denote the set of maximized (minimized) criteria.

These two alternatives V^+ and V^- indicate the most preferable alternative (Ideal solution) and the least preferable alternative (Negative ideal solution or anti ideal solution), respectively.

One calculates the separation values of each alternative ($i \in \{1, \ldots, m\}$) such that, S_i^+ represents the distance from the Ideal solution and S_i^- is the one from the Negative ideal solution:

$$S_i^+ = \sqrt{\sum_{j=1}^{n} \left(\overline{v_{ij}} - v_j^+\right)^2 + \sum_{j=1}^{n} \left(\underline{v_{ij}} - v_j^+\right)^2}$$

and

$$S_i^- = \sqrt{\sum_{j=1}^{n} \left(\overline{v_{ij}} - v_j^-\right)^2 + \sum_{j=1}^{n} \left(\underline{v_{ij}} - v_j^-\right)^2} \tag{28.4}$$

Finely, one computes the quantity P_i which measures the relative proximity to the ideal point:

$$P_i = \frac{S_i^-}{S_i^- + S_i^+} \tag{28.5}$$

Now, the set of alternatives A can be preference ranked according to the descending order of P_i. The best alternative is the one defining the maximum of the quantity P_i.

In the following sections we focus upon the group decision makers versions, we review firstly the extensions of TOPSIS by some authors in the GDM (interval data), after we present our contribution in this area.

28.2.2 *TOPSIS Methods for Group Decision Makers: Interval Data*

In this section, two different versions of TOPSIS method for GDM are reviewed, the first one uses Jahanshahloo et al. [8, 11] procedure for one DM: interval data. Indeed, the proposition of Jahanshahloo et al. [8, 11] uses a new decision problem computed from all the GDM decision problems, such that: from the K decision matrices given by the K decision makers, $k \in \{1, \ldots, K\}$,

$$\underline{a_{ij}} = \frac{1}{K} \sum_{k=1}^{K} \underline{a_{ij}^{(k)}}$$

and

$$\overline{a_{ij}} = \frac{1}{K} \sum_{k=1}^{K} \overline{a_{ij}^{(k)}} \tag{28.6}$$

The weights vector is calculated by the8 same manner:

$$w_j = \frac{1}{K} \sum_{k=1}^{K} w_j^{(k)} \tag{28.7}$$

Upon this new data, on applies the procedure of Jahanshahloo et al. for one DM: interval data.

The second procedure, proposed by Zavadskas et al. [18], uses the following steps: First, a formula of normalization for each DM's data $k \in \{1, \ldots, K\}$ is applied:

$$\underline{r_{ij}^{(k)}} = \frac{\underline{a_{ij}^{(k)}}}{\max_i \overline{(a_{ij}^{(k)})}}$$

and

$$\overline{r_{ij}^{(k)}} = \frac{\overline{a_{ij}^{(k)}}}{\max_i \overline{(a_{ij}^{(k)})}} \tag{28.8}$$

for $i \in \{1, \ldots, m\}$, $j \in \{1, \ldots, n\}$. After, one defines the Ideal and the anti ideal solutions for each DM $k \in \{1, \ldots, K\}$, such that:

$$R^{(k+)} = (< max_{i\in\{1,\ldots,m\}}(\overline{r_{ij}^{(k)}})/j \in J >, < min_{i\in\{1,\ldots,m\}}(\underline{r_{ij}^{(k)}})/j \in J' >)$$

and

$$R^{(k-)} = (< min_{i\in\{1,\ldots,m\}}(\underline{r_{ij}^{(k)}})/j \in J >, < max_{i\in\{1,\ldots,m\}}(\overline{r_{ij}^{(k)}})/j \in J' >) \tag{28.9}$$

The separation measures for the K DMs are calculated by:

$$S_i^{(k+)} = \sqrt{\frac{1}{2}\sum_{j=1}^{n}\left(\underline{r_{ij}^{(k)}} - r_j^{(k+)}\right)^2 + \sum_{j=1}^{n}\left(\overline{r_{ij}^{(k)}} - r_j^{(k+)}\right)^2}$$

and

$$S_i^{(k-)} = \sqrt{\frac{1}{2}\sum_{j=1}^{n}\left(\overline{r_{ij}^{(k)}} - r_j^{(k-)}\right)^2 + \sum_{j=1}^{n}\left(\underline{r_{ij}^{(k)}} - r_j^{(k-)}\right)^2} \tag{28.10}$$

After an aggregation of K separation measures is calculated, such that:

$$S_i^{+} = \frac{\sum_{k=1}^{K} S_i^{(k+)}}{K}$$

and

$$S_i^{-} = \frac{\sum_{k=1}^{K} S_i^{(k-)}}{K} \tag{28.11}$$

P_i is computed by the same quotient as in all the other extensions of TOPSIS and the best alternative(s) is the one which defines the maximum value of P_i.

At this step one cannot pass without discussing the use of 1/2 in the above formulas, in fact there is no need to this quantity because it will be easily simplified in the computation of P_i as well as the parameter K considered in the aggregation formula of K separation measures.

In the following section, two new procedures, inspired from those of Jahanshahloo et al. [8, 11] and Zavadskas et al. [18], are proposed. Two comparison studies are carried out in order to show the most reliable one. In addition, a discussion about the use of another form of normalization upon the first procedure is given.

28.3 TOPSIS Methods for Group Decision Makers Using Nadir Point: Interval Data

In order to study the impact of using another reference point, two different procedures are proposed in which the Nadir solution is usd at the place of the anti ideal solution. The first one is based upon the procedure of Zavadskas et al. [18], such that, from the normalized matrix $\overline{R}$, $\underline{R}$, for each DM's data $k \in \{1, \ldots, K\}$, one computes the local Ideal solutions and the local Nadir solutions.
The Nadir solution is defined by the minimums lower bounds of the intervals (alternatives) forming the Ideal solution. The formula of the Nadir point is given by:

$$A^{(nad,k)} = (< min_{i\in\{1,\ldots,m\}}(\underline{R'^{(k)}_{ij}})/j \in J >, < max_{i\in\{1,\ldots,m\}}(\overline{R'^{(k)}_{ij}})/j \in J' >) \tag{28.12}$$

where $R'^{(k)}$ represents the set of pareto optimal solutions (intervals between the Ideal solution and the Nadir solution).

Indeed, the set of alternatives after normalization should be decomposed into classes, ordered from the first to the last one. The first class contains only intervals (alternatives) between the first local Ideal solution and the first local Nadir solution defined from the whole set of alternatives, the second class is defined by the same manner from the whole set of alternatives minus the alternatives of the first class, and so on until putting all the alternatives into classes for each DM.

The most occurred partition is the one used by each DM. In each class, one computes the separation values:

$$S_i^{(k+)} = \sqrt{\sum_{j=1}^{n} w_j(\underline{r_{ij}^{(k)}} - r_j^{(k+)})^2 + \sum_{j=1}^{n} w_j(\overline{r_{ij}^{(k)}} - r_j^{(k+)})^2}$$

and

$$S_i^{(k-)} = \sqrt{\sum_{j=1}^{n} w_j(\overline{r_{ij}^{(k)}} - r_j^{(k-)})^2 + \sum_{j=1}^{n} w_j(\underline{r_{ij}^{(k)}} - r_j^{(k-)})^2} \tag{28.13}$$

where $r_j^{(k-)}$ (resp. $r_j^{(k+)}$) are the components of the local anti ideal solution (resp. the local Ideal solution). For the problems which have the same most occurred partition, the components of the local Nadir solution are the same of the local anti ideal solution.

After, The overall separation values (of all the group) are computed in each class, are given by:

$$S_i^+ = \sum_{k=1}^{K} S_i^{(k+)}$$

and

$$S_i^- = \sum_{k=1}^{K} S_i^{(k-)} \tag{28.14}$$

In each class a sub-ranking is established by computing the quantity P_i. P_i keeps the same formula and signification as in all the other extensions. The union of all the sub-rankings forms the overall ranking.

A second procedure can be also defined using the procedure of Jahanshahloo et al. [13, 18] in the computation of the overall matrix and weights, so from the K decision matrices given by the K decision makers:

$$\underline{a_{ij}} = \frac{1}{K}\sum_{k=1}^{K} \underline{a_{ij}^{(k)}}$$

and

$$\overline{a_{ij}} = \frac{1}{K}\sum_{k=1}^{K} \overline{a_{ij}^{(k)}}$$

The weight vector is calculated by the same manner, such that:

$$w_j = \frac{1}{K}\sum_{k=1}^{K} w_j^{(k)}.$$

where $k \in \{1, \dots, K\}$.

After, one decomposes the fully set of alternatives into classes and computes the separation values in each class by the same manner as in the first procedure. The best alternative(s) of the problem is defined by establishing a sub-ranking regarding the quantity P_i in each class and ordering the classes from the first one to the last one. Note that the components of the local Nadir solution are the same of the local anti ideal solution.

Upon the same hypothesis of the multi-objective [15], we propose the use of the Nadir solution with GDM interval data. Indeed, in each class the alternatives of pareto optimal area is considered in order to favorite balanced alternatives and weaken the impact of the most no desirable solution (anti ideal solution). To do that, an adaptation of TOPSIS method is proposed by taking some notions from the interval data procedure for one DM of Jahanshahlou et al. [8, 11] and using the results of Deng et al. [4] about the role of weights in the separation values.

In order to show the reliability of the two proposed procedures a simulation study is carried out in which a smart interval data is generated according to two decision makers (K = 2). Indeed, some specific computer instructions are used in order to generate data representing the feature of real GDM for interval data cases. After resolving separately the decision problems of each pair of instances using the procedure of one DM (listed above), we keep only the data (pair of decision problems) that provides the same best alternative(s).
For each number fixed of alternatives (from 2 to 20) and criteria (from 2 to 20), 10000 pair of instances are generated. The commands used to generate the smart random data are given below:

Begin

- ***First decision maker*** $(k = 1)$
  - $a = randi([2, 10], [m, n]);$ (Definition of $\overline{a_{ij}}$)
  - for $i = 1 : m$
    - for $j = 1 : n$
    - $f(i, j) = a(i, j) - randi((a(i, j) - 1));$ (Definition of $\underline{a_{ij}}$)
    - end
    - end.
  - $w = randi([2, 10], [1, n]);$
- ***Second decision maker*** $(k = 2)$
  - $r1 = randi([0, 1], m, n);$
  - $r = randi([-1, 0], m, n);$
  - $a1 = a + r + r1;$
  - $f1 = f + r + r1;$
  - $s1 = randi([0, 1], 1, n);$
  - $s = randi([-1, 0], 1, n);$
  - $w1 = w + s + s1;$

End

where $randi$([2,10],[m, n]) is a random function allows to generate a (m,n)matrix where the components varied in [2,10].

Using the chosen data, we resolve each pair of instances by four different procedures:

+ Procedure of Zavadskas et al. (P1).
+ The modified procedure using one calculated matrix (P2).
+ Procedure of Jahanshahloo et al. (P3).
+ The modified procedure using the K matrices separately (P4).

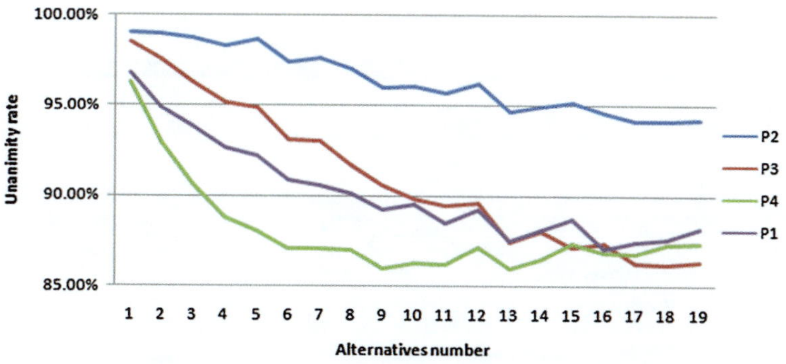

Fig. 28.1 Procedures comparison

Logically, the most reliable procedure is the one which can keep the same best alternative(s) as the two DMs, that is why, for each procedure, the rate of unanimity or the rate of indication of the same best alternative(s) is computed. The first graph (Fig. 28.1), show the results by procedure. Notice that, what ever the range of data, in both cases (new and old procedures) the rate is not less than 85 per cent. However, the most reliable one is the modified procedure which uses one computed matrix. Indeed, it gives the high rate of unanimity. To simplify the presentation, we compute the average of instances by alternatives eliminating by this the criteria axis. Thus the abscissa axis (resp. ordinate axis) represents the alternatives (resp. average of instances).

In addition, in the case of the procedure which uses the K decision problems separately (P4), a form of normalization different from the one used by Zavadskas et al. [18] is used ($\max_i \overline{(a_{ij}^{(k)})}$). In fact, the replacement of this normalization by $\frac{1}{\sqrt{\sum_{i=1}^{m} \underline{(a_{ij}^{(k)})}^2 + \sum_{i=1}^{m} \overline{(a_{ij}^{(k)})}^2}}$ gives more interesting results. The graph 2 (Fig. 28.2) shows that the results of this procedure became better than the modified procedure (P2) which uses one computed matrix, so the divergence of this procedure (P4) was caused by the form of normalization used. The abscissa axis (resp. ordinate axis) represents the alternatives (resp. average of instances).

Clearly, the comparison between the procedures shows that the use of Nadir point can provide more unanimity, their results are the most close to those of one decision maker. At the end, the consideration of the relative pareto optimal area in each class the alternatives insures more reliability to the results of TOPSIS GDM procedures.

28.4 Conclusion

In this paper, which represents the essential of the wor(k)developed in order to study the behavior of TOPSIS method when using Nadir point for group decision makers and interval data area, two new extended versions of TOPSIS method are proposed. These new procedures constitute an improvement of some past contributions.

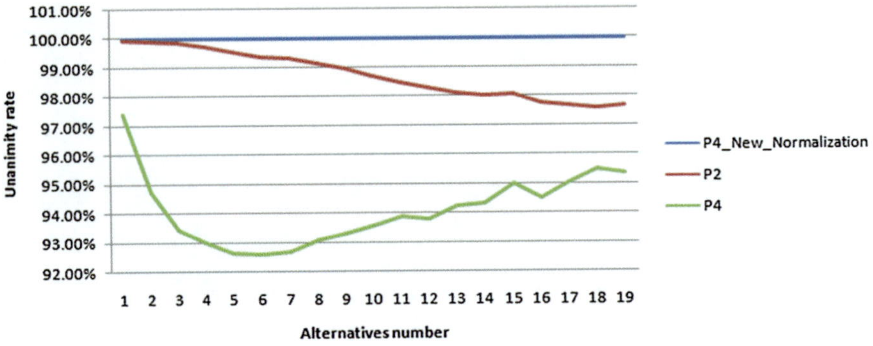

Fig. 28.2 Normalization forms

The reliability of These procedures due mainly to the use of Nadir solution. Theoretical discussions and simulation studies are carried out along this paper.

As a perspective, it will be wise to apply these procedures upon real life cases in order to provide more reliability to obtained results.

References

1. V. Belton, T. Stewart, *Multiple Criteria Decision Analysis: An Integrated Approach* (Springer Science & Business Media, Berlin, 2002)
2. D. Bouyssou, B. Roy, Aide à la décision multicritère : méthodes et cas. Economica (1993)
3. J.P. Brans, B. Mareschal. *PROMETHEE-Gaia : une méthodologie d'aide à la décision en présence de critères multiples*. Ellipes (2001)
4. H. Deng, C.H. Yeh, R.J. Willis, Inter-company comparison using modified TOPSIS with objective weights. Comput. Oper. Res. **27**, 963–973 (2000)
5. J.S. Dyer, Remarks on the analytic hierarchy process. Manag. Sci. **36**, 249–258 (1990a)
6. J.S. Dyer, A clarification of remarks on the analytic hierarchy process. Manag. Sci. **36**, 274–275 (1990b)
7. C.L. Hwang, K. Yoon, *Multiple Attribute Decision Making: Methods and Applications* (Springer, New York, 1981)
8. G.R. Jahanshahloo, F.H. Lofti, M. Izadikhah, An algorithmic method to extend topsis for decision making problems with interval data. Appl. Math. Comput. **175**, 1375–1384 (2006)
9. W.D. Keyser, P. Peeters, A note on the use of promethee mcdm methods. Eur. J. Oper. Res. **0**, 457–461 (1996)
10. S. Opricovic, G.-H. Tzeng, Compromise solution by mcdm methods: A comparative analysis of vikor and topsis. Eur. J. Oper. Res. **156**, 445–455 (2004)
11. E. Roszkowska, Multi-criteria decision making models by applying the topsis method to crisp and interval data. Mult. Criteria Decision Mak. **6**, 200–230 (2011)
12. E. Roszkowska, T. Wachowicz, Negotiation support with fuzzy topsis. Group Decision Negot (2012)
13. E. Roszkowska, T. Wachowicz, Application of fuzzy topsis to scoring the negotiation offers in ill-structured negotiation problems. Eur. J. Oper. Res. **242**(3), 920–932 (2015)
14. H.S. Shih, H.J. Shyur, E.S. Lee, An extension of topsis for group decision making. Math. Comput. Model. **45**, 801–813 (2007)

15. R. Steuer, *Multiple Criteria Optimization : Theory, Computation and Applications* (Wiley, New-York, 1985)
16. E. Triantaphyllou, *MCDM Decision Making Methods: A Comparative Study*. Ellipses (1989)
17. E. Triantaphyllou, *MCDM Decision Making Methods: A Comparative Study* (Kluwer Academic Publishers, Dordrecht, 2000)
18. E.K. Zavadskas, Z. Turskis, J. Tamosaitiene, Construction risk assessment of small scale objects by applying the topsis method with attributes values determined at intervals. in *The 8th International Conference Reliability and Statistic in Transportation and Communication* (2008)